Process Gas Chromatography

Process Gas Chromatography
Advanced Design and Troubleshooting

A Companion Volume of Process Gas Chromatographs (2020)

TONY WATERS
California, USA

This edition first published 2025
© 2025 John Wiley & Sons Ltd

All rights reserved, including rights for text and data mining and training of artificial technologies or similar technologies. No part of this publication may be reproduced, stored in a retrieval system, or transmitted, in any form or by any means, electronic, mechanical, photocopying, recording or otherwise, except as permitted by law. Advice on how to obtain permission to reuse material from this title is available at http://www.wiley.com/go/permissions.

The right of Tony Waters to be identified as the author of this work has been asserted in accordance with law.

Registered Offices
John Wiley & Sons, Inc., 111 River Street, Hoboken, NJ 07030, USA
John Wiley & Sons Ltd, New Era House, 8 Oldlands Way, Bognor Regis, West Sussex, PO22 9NQ, UK

For details of our global editorial offices, customer services, and more information about Wiley products visit us at www.wiley.com.

Wiley also publishes its books in a variety of electronic formats and by print-on-demand. Some content that appears in standard print versions of this book may not be available in other formats.

Trademarks: Wiley and the Wiley logo are trademarks or registered trademarks of John Wiley & Sons, Inc. and/or its affiliates in the United States and other countries and may not be used without written permission. All other trademarks are the property of their respective owners. John Wiley & Sons, Inc. is not associated with any product or vendor mentioned in this book.

Limit of Liability/Disclaimer of Warranty
In view of ongoing research, equipment modifications, changes in governmental regulations, and the constant flow of information relating to the use of experimental reagents, equipment, and devices, the reader is urged to review and evaluate the information provided in the package insert or instructions for each chemical, piece of equipment, reagent, or device for, among other things, any changes in the instructions or indication of usage and for added warnings and precautions. While the publisher and authors have used their best efforts in preparing this work, they make no representations or warranties with respect to the accuracy or completeness of the contents of this work and specifically disclaim all warranties, including without limitation any implied warranties of merchantability or fitness for a particular purpose. No warranty may be created or extended by sales representatives, written sales materials or promotional statements for this work. The fact that an organization, website, or product is referred to in this work as a citation and/or potential source of further information does not mean that the publisher and authors endorse the information or services the organization, website, or product may provide or recommendations it may make. This work is sold with the understanding that the publisher is not engaged in rendering professional services. The advice and strategies contained herein may not be suitable for your situation. You should consult with a specialist where appropriate. Further, readers should be aware that websites listed in this work may have changed or disappeared between when this work was written and when it is read. Neither the publisher nor authors shall be liable for any loss of profit or any other commercial damages, including but not limited to special, incidental, consequential, or other damages.

Library of Congress Cataloging-in-Publication Data applied for

Hardback ISBN: 9781119791478

Cover Design: Wiley
Cover Image: Courtesy of ABB Inc.

Set in 11/13pt STIXTwoText by Straive, Pondicherry, India

SKY10093090_120524

For Sonia

*"I pray you choose Joy.
Anything else is a waste of time."*
Rev. Dr. Sonia E. Waters (1972–2023)

Process Gas Chromatographs at the INEOS Olefin Plant in Cologne, Germany.
Photo © INEOS in Cologne 2019.

*"We cannot teach people anything;
we can only help them discover it within themselves"*

Attributed to Galileo Galilei 1564-1642

Contents

Contributors xxiii
Preface to the first book xxviii
Preface to the second book xxx
Acknowledgments xxxi

1 Fundamentals 1
  Introduction 1
  Gas chromatography 1
    A unique analytical technique 1
    A basic gas chromatograph 2
    The columns 3
    The detector 3
    The chromatogram 4
  The science of GC 4
    The basic science 5
  The gas chromatograph 7
    The basic instrument 7
    The process instrument 7
  The contents 8
  Becoming a PGC expert 10
  Knowledge Gained 11
  Did You Get It? 12
    Self-assessment quiz: SAQ 01 12
  References 13
    Further reading 13
    Cited 13
    Figures 13
    New technical terms 13

2 Measures of perfection 15
  Performance indicators 15
  Chromatogram data 16
    Two fundamental measurements 16

Contents

Significance of the air peak	17
Time in each phase	17
Calculated performance indicators	19
Average carrier gas velocity	19
Retention factor	19
Separation	21
Resolution	21
Practical aspects of theory	23
Plate theory	23
SCI-FILE: On Plate Theory	24
Ideal peak shape	24
Peak width	25
Measures of column efficiency	25
Plate height	25
Development of peak shape	26
Peak shape at elution	27
Chromatogram peak shape	27
Column efficiency	28
Effective plate number	28
Using the theory	29
Resolution on longer columns	29
Estimating plate number	30
Estimating plate height	32
Knowledge Gained	33
Did you get it?	34
Self-assessment quiz: SAQ 02	34
On Practice	35
On Theory	35
References	35
Cited	35
Figures	35
Symbols	35
Equations	36
New technical terms	38

3 Column technology — 39

Introduction	39
Column types	40
Two types of column	40

Packed columns	40
Capillary columns	41
Mechanism of retention	42
Mutual affinity	42
SCI-FILE: On Mutual Affinity	43
Molecular forces	43
Dispersion forces	43
Polar forces	44
Ionic forces	45
Nomenclature	45
Summary	46
Packed columns for GSC	46
Solid-phase columns	46
Active solids	46
Operation and maintenance	47
Pros and cons of GSC	48
Packed columns for GLC	49
Solid support	49
Coating the support	50
Pros and cons of GLC	51
Packed column technology	51
Tubing	52
Mesh size	52
Packing the tube	54
Conditioning the column	54
Pros and cons of packed columns	55
Micropacked columns	56
Typical operating conditions	57
Capillary column technology	57
Capillary tubing	57
Wall-coated open tubular (WCOT) columns	60
Porous-layer open tubular (PLOT) columns	60
Support-coated open tubular (SCOT)	60
Phase ratio	61
Carrier gas flow rate	63
Separating power	64
Pros and cons of capillary columns	64
Packed or capillary? A comparison	64
Acknowledgment	64

Design of equipment	65
Efficiency and resolution	65
Sample size	67
Feed volume	67
Sample capacity	68
Measurement sensitivity	71
Analyte rangeability	71
Column bleed	72
Knowledge Gained	74
Did you get it?	77
Self-assessment quiz: SAQ 03	77
References	77
Further reading	77
Cited	78
Tables	79
Figures	79
Symbols	80
Equations	80
New technical terms	81
4 The stationary phase	**82**
Liquid phases	82
The need for variety	82
Column identification	83
Liquid phase polarity	83
Solute affinity	85
Polar or Nonpolar?	86
Silicone liquid phases	87
Methyl silicones	87
Phenyl silicones	88
More possibilities	89
Nonpolar silicones	89
Fluoro silicones	90
Cyano silicones	91
Multifunctional silicones	91
Bonded phases	92
A precaution	92
Non-silicone liquid phases	92

Dialkyl esters	92
Polyesters	93
Waxy polymers	95
Nitrile ethers	95
Column temperature effects	96
Liquid phase temperature rating	96
Liquid phase viscosity	96
Choosing a liquid phase	96
The practical method	96
Using retention data	98
SCI-FILE: On Retention Data	98
Retention index	98
McReynolds constants	99
Using McReynolds	100
System constants	102
Making a choice	102
Are McReynolds constants useful?	102
Adsorption columns	103
Distribution in GLC and GSC	103
Adsorption isotherms	104
Nonideal peak shape	104
Activation	105
Molecular sieves	106
Carbon	106
Silica gel or porous silica	107
Alumina	108
Porous polymers	108
Knowledge Gained	110
Did you get it?	113
Self-assessment quiz: SAQ 04	113
References	113
Further reading	113
Cited	114
Tables	117
Figures	117
Symbols	117
Equations	118
New technical terms	118

5 PGC column design — 119

- The design objective — 119
- Design strategy — 120
  - It's always about resolution — 120
- SCI-FILE: On Resolution — 121
  - Theory — 121
  - The resolution equation — 121
  - Endnote — 122
- Achieving resolution — 123
  - The resolution equation — 123
  - Interaction between factors — 123
  - The column length — 124
- A deeper understanding — 125
- SCI-FILE: On Distribution — 125
  - Retention volume — 125
  - Distribution constant — 126
  - Definition of retention factor — 127
  - Definition of separation factor — 128
- Optimizing liquid phase performance — 128
  - Adjusting the separation factor — 128
  - Adjusting the retention factor — 128
- SCI-FILE: On Retention Factor — 129
  - Optimum resolution — 129
  - Optimum analysis time — 130
  - Optimum retention factor — 130
- Exploring the power region — 131
- Optimizing retention factor — 133
  - Ineffective variables — 133
  - Effective variables — 133
  - Optimizing the column temperature — 134
  - Temperature programming — 134
  - Optimizing the column phase ratio — 134
- A worked example — 135
  - Evaluation of options — 136
  - How to optimize liquid phase performance — 137
  - Reducing the column length — 139
  - Getting a faster analysis — 142
- Knowledge Gained — 142
- Did you get it? — 144
  - Self-assessment quiz: SAQ 05 — 144
- References — 144

	Further reading	144
	Cited	144
	Figures	145
	Symbols	145
	Equations	146
	New technical terms	148

6 Optimizing performance — 149

Caution	149
Optimum flow rate	150
Experimental	150
The optimization procedure	151
Relation to theory	153
SCI-FILE: On Rate Theory	153
Too much theory?	153
Rate theory	154
What we learn	154
Carrier gas velocity	154
Solute band velocity	155
The van Deemter equation	155
The A-term	155
What we learn	156
The B-term	156
What we learn	156
The C-term	156
What we learn	157
Extended van Deemter equation	157
The Golay equation	158
What we learn	158
Other effects	158
Getting a fast analysis	159
A worked example	160
Introduction	160
Initial assumptions	160
Step #1: Evaluate the existing column	161
Summary of chromatogram 6.4a	164
Step #2: Use a longer column	164
Summary of chromatogram 6.4b	165
Step #3: Regain the analysis time	166
Summary of chromatogram 6.4c	168

Step #4: Faster analysis with the same resolution	168
Summary of chromatogram 6.4d	169
Time efficiency	170
Designing for fast analysis	170
The efficiency curve	170
Narrow column diameter	171
Choice of carrier gas	171
Liquid phase loading	172
Liquid phase viscosity	173
Knowledge Gained	173
Did you get it?	174
Self-assessment quiz: SAQ 06	174
On practice	174
On theory	174
References	175
Further reading	175
Cited	175
Tables	175
Figures	176
Symbols	176
Equations	177
New technical terms	178

7 Extracolumn broadening — 179

Introduction	179
Sample injection	180
Sample volume	181
Solubility limit	181
Peak width	181
Overall effect	182
Feed volume	183
Practical outcomes	184
Sample size calculations	185
Duration of injection	185
An example	186
Flow path geometry	187
Open tubing	188
Mixing chambers	188
Dead volumes	188
Detectors	189

The theory	190
SCI-FILE: On Extracolumn Variance	190
Introduction	190
Acceptable dispersion	190
Sources of variance	191
Warning	191
Sample injection profile	191
Plug injection	191
Exponential injection	191
Gaussian injection	191
What we learn	192
Dispersion in open tubes	192
Laminar flow	192
What we learn	193
Dispersion by mixing	193
What we learn	194
Dispersion in dead volumes	194
What we learn	194
Dispersion in detector	194
Mass flow detectors	194
Concentration detectors	194
What we learn	194
Response speed	195
What we learn	195
Practical limits	195
More detail	196
Worked examples	196
Sample injection	196
Open tubing	197
Knowledge Gained	198
Did you get it?	199
Self-assessment quiz: SAQ 07	199
On practice	199
On theory	200
References	200
Further reading	200
Cited	200
Table	201
Figure	201
Symbols	201

	Subscripts	201
	Equations	202
	New technical terms	203
8	**Evaluating peak shape**	**204**
	Real chromatogram peaks	204
	Three peak shapes	204
	Normal asymmetric peaks	206
	Measuring peak asymmetry	206
	What causes asymmetry?	207
	Limited isotherm linearity	207
	Curved isotherms	209
	A second retention mechanism	209
	Column overload	211
	SCI-FILE: On Asymmetric Peaks	212
	Natural deviations from perfection	212
	Poisson distribution	212
	Transformation to the time domain	212
	Thermal peak distortion	213
	A second retention mechanism	214
	Problems with asymmetric peaks	214
	Retention time shift	215
	Loss of resolution	216
	Tailing peaks	216
	All peaks are tailing	217
	Some peaks are tailing	217
	Offscale peaks are tailing	217
	Diagnosis of deviant peak shapes	218
	Reading the chromatogram	218
	Overlapping peaks	219
	Offscale peaks	221
	Partial peaks	222
	Knowledge Gained	225
	Did you get it?	226
	Self-assessment quiz: SAQ 08	226
	On practice	226
	On theory	227
	References	227
	Further reading	227

	Cited	227
	Table	228
	Figures	228
	Symbols	228
	Equations	229
	New technical terms	229

9 Columns in series — 230

The need for multiple columns	230
To minimize analysis time	230
To do the housekeeping	231
To regroup a set of components	231
To enhance the separation	231
To separate trace amounts of analyte	232
To protect a sensitive column or detector	232
History of the technique	232
Development of laboratory chromatographs	232
Development of process chromatographs	233
The power of dissimilar columns	235
Pressures in series columns	237
SCI-FILE: On Pressure Drop	238
Pressure loss in columns	238
Average pressure in a column	238
Average velocity in a column	238
Velocity in series columns	239
Retention time in series columns	239
Retention time in series columns	239
Pressure effect on solubility	240
Example calculations	240
Pressure in series columns	240
Velocity in equal columns	242
Retention time in equal columns	242
Choosing column lengths	243
Column systems	243
Knowledge Gained	244
Did you get it?	245
Self-assessment quiz: SAQ 09	245
References	246
Further reading	246

	Cited	246
	Tables	247
	Figures	247
	Symbols	247
	Subscripts	248
	Equations	248
	Technical terms	249

10 Backflush systems — 250

Introduction — 250
    Backflush functions — 250
    Backflush to vent — 251
    Backflush to measure — 252
    Pressure balance switching — 254
    Setting the backflush time — 256
    Stuttering backflush — 257
    Parallel chromatography — 257
    Regrouping systems — 257

Backflush theory and practice — 259
    Limitations — 259
    Column design and operation — 260
    Using the same stationary phase — 260
    Using different stationary phases — 261

A worked example — 262
    Calculated backflush event timing — 262
    Valley point — 263
    Timed events — 264
    Peak B retention time — 267
    Regrouping effectiveness — 268
    Regrouping column — 269

Knowledge Gained — 269
Did you get it? — 271
    Self-assessment quiz: SAQ 10 — 271
References — 271
    Further reading — 271
    Cited — 271
    Tables — 272
    Figures — 272
    Subscripts — 272
    New technical terms — 273

11 Heartcut systems — 274

- Origin and development — 274
  - Early days — 274
  - Heartcut functions — 275
- Heartcut column system — 276
  - Pressure balance column switching — 277
  - Column system design — 278
- Making the cuts — 279
  - Modes of operation — 279
  - A single cut for one analyte — 279
  - A single cut for two analytes — 281
  - Multiple cuts — 282
  - Avoiding complexity — 283
  - Adding a backflush — 284
- Setting the valve timing — 284
  - The setup procedure — 284
  - Setting multiple cuts — 286
  - General rules of procedure — 286
- Diagnosis — 286
  - Observing the remnant peak — 287
  - Observing the analyte peak — 288
- Similar column systems — 289
  - Trap-and-hold column system — 289
  - Distribution column system — 290
  - Parallel chromatography — 290
  - The real power — 291
- Knowledge Gained — 292
- Did you get it? — 294
  - Self-assessment quiz: SAQ 11 — 294
- References — 295
  - Further reading — 295
  - Cited — 295
  - Figures — 295
  - Symbols — 296
  - New technical terms — 296

12 PGC troubleshooting — 297

- Is there a problem? — 297
  - The usual suspects — 297
  - Why validation? — 298

Choose your champion	298
Benefits	298
Validation strategies	299
Watching process data	300
Alarming the calibration factors	301
Statistical quality control	301
Alarming the validation fluid analysis	302
Comparing PGC and laboratory analyses	303
Comparing redundant PGC analyzers	303
Comparing PGC with another analyzer type	303
Troubleshooting	303
Two kinds of problem	303
Evaluating the baseline	304
What is a baseline?	304
Checking the baseline	306
Prime suspect – the detector	306
Concentration-sensing detectors	307
Mass flow-sensing detectors	307
Another thought	307
A logical approach	307
The detector is working but responding to external influences	308
The detector is not working correctly	309
The detector is responding to molecules	309
The detector is not generating the anomaly	310
Working from the symptoms	310
Diagnosis	312
Diagnosing baseline drift	312
Diagnosing wander	315
Diagnosing cycles	317
Diagnosing noise	319
Overall effect	322
Knowledge Gained	323
Did you get it?	326
Self-assessment quiz: SAQ 12	326
References	326
Further reading	326
Cited	326
Tables	327
Figures	327
New technical terms	327

13 Troubleshooting chromatograms — 329

- Reading the chromatogram — 329
  - Your ultimate skill — 329
- Diagnosing chromatographic faults — 330
  - Noticing a problem — 330
  - Recognizing a malfunction — 331
  - Seeking the cause — 332
  - Identifying artifacts — 333
  - Confirming the diagnosis — 335
- SCI-FILE: On Diagnosis — 335
- Worked example — 335
  - Checking the history — 335
  - Finding the valve actions — 336
  - Studying the shape — 336
  - Estimating elapsed times — 336
  - Seeking the cause — 336
  - Confirming the diagnosis — 337
  - Other observations — 337
  - Fixing a chromatographic fault — 337
- Key properties of peaks — 338
  - What to know about peaks — 338
  - Retention time pattern — 338
  - Retention time shift — 339
  - Peak width — 339
  - Peak height — 340
  - Unusual peak shape — 341
  - Unexpected peaks — 341
  - Disappearing peaks — 348
  - Negative peaks — 349
  - Diagnosing peaks: An example — 349
- Key properties of spikes — 352
  - What we know about spikes — 352
- Key properties of bumps — 352
  - What we know about bumps — 352
- Key properties of steps — 356
  - What to know about steps — 356
- Epilog — 359
  - Last words — 359
  - Chromatogram troubleshooting — 359
- Knowledge Gained — 359

Did you get it?	361
Self-assessment quiz: SAQ 13	361
References	362
Further reading	362
Vendor troubleshooting guides	362
Cited	363
Tables	363
Figures	363
New technical terms	364
Answers to self-assessment questions	365
Subject index of SCI-FILEs	378
Glossary of terms	379
Index	417

Contributors

An international team of expert chromatographers has peer reviewed the technical content of this text. This **Editorial Advisory Board** comprised the experienced analyzer engineers listed below. Together with the author, this team has accumulated **487 years** of practical experience with process gas chromatographs. We gratefully acknowledge their contributions.

Massimo Baldizzone, PhD

Area Market Manager
ABB Measurement
and Analytics
Milan, Italy.

Massimo holds both Bachelor's and Master's degrees in Chemistry, as well as a PhD in the same field. Additionally, he has earned an Executive Master in Marketing and Sales (EMMS) from Bocconi and ESADE School of Management. Throughout his career, he served as Technical Sales Support for ABB Lewisburg Process Analytics. Currently, he holds the position of Area Market Manager for the Process Gas Chromatographs product line at ABB Measurement and Analytics.

Massimo has 15 years of experience with process gas chromatographs.

Jerry Clemons, PhD

Process Gas
Chromatograph Consultant

Formerly, General Manager
ABB Process Analytics
Ronceverte, West
Virginia USA.

Jerry has worked with gas chromatographs during his entire career starting at Virginia Polytechnic University where he earned his PhD with Dr Harold McNair.

He has held many engineering and management positions at ABB Process Analytics and its predecessors, always focused on their process gas chromatographs. Now retired from active duty, Jerry continues to provide his technical expertise as a consultant to that company.

Jerry has 50 years of experience working with process gas chromatographs.

R. Aaron Eidt, BSc

Process Analyzer Consultant
PEAK PERFORMANCE
Analytical Consulting Ltd.
Delta, British Columbia, Canada.

Formerly, Analyzer and PGC
Manager Dow Chemical
Canada ULC
Fort Saskatchewan,
AB, Canada.

Aaron is a chemist with 25 years of experience developing GC methods for research and industrial chromatographs at Dow Chemical Canada. Aaron specialized in process analyzer validation, troubleshooting, and performance improvement. For several years, he led the Dow Global Process Chromatography Technology Network. Since retiring from Dow, Aaron has had process analyzer consulting engagements with the Sadara Chemical Company in Jubail, Saudi Arabia, and most currently with MEGlobal.

Aaron has 35 years of experience practicing industrial gas chromatography.

Michael Hoffman

Business Development Manager Siemens Industry, Inc.
Gas Chromatography Systems MAXUM LLC
Houston, Texas, USA

Michael started the industry in 1979 at Phillips 66, Bartlesville, OK. The journey continued with Standard Oil Chemicals, BP, Innovene, and INEOS before Siemens, soon to be Valmet.

Michael initially worked with laboratory chromatographs but transitioned to process chromatographs in the early 1980s. Primarily focusing on online analyzer reliability, the journey expanded to advanced process control, materials handling, and analyzer data management technologies. Michael joined Siemens in 2007 and is now responsible for the marketing and technical support of analytical solutions, data communications, process GC applications, and sample system designs.

Michael has 41 years of experience working with laboratory and process gas chromatographs.

Dirk Horst

Freelance Trainer and Consultant Process Analyzer and Custody Transfer Systems

Formerly Global QMI Engineer and Consultant Shell Global Solutions Team The Hague, The Netherlands

Dirk has a long experience with process analyzers, including startup assignments at Shell jobsites in India, Germany, Nigeria, and Russia. He is also well known for his many classroom and practical training programs for analyzer maintenance technicians. Dirk has 34 years of experience working with process gas chromatographs.

Junji Koyama, ME

Process Gas Chromatograph Application Specialist
YPHQ System Analyzer Center
Yokogawa Electric Corporation Mitaka,
Tokyo, Japan.

Junji-san has been a leader of the PGC application development team at Yokogawa, holding engineering and management roles in application development and column design. Additionally, Junji-san looks after Yokogawa's global application labs while also providing technical support to sales and service engineers.

Junji-san has 16 years of experience working with process gas chromatographs.

Dr. Daniel Kuehne

Process Gas Chromatograph Consultant
Head of Global GC Proposals and FEED Support
Gas Chromatography Systems MAXUM GmbH;
Part of Valmet Group
Karlsruhe, Germany.

Daniel studied Chemistry at the University of Bremen, earning his diploma and completing a doctorate thesis in Analytical Chemistry. He joined Siemens in 2005 as a method developer for process GCs. He stayed in method development for 11 years, the last 5 years as head of the PGC method developer team. In 2016, he joined the technical proposals team for PGC and worked as a technical consultant for sales and customers. Daniel has been in his current role as head of the global team for GC proposals and FEED support since 2022.

Daniel has 19 years of experience working with process gas chromatographs.

James Leonard, PhD

Senior Technical Associate
Corporate Analytical Division
Eastman Chemical
Company Kingsport,
Tennessee, USA.

James received his PhD in Analytical Chemistry from The Ohio State University. He has 24 years of experience working in the field of Process Analytics at Eastman Chemical. During this time, James has designed, installed, and commissioned analyzer systems incorporating modern on-line techniques throughout the world. He has presented lectures on process analytics at university and other organizations to promote the use of on-line technologies to improve process control and reduce waste.

James has 20 years of experience working with process gas chromatographs.

Harald Mahler

Business Development
Professional Gas
Chromatography
Systems MAXUM GmbH
Karlsruhe, Germany.

Harald Mahler studied chemistry at Reutlingen University. Since 1989, he gained experience in process analytics from various engineering and management positions within Siemens Process Analytics; these include assignments in the applications and method development laboratory, project management, and industry marketing and product management. Currently, he is Global Sales and Business Development Manager for process analytics within Gas Chromatography Systems MAXUM GmbH. Major working areas are the (petro)chemical and oil and gas industries including decarbonization markets.

Harald has presented technical papers at several international symposia and conferences in Europe and North America. He has also authored articles in technical magazines and books and acted as session chair at several conferences.

Harald has 35 years of experience working with process gas chromatographs and other process analyzers.

Gen Matsuno, ME

Manager, Analyzer
Marketing
Sensing Center
Yokogawa Electric
Corporation Mitaka,
Tokyo, Japan.

Matsuno-san was the leader of the Yokogawa GC8000 PGC development team. In addition to his experience of designing process gas chromatographs, he has five years of experience as a laboratory GC user.

Gen-san has 17 years of experience working with process gas chromatographs.

Takashi Matsuura, BE

Senior Field Engineer
Nippon Swagelok FST, Inc.
Yokohama, Japan

Formerly, Manager of
Process GC Development
Yokogawa Electric
Corporation.

Taka designed the Yokogawa GC1000 PGC oven and was the leader of the engineering team that developed the Yokogawa GC1000 Mk2 PGC. He wrote the specification for the Yokogawa GC8000 PGC.

Taka has over 25 years of experience working with process gas chromatographs.

Phillip McKay

Process Gas Analyser Specialist
Australia and Pacific Region
Support Services Manager at Integrated Analytical Systems
Service Manager at Tresco International (Aust) Pty. Ltd.

Phil came into gas chromatography with a strong electronics background. In the early 1980s, he joined Beckman Instruments as a service engineer specializing in process gas chromatographs and a wide range of process gas analyzers. He has extensive practical experience of process analyzers including sample system design and commissioning; troubleshooting and repair; and programming and data communications interfacing. Phil also teaches introductory and advanced training courses on the operation and maintenance of process gas chromatographs.

Phil has over 40 years of experience with process gas chromatographs.

Suru Patel, PhD

Process Analyzer Consultant
Patex Controls Ltd.
Calgary, Alberta, Canada.

Formerly, Distinguished Engineering Associate for Process Analyzers
Exxon Chemical Company Sarnia, Canada and Singapore.

In addition to his process analyzer engineering work, Suru developed PGC training courses for process analyzer technicians and PGC data users. Previously, for several years, Suru was a PGC Applications Engineer at Servomex Company in the UK and was the Lead Analyzer Engineer in Houston for Exxon's Singapore Chemical Complex Project. He was also the development engineer for a new flame ionization detector.

Suru has 40 years of experience working with process gas chromatographs.

Eric Schmidt, PhD

R&D Fellow
The Dow Chemical Company
Analytical Sciences
Freeport, Texas, USA.

Eric Schmidt received his PhD in Analytical Chemistry from The University of Texas at Austin. He has worked at the Dow Chemical Company in Freeport, TX, for over 25 years as a Research Scientist where he spends his time developing new on-line process measurements for R&D and manufacturing. He is currently leading the On-line Chromatography Strategic Capability Team at Dow.

Eric has 20 years of experience working with process gas chromatographs.

Ivan Rybár, PhD

Head of the Electrical and Automation Maintenance
Slovnaft Refinery MaO, a.s.
Bratislava, Slovakia.

Formerly, Research and Teaching Assistant,
Department of Analytical Chemistry Comenius University
Bratislava, Slovakia.

For 15 years, Ivan has been responsible for the reliability of process analyzers at the Slovnaft refinery, including managing the maintenance of existing systems and designing new analytical equipment. During his career, he worked on more than 120 projects. As a supervisor for process analyzers, he creates work procedures and provides consultation and support to his team.

He was awarded the accolade "Slovnaft Star" twice, in 2013 and 2015.

Previously, he worked as an analyzer engineer for several companies providing engineering services to industrial plants, including the selection of analyzers and the design of entire sampling systems.

During his time at the university, Ivan developed new methods in liquid chromatography, taught several postgraduate courses, and published four scientific papers in this field.

Ivan has 20 years of experience working with process gas chromatographs.

Preface to the first book

Welcome to the world of Process Gas Chromatography!

This book (Waters 2020) focuses on the **Process Gas Chromatograph** (PGC). There are dozens of fine books on the science of gas chromatography but few on the technology of the process instrument. I found only two previous books dedicated to online gas chromatographs (Huskings 1977; Annino and Villalobos 1992).

Process gas chromatographs are complex instruments and the people that design and operate them need special knowledge and unique skills. With that in mind, I designed the book to serve the needs of the journeyman analyzer technician, the process instrument engineer, and the process analyzer specialist.

PGC is a practical technology and this is a practical book. It's an effective classroom training manual for those currently learning the art and a handy reference manual for those already practicing it.

Chapters are deliberately compact, suitable for a weekly reading program or as focused lessons in an educational course. Each chapter ends with a summary of knowledge gained and a self-assessment quiz with answers provided. In addition, there are nine optional test questions for students; three easy, three moderate, and three challenging.

Why is such a book necessary?

Anyone working in the fluid processing industries knows that their knowledge base is in full flight. Due to staffing reductions and mass retirements, our industry is losing decades of hard-won experience.

Walter Jennings and Colin Poole recently expressed this situation rather well (Jennings and Poole 2012, 16):

> *This [automation of gas chromatographs] has led to a continuing decline in the expertise of the average practicing chromatographer from the mid-1980s to the present time. This can be perilous, because everything from column selection to trouble-shooting skills is based on a fundamental knowledge of chromatographic principles, the absence of which degrades the quality and usefulness of the information acquired by these instruments. To address these problems requires a massive educational effort before the knowledge is lost and the usefulness of gas chromatography to decision makers is called into question.*

There can be no clearer call to justify this book. While the authors were writing to laboratory chemists, those working on process gas chromatographs also need a *fundamental knowledge of chromatographic principles* presented in a way that facilitates a *massive educational effort*. This book sets out to satisfy those needs. It's primarily written for process analyzer engineers and technicians but should be helpful to anyone using or maintaining a process gas chromatograph.

To succeed in its mission, a book needs to so excite the reader that they want to read more. It should be so useful that they immediately return to it when they need information. Yet the average book on gas chromatography is abysmally boring and poses an intellectual challenge even to post-doctoral scientists, let alone the lonely guy faced with fixing a broken process chromatograph at midnight.

This text teaches the fundamental knowledge of process gas chromatography by encouraging the reader to think critically about what is happening in the instrument, mostly without recourse to analogy or math. It also describes some practical procedures for design or troubleshooting.

So, here you have it. A clear yet detailed book that is ideal for classroom instruction, private study, or distance learning. Focused chapters unfold the technology of a process gas chromatograph to an engineer or technician who may have no previous experience of the technique. The content is basic, yet thorough, so it should meet the needs of many readers.

I'm glad that you're here. I hope you enjoy the book!

Tony Waters
Atascadero, California
January 2020

Preface to the second book

Welcome back to the world of Process Gas Chromatography!

This second book about process gas chromatographs picks up from its predecessor, going deeper into PGC application and care. As such, companion volume to the first book takes off where the first one landed. The focus here is on the columns and the chromatograms they generate. We take a detailed look at how to design chromatographic columns, how to make them work together to speed analysis, and how to diagnose symptoms that presage trouble.

There's more than enough theory here, but it's mostly set out in optional SCI-FILES for those who need to know the mathematical foundation for the practical procedures given. This compilation of theoretical and practical knowledge is a highly focused resource for applications engineers who design column systems for industrial applications, for service engineers who maintain them, and for anyone else working with gas chromatographic columns. Analytical chemists in industrial laboratories, pilot plants, and academia will appreciate the detailed information on column design and the automation of routine chromatographic analyses.

Unless you're very familiar with process gas chromatographs, you'll want to own both books. They work together to provide a complete knowledge of the technology without reference to the subtleties of the brand. This second book is designed not only to support an advanced-level course in Process Gas Chromatography but also to serve as a handy reference to anyone needing a quick answer to an immediate problem. It is a great reference source to have by your side when you need it.

Enjoy your work with process gas chromatographs!

Tony Waters
Atascadero, California
August 2024

References Cited

Annino, R., and Villalobos, R. (1992). *Process Gas Chromatography*. Research Triangle Park, NC: Instrument Society of America. ISBN: 1-55617-272-9.

Huskings, D.J. (1977). *Gas Chromatographs as Industrial Process Analyzers*. New York, NY: Pergamon Press, Inc.

Jennings, W.G. and Poole, C.F. (2012). Milestones in the development of gas chromatography. In: *Gas Chromatography* (ed. C.F. Poole), 1–17. Oxford, UK: Elsevier. ISBN: 978-0-12-385540-4.

Waters, T. (2020) *Process Gas Chromatographs: Fundamentals, Design and Implementation*. Chichester, UK: John Wiley & Sons Ltd.

Waters, S.E. (2023). *Letter to the President, Princeton Theological Seminary*. Princeton, NJ.

Acknowledgments

Hearty thanks to friends and associates who contributed material to this text. Their valuable contributions of time and knowledge are much appreciated.

Andy Evans
Analyzer Engineer
Dow Corning
Swansea, Wales, UK.

Jay Brown
General Manager
Ohio Valley Specialty
Company Marietta,
Ohio USA.

Daren Brumley
Senior Technical Instructor
ABB Process Analytics
Skiatook, Oklahoma
USA.

Damian Huff
Senior PGC Applications Engineer
ABB Process Analytics
Bartlesville,
Oklahoma USA.

Andrzej Jopek
Director
Process Analytical Middle East WLL
Bahrain.

Fusan Karaburun
Technical Product Owner
Qmicron B.V.
Enschede,
Netherlands.

Dorothy Kwidama
Technical Product Specialist
Qmicron B.V.
Enschede,
Netherlands.

Bert Laan
Process Analyser Consultant
Olpass Ltd
Hereford,
England UK.

Ryan McSherry
Analytical Solutions
Yokogawa Corporation of America
Newnan,
Georgia USA.

Nicholas Meyer
Manager—Industry and Product Marketing
Yokogawa Corporation of America
Newnan,
Georgia USA.

Martin Pijl
Product Manager Gas Analyzers
Yokogawa Europe B.V.
Amersfoort, Netherlands.

Wouter Pronk
Business Development & Trainer
B.E.S.T. Fluidsysteme GmbH Neuss,
Germany.

Asad Tahir
Director of Product Management
Emerson Process Management
Houston, Texas
USA.

John Wasson
President
Wasson-ECE Instrumentation
Fort Collins,
Colorado USA.

Henk van Well
Manager Analytisch Technischer Service
INEOS Manufacturing Deutschland GmbH Köln,
Deutschland.

Dr. Martin Wieser
Process Analyzer Engineer
BASF SE
Ludwigshafen,
Germany.

A Modern Process Gas Chromatograph.
Source: Reproduced with permission from Yokogawa Electric Corporation.

1

Fundamentals

"This is our second book on the process gas chromatograph. Written as a companion volume, this book doesn't supersede its predecessor; it builds on that firm foundation to examine the special nature of process gas chromatography rather than the process instrument itself."

Introduction

Academic theories of gas chromatography have been hugely successful in explaining the chromatographic process and predicting ways to improve its performance. Many books and scientific papers are available to those looking for a rigorous dissertation on the subject, but they are often inscrutable. This book takes a pragmatic approach without disregarding well-established theory. As optional reading, separate **SCI-FILEs** located at key moments throughout the book capture a simplified and accessible explanation of the relevant science. There's a list of these technical asides in Appendix B.

We assume the reader has a good understanding of PGC hardware gained from practical experience or from studying our previous book. That said, this chapter has the simple goal of introducing some of the concepts and terminology used herein.

Gas chromatography

A unique analytical technique

Gas chromatography[1] is a powerful method of chemical analysis that has been adapted for use in industrial processes to measure the concentration of selected chemical substances in a small sample of the process fluid.

---

[1] The bold font used in the text indicates a technical term further defined in the Glossary of terms (pp 379–416).

Process Gas Chromatography: Advanced Design and Troubleshooting, First Edition. Tony Waters.
© 2025 John Wiley & Sons Ltd. Published 2025 by John Wiley & Sons Ltd.

The chemicals present in the sample are known as **components**. Of these, the ones actually measured are the **analytes**. An analyte is usually a single chemical substance but sometimes can be several components measured together as a group.

Unlike other process stream analyzers, a PGC doesn't try to measure a desired analyte in the presence of the other substances in the sample. It's always difficult to do that. Instead, it separates each target analyte from everything else and then measures it independently. The two-stage procedure of separation-before-measurement is why the PGC is so versatile. It allows a PGC to measure almost anything in any process gas or volatile liquid.

A basic gas chromatograph

Figure 1.1 shows the main parts of a basic gas chromatograph. To start an analysis, the sample valve injects a discrete sample of the process fluid into a stream of inert **carrier gas**, typically hydrogen, helium, or nitrogen. The injected sample can be a gas or a volatile liquid that immediately vaporizes. The flowing carrier gas immediately carries the vapor sample into the **column**, a narrow tube containing an active solid or an immobilized liquid. As they pass through the column, the analytes separate from the other components and from each other, arriving at the end of the column at different times. In practice, most PGCs need two or more columns to fully separate the analytes. Then, each analyte passes in turn through a **detector**, which senses its presence and outputs an electronic signal proportional to its concentration in the original sample.

When expressed in more formal terminology:

➢ The carrier gas is the **mobile phase**.
   It's the driving force that powers the chromatography.

➢ The active solid or immobilized liquid is the **stationary phase**.
   It's the agent that creates the separation.

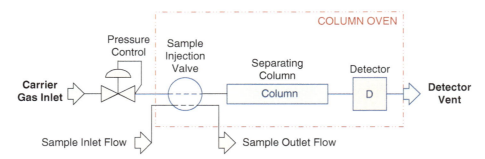

Figure 1.1 A Basic Gas Chromatograph.

When actuated, the sample injection valve transfers a miniscule aliquot of the sample fluid into the flowing carrier gas. The carrier transports the injected sample though one or more columns where separation of the desired analytes occurs. The detector senses the component molecules exiting the columns and generates an electronic measurement signal.

A mobile phase moving in contact with a stationary phase is the essence of chromatography. In gas chromatography, the mobile phase and the sample (after injection) are both gases. In another kind of chromatography, the mobile phase and the sample are both liquids. That's **liquid chromatography**, a technique rarely used in process analyzers and not further discussed herein.

The columns

As the carrier gas carries the analytes through the column, they experience an affinity for the stationary phase which delays their progress more or less than the other components. A careful choice of stationary phases can combine these delays to accomplish a complete time-separation of the desired analytes.

PGC **column systems** are custom designed to isolate the desired analytes from all the other components in the sample. An applications engineer chooses the stationary phases and an arrangement of columns to accomplish that. The special techniques of column design, optimization, and troubleshooting are the core subjects of this book.

Gas chromatographs employ two types of column:

➤ A **packed column** is typically a ⅛″ or ¹⁄₁₆″ o.d. tube closely packed with small particles. In a solid-phase column the particles themselves are the adsorbent solid that performs the separation. In a liquid-phase column the particles are inert but they have a thin coating of the selected liquid phase.

➤ A **capillary column** is typically a long 0.53 mm (or smaller) i.d. tube that has an internal wall coating. Most capillary columns have the selected liquid phase coated or chemically bonded to their inner walls. A few have a finely powdered solid adsorbent on their walls.

The detector

After a predictable delay, each analyte emerges from the column system as pure vapor diluted with carrier gas, an easy mixture to measure. The carrier gas carries it into a detector which responds with an electronic signal to the data processor.

As each component emerges from the column system the detector signal gradually increases from its quiescent **baseline** level, transits through a maximum, and then decays back to the baseline forming a nearly symmetrical **peak** shape over time. Analyte peaks are typically between 1 and 10 seconds wide and separated in time from all other peaks.

It's normal to refer to the bands of solute molecules migrating along the column as "peaks" even before they reach a detector.

A PGC uses the elapsed time between sample injection and peak apex to identify each analyte peak. It also measures the area of each analyte peak to determine its concentration in the original sample. The concentration is usually a linear function of the peak area, a relationship established by calibration.

The chromatogram

The detector signal viewed as a function of time is a **chromatogram**. The chromatogram contains all the analytic information generated by a PGC and a display or printout is indispensable for design, maintenance, or troubleshooting work. In normal operation, though, the PGC processes the detector signal automatically and displays a visible chromatogram only on demand.

Figure 1.2 shows a chromatogram as displayed on a typical PGC. This one shows the separation of small amounts of aromatics[2] and higher paraffins in natural gas. The C_{11} and higher components are backflushed[3] and elute first as a partly regrouped composite peak.

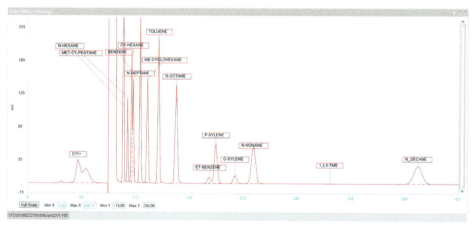

Figure 1.2 A *Typical* PGC Chromatogram.
Source: Reproduced with permission from ABB, Inc.

In this chromatogram to measure impurities in natural gas, the C_{11} and later components are backflushed and elute first as a "$C_{11}+$" composite peak. The dashed lines show the limits of peak area measured by the PGC processor. The cyclohexane peak has not fully separated from the benzene peak, so the processor drops a perpendicular between them to get a good estimate of their individual concentrations.

The science of GC

The theory of gas chromatography is complex and its technology intricate, but once you get past all the equations, the simplicity of the core science may come as a shock. We introduce it here and define some of the terms we use.

---

[2] The Glossary of terms (pp 379–416) explains the names of many chemicals mentioned in the text.
[3] See Chapter 10 for a detailed explanation of backflush.

The basic science

The two forms of gas chromatography use different stationary phases, solid or liquid. Their mechanisms of separation are different. Solid phases delay components by adsorbing their molecules onto an active surface, whereas liquid phases delay molecules by dissolving them. Yet the result is much the same: the molecules can't move while dissolved or adsorbed by the stationary phase. Solid phases have a strong affinity for gas molecules, which makes them suitable for separating low-density gases such as hydrogen, oxygen, nitrogen, carbon oxides, and lighter hydrocarbons. Liquid-phase columns separate most other components, so the majority of PGC columns have liquid stationary phases. Let's see how those work.

When in the vapor phase most substances will dissolve in a liquid, but to different degrees. Two technical terms apply to the resulting solution:

- The substance dissolving in the liquid is the **solute**.
  It's a general term for any component of the analyzed sample retained by the stationary phase.

- The liquid dissolving the solute is the **solvent**.
  We usually just call it the liquid phase.

We rarely use the word *solvent* in process gas chromatography because it can be confusing. The confusion arises from the common laboratory practice of dissolving samples in a solvent before injecting them into a gas chromatograph. This creates a huge **solvent peak** on the chromatogram. PGCs don't dilute their samples, so solvent peaks don't occur.

Let's examine what happens when a solute contacts a stationary phase. When a small amount of sample gas enters an enclosed space containing a gas and a liquid, solute molecules distribute themselves between the gas phase and the liquid phase. This happens because molecules possess kinetic energy and move rapidly and randomly in every direction. Their motion brings them into contact with the liquid surface where some of them dissolve. Then, as their number in the liquid phase increases, they start to reenter the gas phase. This rapid exchange inexorably leads to a situation where their rates of arrival and departure at the liquid surface become equal, forming a **dynamic equilibrium**. It then seems like nothing is happening because the number of solute molecules in each phase remains constant. Yet solute molecules continue to move rapidly between the phases.

At equilibrium, the solute concentration in the gas phase and the solute concentration in the liquid phase are both constant, but not identical. Their values depend only on the temperature and the inherent **solubility** of the solute in the selected liquid.

Then:

- The ratio of solute concentration in the liquid phase to solute concentration in the gas phase is the **distribution constant**.

- Increasing the temperature reduces the distribution constant, and vice versa.

The distribution constant[4] is the fundamental science of chromatography. In a column, it determines the number of dissolved solute molecules left behind when the gas phase moves, carrying with it the undissolved solute molecules. We shall see that the position of each peak on the chromatogram is a simple function of its distribution constant.

The science of chromatography comes down to this simple concept: *gases dissolve in liquids to different extents.*

That's it. Chromatographic separation occurs because different sample components each have their own unique solubility in the liquid phase. As the carrier gas moves, the more soluble solute molecules tend to stay in the liquid phase, delaying them longer than those having lower solubility. When their solubility difference is sufficient the solutes separate from each other.

> *No chemistry is involved. Interaction with the stationary phase is a simple physical process of dissolving in a liquid or adsorbing onto a solid surface.*

If that was all, this book would be very short and gas chromatography would be simple, as indeed it was in the early days. Back then, two incompatible needs pulled process and laboratory chromatography apart, dragging them both into realms of increased complexity:

> In the laboratory, sample composition is often unknown, so there's a need to separate *and identify* many components – sometimes hundreds of them. The eventual solution was to use a capillary column and continually raise its temperature during analysis to separate dozens of components, a procedure known as **temperature programming**. Then, a mass spectrometric detector can identify and measure the plethora of peaks.

> In the process plant, the stream compositions are known and the PGC measures only a few designated analytes. Barring simple mistakes, there's rarely an issue with peak identity but the process control systems need fast results – sometimes in less than one minute. The eventual solution for the fastest separation was to use multiple columns and special valves to direct each component peak into the desired column.

The employment of multiple columns is a distinctive hallmark of process gas chromatographs. Although a few multiple-column arrangements have migrated into the laboratory, it's still uncommon to find more than one column in a laboratory gas chromatograph. In contrast, virtually all PGCs use multiple columns to separate and measure a few known analytes in the shortest time with the highest reliability and lowest cost.

Temperature programming is a distinctive hallmark of laboratory gas chromatographs as most use it to separate and identify large numbers of analytes. The technique is available in modern PGCs but when designed for

---
[4] For a detailed discussion on the distribution constant see the **SCI-FILE** *On Distribution* in Chapter 5, pages 119–148.

autonomous operation in hazardous environments it tends to be costly and demand more maintenance than an isothermal analyzer. Therefore, PGCs employ temperature programming only when it confers a significant advantage.

The gas chromatograph

The basic instrument

A **gas chromatograph** is an analytical instrument that uses the techniques of gas chromatography to measure the concentration of selected chemical compounds in a small sample containing a mixture of compounds.

The essential hardware devices found in any gas chromatograph include:

- One or more temperature-controlled zones.
- A carrier gas supply and pressure control system.
- A sample injector that injects a repeatable volume of sample into the flowing carrier gas.
- One or more separating columns.
- One or more detectors.
- A processor to control operations and calculate results.

While all gas chromatographs have these basic functions, there are large variations in their design and fabrication.

The process instrument

A gas chromatograph working at an industrial processing plant is strikingly different from a gas chromatograph sitting on a laboratory bench. The main reasons for these differences are:

- The PGC operates in a potentially hot, cold, dusty, wet, windy, corrosive, or hazardous environment.
- The PGC operates continuously 24 hours per day, 7 days per week.
- The PGC must operate reliably with almost no human intervention – perhaps only one calibration check each month.
- The PGC applications engineer knows in advance the components and concentrations expected in the sample.
- The PGC can focus on measuring just a few of the components in a sample – the ones needed for process control.
- The PGC suffers a fanatical quest to minimize the analysis time, to ensure that its measurements are valid for process control.

For all the above reasons, a process gas chromatograph may include some devices and properties not shown in Figure 1.2, such as:

> External devices to condition the incoming process sample to make it compatible with the PGC; that is, a **sample conditioning** system.

> Multiple columns with special devices to switch analyte molecules from one column to another, thus maximizing the rate that separated components arrive at the detector. Laboratory instruments rarely need this additional complexity.

> **Housekeeping columns** that allow strongly retained components to quickly exit the column system. These are also becoming popular in laboratory chromatographs used frequently for routine analyses. Those used only a few times each day have plenty of time to recover between sample injections.

> Robust column systems and stable devices, all designed to operate for a long time without adjustment.

> Hardened microprocessor systems to capture and process the detector signal and to schedule timed events.

> Self-diagnostics and alarm generation and optional software for centralized maintenance.

> Optional applications software for autonomous validation and statistical quality control.

> Protected and suitable for continuous operation in a dirty, corrosive, and potentially hazardous environment.

> Housed in an analyzer shelter or building to protect the analyzers and the workers from the plant environment.

For further information, refer to our companion book on the hardware of process gas chromatographs (Waters 2020). The present volume focuses on the design, optimization, and troubleshooting of PGC column systems.

The contents

Here's a brief review of the information this book contains.

Chapter 2 illustrates a few simple chromatogram measurements we can make to evaluate and optimize column performance.

When looking at a chromatogram an inexperienced observer sees a range of mountain peaks on a flat plain. There's so much more to see. From a few simple measurements we learn what's going on in the columns and find ways to optimize their performance. The chapter also introduces the use of plate theory to explain and then quantify peak shape, separation, and resolution.

Chapter 3 is a detailed review of PGC column technology, comparing and contrasting the structure and properties of the different kinds of columns.

This practical chapter describes how packed and capillary columns are made and how they work. It examines in fine detail the molecular forces involved in various mechanisms of separation and defines what we mean by the *affinity* between a solute and a stationary phase.

Chapter 4 examines the choice of PGC stationary phases and tabulates the properties of many solid and liquid phases.

This is a detailed review of the liquid and solid stationary phases typically used in process gas chromatography. It provides practical guidance on the use of Kovâts indexes and McReynolds constants for selecting a suitable liquid phase and gives the polarity factors for many liquids.

Chapter 5 recognizes the need to tune the stationary phase to deliver the best resolution between analyte peaks in the shortest time.

It expounds the theories of distribution and resolution and examines the practical consequences, deducing the optimum conditions for maximum resolution of peaks. This leads into a practical example that shows how to obtain improved resolution in less analysis time.

Chapter 6 introduces the rate theory of gas chromatography and explains its use for optimizing the operating conditions of a single column.

You'll learn how to determine the optimum carrier gas flow rate and why it's not used in process chromatographs. A practical example illustrates the optimization of flow and column length, achieving the same or better resolution in less time by using a longer column!

Chapter 7 discusses the causes and remedies for asymmetric peaks.

Real peaks are slightly skewed, they rise to their apex a little faster than they decline to their baseline. This minor asymmetry isn't a problem. But some peaks are grossly misshapen and we need to know why. Here, we discover the reasons for tailing and fronting (or "leading") peaks and what we can do to minimize the problem: the handy troubleshooting table given here might be helpful.

Chapter 8 identifies causes for peak broadening in the analyte flow paths external to the column.

Peaks become wider as they travel through devices such as the sample injector, column valve, detector, or even connecting tubing. If significant, this broadening can spoil a perfect resolution. The text gives quantitative methods for evaluating these effects and suggests some ways to minimize their influence on column performance.

Chapter 9 examines the effect of the pressure gradient in series columns.

Most process gas chromatographs employ multiple column arrangements to achieve the necessary separation in the shortest possible time and this results in columns working at different pressures. While the mass flow rate is constant within columns in series, the volumetric flow varies with the pressure. The chapter develops quantitative methods for determining the retention times of peaks in each column and gives some examples for the calculation of column lengths.

Chapter 10 examines the backflush column system in detail, giving both theoretical and practical outcomes.

Neary all PGC column systems include a backflush function, so it's essential to understand its function and its quirks. The text describes backflush column systems using valve and valveless arrangements, then looks at why practical outcomes differ from the theoretical ideal. The discussion includes the optimization of column lengths.

Chapter 11 examines the heartcut column system in detail, giving both theoretical and practical outcomes.

PGCs always use the heartcut function to measure low concentrations of impurities in otherwise pure samples. The chapter explains why heartcut is necessary and then describes heartcut column systems that use both valve and valveless column switching. The discussion includes the column design features necessary for successful operation of single and multiple heartcuts.

Chapter 12 describes useful troubleshooting techniques for diagnosing baseline problems.

When only carrier gas is emerging from the column, the chromatogram should display a flat baseline with no change over time. If not, any visible disturbance is a potential symptom of trouble. The discussion includes diagnostic techniques for finding the root cause of drifting, wandering, cycling, or noisy baselines. The extensive troubleshooting tables may be helpful.

Chapter 13 describes useful troubleshooting techniques for diagnosing chromatographic problems.

After obtaining a flat and smooth baseline any symptoms that remain must be due to the cyclic operation of the PGC analysis. The often subtle symptoms of incipient failure may escape the notice of a casual observer but are bright red flags to an experienced troubleshooter. They include unexpected peaks, spikes, bumps, or steps in the baseline. The extensive troubleshooting tables may be helpful.

The Glossary of terms is an extensive dictionary of the technical terms and expressions used in gas chromatography and a useful resource. It also includes the name, formula, and molar mass of many chemicals found on chromatograms. If you're uncertain about the meaning of an English word or phrase, look it up in the Glossary!

Becoming a PGC expert

In contrast to holistic methods of analysis, measuring an isolated analyte is simple and rarely troublesome. But the chromatographic process used to secure that isolation is not so easy to understand. It's often arcane and hard to grasp. We intend to demystify that process by describing many practical ways to evaluate and optimize column performance or diagnose column malfunction.

It turns out that most of the evidence you need is plainly visible on the chromatogram and the secret skill of the expert is simply knowing how to decipher the information it contains.

The secret skill of the PGC expert is their ability to read a chromatogram and understand what it's telling them.

The chromatogram readout is a vital design and troubleshooting tool; so much so that it's difficult to overstate its usefulness. Discounting calibration errors and electrical failures that are easy to fix, all faults are visible on a chromatogram, either directly or by comparing the current chromatogram with a previous one.

If you aspire to be an expert PGC applications engineer or troubleshooter, learn to read the chromatogram. You'll find the knowledge you need on the following pages.

Knowledge Gained

- *Chromatography uses a fluid mobile phase passing over a liquid or solid stationary phase.*

- *In gas chromatography, the mobile phase is gas; in liquid chromatography, the mobile phase is liquid.*

- *Chromatographic analyzers separate the desired analytes and then measure them one by one.*

- *Other analyzers attempt to measure the analyte molecules in the presence of other molecules.*

- *Gas chromatographs inject a tiny volume of sample into the flowing carrier gas.*

- *The sample must be a gas or a volatile liquid that quickly vaporizes and enters the column as a vapor.*

- *Process gas chromatographs (PGC) use an automatic injection valve to inject the sample.*

- *The carrier gas carries the sample into the column where it contacts the stationary phase.*

- *The stationary phase may be a solid adsorbent or an immobilized nonvolatile liquid.*

- *Contact with the stationary phase delays some peaks more than others, so separation occurs.*

- *Special routing devices may direct the peaks into different columns to finish the desired separation.*

- *PGCs can use either packed columns or capillary (open tubular) columns.*

- *The separation process takes time; typically 1 to 10 min, sometimes longer.*

- *The carrier gas elutes peaks from the column into a chosen detector for measurement.*

- *The detector responds to a property of analyte molecules that differs from carrier gas molecules.*

- *The detector output signal forms a chromatogram display when plotted against elapsed time.*

- *Analyte molecules cluster together at different times to form separate chromatogram peaks.*

- *The PGC typically measures peak area to compute the concentration of an analyte.*

- *The chromatogram is a most valuable source of information, but one must learn to read it.*

- *To the expert troubleshooter, all chromatographic faults are visible on the chromatograms.*

Did You Get It?

Self-assessment quiz: SAQ 01

Q1. In gas–liquid chromatography, what are the gas and liquid doing?
Select the one correct answer:
- **A.** They are both moving.
- **B.** The gas is moving and the liquid is stationary.
- **C.** The gas is stationary and the liquid is moving.
- **D.** They are both stationary.

Q2. In a gas chromatograph, which of the gases listed below would be suitable as the carrier gas?
Select all the correct answers:
- **A.** Oxygen
- **B.** Nitrogen
- **C.** Hydrogen
- **D.** Helium

Q3. In a gas–liquid chromatograph, which one of the materials listed below might be the stationary phase?
Select the one correct answer:
- **A.** An inert gas
- **B.** A granular adsorbent solid
- **C.** A volatile liquid
- **D.** A nonvolatile liquid

Q4. In a gas–liquid chromatograph, what really causes the separation?
Select the one best answer:
- **A.** The carrier gas causes the separation.
- **B.** The mobile phase causes the separation.
- **C.** The stationary phase causes the separation.
- **D.** The chromatogram causes the separation.

Q5. In a gas chromatograph, what do the columns do?
Select the one best answer:
- **A.** They convert all the components into peaks.
- **B.** They separate the analytes from all other components and from each other.
- **C.** They separate all the components of the sample.
- **D.** They allow only measured components to enter each detector.

Q6. In a gas chromatograph, what does a detector do?
Select the one best answer:
- **A.** It provides a continuous flat baseline as a reference for measuring the peaks.
- **B.** It generates a signal proportional to the instantaneous number of component molecules leaving the column.
- **C.** It measures the area of each peak.
- **D.** It converts each component of the sample to a concentration.

Q7. Why is the chromatogram so important?
Select the one best answer:
- **A.** It shows the baseline used for measuring the peaks.
- **B.** It shows the shape of each peak.
- **C.** It shows the separation between peaks.
- **D.** All of the above.

Check your SAQ answers with those given at the end of the book.

References

Further reading

The companion book (Waters 2020) establishes the basic functions and hardware of the PGC and is essential reading to those who desire a full understanding of the equipment.

The 26-page encyclopedia article by Clemons (2011) is an excellent review of process gas chromatographs and their applications in industry. It's highly recommended for those seeking a wide overview of the techniques and applications of the PGC.

The delightful book by Ettre (2008) is an interesting historical account of the development of chromatographic science by one who was there.

The two earlier books on process gas chromatographs by Huskings (1977) and Annino and Villalobos (1992) may be of interest for their historical perspectives.

Cited

Annino, R. and Villalobos, R. (1992). *Process Gas Chromatography: Fundamentals and Applications*. Research Triangle Park, NC: Instrument Society of America. ISBN: 1-55617-272-9.

Clemons, J.M. (2011). Chromatography in process analysis. In: *Encyclopedia of Analytical Chemistry* (ed. R.A. Meyers). Chichester, UK: John Wiley & Sons Ltd.

Ettre, L.S. (2008). *Chapters in the Evolution of Chromatography* (ed. J.V. Hinshaw). London, UK: Imperial College Press. ISBN 1-86094-943-6.

Huskings, D.J. (1977). *Gas Chromatographs as industrial process analyzers*. New York, NY: Pergamon Press, Inc.

Waters, T. (2020). *Process Gas Chromatographs: Fundamentals, Design and Implementation*. Chichester, UK: John Wiley & Sons Ltd.

Figures

1.1 A Basic Gas Chromatograph.
1.2 A *Typical* PGC Chromatogram.

New technical terms

When first introduced, these words and phrases were in bold type. You should now know the meaning of these technical terms. If still in doubt, consult the Glossary at the end of the book:

analyte	**housekeeping column**
baseline	**liquid chromatography**
capillary column	**mobile phase**
carrier gas	**packed column**
chromatogram	**peak**
column	**sample conditioning**
column system	**solubility**

component
detector
distribution constant
dynamic equilibrium
gas chromatograph
gas chromatography

solute
solvent
solvent peak
stationary phase
temperature programming

2

Measures of perfection

"Never underestimate the chromatogram. The information you need to evaluate column performance or diagnose faults is all hiding there in plain sight. You just need to read it. This chapter examines some ways to evaluate column efficiency. Later, we'll use the chromatogram to diagnose a host of instrument faults. It's the best troubleshooting tool you'll ever have."

Performance indicators

This chapter introduces some useful column performance indicators derived from two simple chromatogram measurements. There are lots of equations here, but the math is not difficult.

The main text discusses the five equations you will need in your PGC toolbox. These are the sharp implements of practical chromatography. They are simple ratios of chromatogram measurements that are easy to calculate and tell you how well your columns are doing: *retention factor*, *separation factor*, and *resolution*. And after a little theory, *plate number* and *plate height*.

The theory behind this practical work is in the **SCI-FILE:** *On Plate Theory* and is optional reading.

Since the majority of PGC columns have a liquid stationary phase, we use gas–liquid chromatography as a common example of theory and practice. Solid stationary phases work in a different way, but the outcome is much the same. A later chapter will examine the properties of solid adsorbent columns.

Also, to keep it simple at this point, we start by looking at a single column. Or maybe it's several columns in series that use the same stationary phase and act like a single column. It's a little more difficult to assess the performance of series columns that use different stationary phases, so we defer that discussion until later.

With those disclaimers, let's get back to the key question: How can you tell whether your column is working perfectly – or not? It will take just two easy chromatogram measurements to find out.

Process Gas Chromatography: Advanced Design and Troubleshooting, First Edition. Tony Waters.
© 2025 John Wiley & Sons Ltd. Published 2025 by John Wiley & Sons Ltd.

Chromatogram data

Two fundamental measurements

Start by measuring the position and width of a peak on a chromatogram. That's all the data we need to evaluate column performance.

> *The theory of chromatography is quite complex, so it's good to remember that the only real data is the measurement of peak position and peak width on a chromatogram. There are reams of theory to explain why peaks are where they are and as wide as they are, but from a practical perspective all you need to do is measure them.*

A microcomputer PGC may calculate peak measurements automatically, and we'll certainly take advantage of that luxury when troubleshooting. But you'll need to know what your computer is doing, and the best way to learn is by doing it manually.

To prepare for a manual measurement, obtain a printout or chart record of your chromatogram. Then draw tangents at both sides of the peak and extend them downward to intersect an extended baseline drawn across the base of the peak, as shown in Figure 2.1. Chromatographers call this procedure **triangulating** the peak.

Then, make the following measurements:

> Measure the **holdup time** (t_M) from the injection time mark to the apex of the air peak.

> Measure the **retention time** (t_R) of each component peak from the injection time mark to the intersection of its tangent lines.

> Measure the **base width** (w_b) of each component peak between the intersections of the tangent lines with the extended baseline.

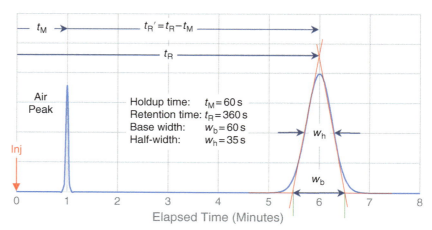

Figure 2.1 Typical Chromatogram Measurements.

You can make simple measurements from a chromatogram to discover how well the columns are performing. The data most often collected are shown in this illustrative chromatogram, and are further discussed in the text. Obtaining such data is an essential prerequisite for optimizing or troubleshooting column performance.

Some chromatographers prefer to measure the peak **width at half height** (w_h). Simply measure the horizontal peak width at an elevation of half the peak height. This is often easier to do than triangulating the base width and is perhaps a more accurate method, since no triangulation errors occur. The two width measurements are related and either of them is valid for evaluating column performance.

For clarity of display, the single component peak in Figure 2.1 is very wide. You may have many component peaks on your chromatogram, and they are likely much narrower. It's difficult to measure the width of a narrow peak. To obtain a good measurement, you might have to expand the time base on a computer display or increase the chart speed on a recorder.

Most workers like to make chromatogram measurements in millimeters, since it's a convenient way to get a precise measurement on a chart or printout. It works well in practice because most performance parameters are a ratio of two values and the units cancel out. Yet it's essential to remember that chromatogram measurements of peak position and peak width represent time durations, whether you express them in millimeters, chart divisions, digital counts, or seconds.

The distinction is important because some column variables are true lengths measured in real millimeters. On a chromatogram, a measurement in millimeters is not really a length; it's just a surrogate for time. If you mix the two units of measure, your calculations won't work.

Significance of the air peak

The **air peak** is a valuable indicator of liquid-phase column performance, but not applicable to solid-phase columns. Since air doesn't dissolve in column liquids to any significant extent, the air peak remains in the gas phase all the way through the column – traveling at full gas velocity.

If using a detector that doesn't respond to air, another **unretained peak** can act as a surrogate. For instance, methane often serves as an adequate "air peak" on a flame ionization detector.

The significance of the air peak is that it travels at the same speed as the carrier gas. Its position on the chromatogram tells us the elapsed time for the carrier gas to travel from one end of the column to the other end. It also gives us a way to measure the average carrier gas velocity in the column, an important variable used in optimization (Chapter 6).

Time in each phase

The holdup time is the time that each component molecule spends in the gas phase. Recall that component molecules can move along the column only when they are in the carrier gas. So, to get through the column, every component molecule must spend the same time in the carrier gas as the air peak does.

18 Measures of perfection

> *Each sample molecule that passes through a column spends the same amount of time in the carrier gas as every other sample molecule does – whatever its chemical composition!*

After the air peak, the additional retention time taken by a component molecule to reach the detector is the time it was stationary in the liquid phase. By random chance, identical molecules don't stay in the liquid for identical times; they reach the detector at different times. This is the root cause of peak shape. The time measured from the air peak to the apex of a component peak is therefore the average time[1] those molecules spent in the liquid phase, not moving. We call that the **adjusted retention time**.

Thus, the retention time (t_R) of a peak is the sum of two times: the time that the peak spends in the gas phase (t_M) plus the time it spends in the liquid phase (t_R'):

$$t_R = t_M + t_R' \tag{2.1}$$

Figure 2.2 illustrates the point rather well; the air peak neatly divides the chromatogram into two independent time zones and makes them easy to measure:

➤ From injection to the air peak is the time that all components spend in the *gas phase*. Some workers call this time the "dead time" of the column because it doesn't contribute to separation. It's formal name is the holdup time.

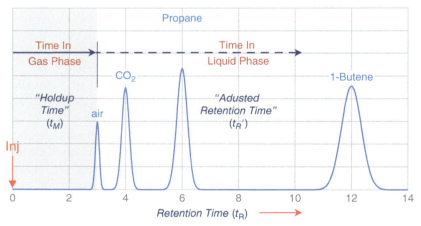

Figure 2.2 Significance of an Air Peak.

An air peak tells us how long the carrier gas takes to get through the column, which is also the time that each component spends in the gas phase. The additional retention time for each peak is the time that the peak spends in the liquid phase. For example, on this chromatogram, the propane peak spends three minutes in the gas phase and three minutes in the liquid phase for a total retention time of six minutes.

---

[1] This assumes the peak is symmetric, which is close enough for our present discussion. Real peaks rarely enjoy perfect symmetry and we have some work to do before we can explain why.

> From the air peak to the apex of each peak is the average time that each component spends in the *liquid phase*. It's formal name is the adjusted retention time.

The air peak location is also useful for troubleshooting. Since none of the component molecules can travel faster than the carrier gas, any peak that appears on the chromatogram before the air peak can't be from the current sample injection.

Calculated performance indicators

Having measured peak position and peak width, you can calculate several useful indicators of column performance.

Average carrier gas velocity

Since you know the column length (L), the air-peak time (t_M) allows you to calculate the average **carrier gas velocity** ($\bar{u}$) in m/s:

$$\bar{u} = \frac{L}{t_M} \quad (2.2)$$

We'll come back to the average velocity in Chapter 6, where it helps to optimize the carrier gas flow rate.

Retention factor

The stationary phase is totally responsible for separation, so the peaks should not waste too much time in the mobile phase. Nothing happens there. So an important measure of column performance is the ratio of the time a peak spends in the liquid phase to the time it spends in the gas phase. This parameter is the **retention factor** (k), previously known as the capacity factor. Later, we shall see that the retention factor is very helpful when optimizing the operating conditions of a column.

The measurements made in Figure 2.1 illustrate the data you'll need to determine the retention factor for a selected peak. The air-peak time tells you the time-in-gas-phase (t_M) and reveals the time-in-liquid-phase (t_R'). The retention factor then follows from:

$$k = \frac{t_R - t_M}{t_M} = \frac{t_R'}{t_M} \quad (2.3)$$

In Figure 2.2, for example, the propane peak spends three minutes in the liquid phase and six minutes in the column, so its retention factor is:

$$k = \frac{6-3}{3} = 1.0$$

By similar logic, you should see that the retention factor of the 1-butene peak is $k = 3.0$.

Measures of perfection

When there are two or more component peaks on a chromatogram, a fixed ratio exists between the adjusted retention times of two selected peaks. This ratio is the **separation factor** for the two components, and it tends to remain constant for a given liquid phase regardless of the column operating conditions.

For a specified liquid phase, the separation factor between two peaks is essentially constant, whatever you do. There are a few exceptions, but for most analytes the only way to change the separation factor is to use a different liquid phase.

Figure 2.3 shows the chromatogram measurements you'll need. Again, the chromatogram must include an air peak. Use the holdup time (t_M) to calculate the adjusted retention times (t_R') of the two component peaks. Alternatively, just measure the adjusted retention times directly from the chromatogram, as done in Figure 2.3. The separation factor (α) is then:

$$\alpha = \frac{t_R'(B)}{t_R'(A)} \tag{2.4}$$

For example, inserting the values from Figure 2.3 gives a separation factor of 1.25:

$$\alpha = \frac{300 \text{ s}}{240 \text{ s}} = 1.25$$

The separation factor indicates the ability of the liquid phase to separate the peaks, so it's sometimes called the **selectivity** of that stationary phase for those components.

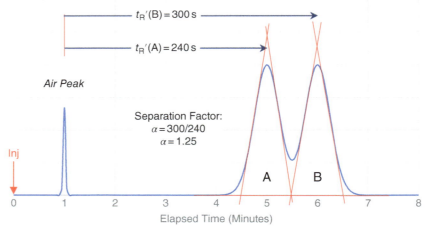

Figure 2.3 Measurements for Separation Factor.

The separation factor is a property of the stationary phase, so the chromatogram measurements needed are the adjusted retention times of the peaks, starting from the air peak. For clarity of display, this illustrative chromatogram shows two very wide peaks; in practice, a separation factor of 1.25 would normally yield a nice baselne separation of the two peaks.

It's useful to know the alpha-value when selecting or troubleshooting a column to separate two components. Generally, the separation is easy when the separation factor is greater than 1.1 and becomes more difficult as it approaches unity. A separation factor of 1.0 indicates that the two components will not separate on that liquid phase whatever you do.

For most peaks, the only way to change the separation factor is to select a different liquid phase. But even a different liquid phase won't change the separation factor unless the molecules of the two peaks are chemically different. For instance, all paraffins have a similar chemical structure, so their separation factors are about the same on all columns. Adjacent paraffins have a separation factor of about 2.0 on any liquid column under any operating conditions.

Separation

As a technical term, **separation** (S) simply means the time between the two peak apexes.

$$S = t_R(B) - t_R(A) \tag{2.5}$$

But people tend to use "separation" in its everyday sense of moving peaks apart from one another. So when two peaks overlap each other, they say it's a poor separation. Because of this potential misunderstanding the chromatography world now shuns the use of *separation* as a variable, but we still find it useful for simple explanations. The separation of the peaks in Figure 2.3 is 60 seconds. This says nothing about how they part or overlap at the baseline. To capture that information we need a different variable, one that includes the effect of peak separation and peak width.

Resolution

The **resolution** of adjacent peaks expresses peak separation relative to peak width and is an effective measure of column performance.

Figure 2.4 shows how this works. Measure the retention time (t_R) of the peaks and their triangulated base widths (w_b). Get their separation (S) from Equation 2.5 as before. Then, figure their average base width ($\overline{w}_b$):

$$\overline{w}_b = \frac{w_b(B) + w_b(A)}{2} \tag{2.6}$$

Alternatively, measure the peak widths at half height (w_h) and get their average width ($\overline{w}_h$) from:

$$\overline{w}_h = \frac{w_h(B) + w_h(A)}{2} \tag{2.7}$$

Finally, calculate their resolution (R_s) from:

$$R_s = \frac{S}{\overline{w}_b} \tag{2.8}$$

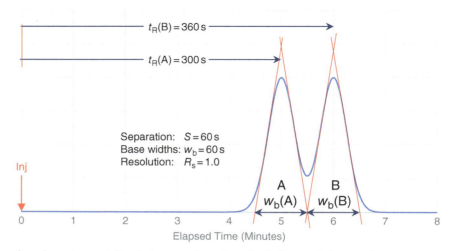

Figure 2.4 Measurements for Resolution.

An adequate resolution between adjacent peaks is the goal of any column system. The figure shows a resolution of 1.0, which is not enough to accurately measure the peaks. A resolution of 1.5 would be adequate to accurately measure adjacent peaks of similar size, but even more resolution is necessary when the two peaks have radically different heights.

To use peak width at half height, refer to the **SCI-FILE:** *On Plate Theory* (below) for this data on the properties of a Gaussian curve:

$$\frac{\overline{w}_h}{\overline{w}_b} = \frac{2.354\,\sigma}{4\,\sigma} = 0.59$$

Substitute into Equation 2.8:

$$R_s = 0.59 \cdot \frac{S}{w_h} \tag{2.9}$$

The peaks in Figure 2.4 have a resolution of 1.0, which always generates an overlap at the baseline: the valley point between the peaks is about 25 % of each peak height. Yes, a computer program can estimate the area belonging to each peak but its digital procedure is subject to error. Always try to get the best possible resolution.

Computer programs can't resolve peaks. Only chromatographic columns can do that.

Well-resolved peaks are essential for accurate measurement.[2] When the resolution is 1.5, the valley between two equal peaks almost touches the baseline. That's defined as a **perfect resolution** and is adequate for their

---

[2] Later, we will find that the exigency to achieve a fast analysis often produces a design with inadequate resolution for accurate measurement. You may have the opportunity to rectify that.

accurate measurement. But peaks that are unequal in size require even more resolution. The resolution necessary for accurate measurement will always depend on the relative size of the peaks.

Practical aspects of theory

Plate theory

There are two prominent theories of chromatography, the **plate theory** described here and the **rate theory** described in Chapter 6.

Waters (2020, 28–36) uses plate theory to explain the formation of peaks. This theory easily shows why the peaks should be symmetrical and grow wider as they stay longer in the column. The theory imagines that the injected sample transits a series of "plates" where equilibrium occurs between the molecules in the mobile phase and the molecules in the stationary phase. Of course, there are no separate plates inside a column, so we call them **theoretical plates.** The basic idea is that each theoretical plate is the equivalent of one equilibrium.

The theoretical plate theory of chromatography arose by analogy with the liquid distillation process. Industrial distillation towers have trays that hold a liquid in equilibrium with a vapor above; the vapor moves up the tower, reaching a new equilibrium with the liquid on each tray. In the laboratory, the apparatus for distillation is much smaller. Chemists say their distillation tower is a *column* and they call each tray a *plate*. Laboratory stills don't always use plates; instead, they use a column containing irregular-shaped glass or ceramic objects (called *saddles* or *Raschig rings*) packed inside a straight vertical tube. Although such columns don't have real plates, chemists rate them by the effective number of vapor–liquid equilibria that occur; thereby specifying the *number of theoretical plates*.

The original plate theory of chromatography used that same terminology, but a later standard simplified the nomenclature from *number of theoretical plates* to **plate number** (Ettre 1993).

The theoretical plate model of chromatography effectively explains the shape of chromatographic peaks. The movement of individual component molecules is a random process; some move with the carrier gas while others stay in the liquid phase. This random motion tends to disperse the narrow band of solute molecules, spreading them apart. After many such movements, the linear position of molecules along the column follows the **normal distribution** of statistics – often called the "bell curve" – that features in the SCI-FILE that follows. Another name for this symmetrical shape is a **Gaussian peak**.

Most chromatogram peaks are close to being Gaussian in shape but are never exactly so. The theories of chromatography always start from the assumption that peaks are Gaussian and modify their conclusions to account for observed deviations from the normal distribution. We shall do the same.

As a practical chromatographer, you're probably anxious to know *why* you have non-Gaussian peaks on your chromatogram. As yet, we can't answer that more-complex question. First, we need to explore the properties of simple Gaussian peaks. Once you understand those, you'll be ready for the more complex discussion on how peaks can become asymmetric.

The following SCI-FILE: *On Plate Theory* is optional reading, but you might find it interesting. The main text continues after the SCI-FILE and employs a few of the equations introduced there to illustrate the practical uses of plate theory to evaluate the performance of chromatographic columns.

SCI-FILE: *On Plate Theory*

Ideal peak shape

A chromatogram peak represents the distribution of the component molecules in time as they pass through the detector. It's obvious that all the molecules don't reach the detector at the same time. Some arrive earlier and some later, due to the cumulated effect of multiple random processes in the column.

Theories of chromatography posit that the arrival times of molecules at the detector are normally distributed. Figure 2.5 is an example of a normal distribution. It's a well-known curve found in many random processes and often called a "Gaussian" distribution in honor of the mathematician Carl Gauss.

Real peaks on a chromatogram don't quite live up to the expectation of being normal, but they come close and that allows us to make deductions about their shape.[3]

A **Gaussian peak** is an ideal symmetrical peak and its central point is the **mean**. The average molecule is also at the center, with an equal number of molecules arriving earlier and later in time. The time coinciding with the apex of the peak is the retention time recorded for that solute.

Since all chromatograms are a plot of detector response versus time, measurements made on the chromatogram to determine the retention time and width of a peak are *time measurements*. This is true even when you measure distances on a printout in millimeters.

Of course, the shape of the peak forms inside the column before the molecules arrive at the detector.

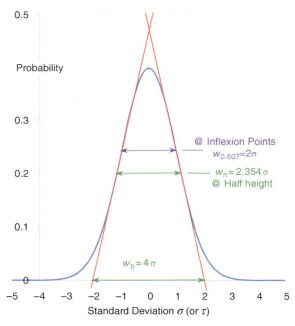

A perfect peak would follow the Gaussian distribution, which has the known characteristics are shown here. Real peaks show slight deviations from perfection that demand a more complex theory.

Figure 2.5 The Gaussian Peak Shape.

As the injected **band** of molecules (that is, an embryo "peak") works its way down the column, the band tends to spread, occupying a small but gradually expanding segment of the column length as it inexorably migrates toward the detector.

---

[3] To study the reasons for asymmetric peaks, refer to Chapters 7 and 8.

Thus, the band of molecules in the column becomes *distributed in space*; although all the molecules have traveled for the same amount of time, some of them get farther along the column than others do. Therefore, when the molecules are inside the column we express the width or position of the band in length units, not time units.

In gas chromatography, it's not practical to measure the location of molecules inside the column; we must wait until they reach the detector. We deduce their behavior in the column by measuring the time they appear on the chromatogram.

Peak width

The **standard deviation** of a normal distribution is a measure of the spread of values around the mean. It's therefore a very good indicator of peak width.

In gas chromatography, we employ two alternative symbols to represent the standard deviation. We use *sigma* (σ) to represent the distribution of molecules occurring in the column, so *sigma* always carries units of length, usually millimeters.

We introduce *tau* (τ) to represent the time distribution of a chromatogram peak, so *tau* always carries units of time, usually seconds.

Really, the two distributions are equivalent; only the units have changed. Even so, failure to appreciate the difference will lead to confusion and perhaps to errors in calculation.

The length and time distributions are related in a simple way. The *distance* a peak travels is the length of the column (L) and the *time* that it travels is its retention time (t_R). Thus, for a specified peak:

$$\frac{\sigma}{\tau} = \frac{L}{t_R} \quad (2.10)$$

From statistical theory, the normal curve has inflexion points at 60.7 % of its height, where its width is $\pm\tau$ from the mean. Figure 2.5 shows that tangent lines drawn at these inflexion points intersect the baseline at $\pm 2\,\tau$ from the mean, a known property of the normal curve.

Thus, the base width (w_b) of an ideal chromatogram peak is:

$$w_b = 4\,\tau \quad (2.11)$$

Note that the standard deviation of a chromatogram peak is one-fourth of its base width, measured in time units.

Also from the standard shape of the normal curve, the peak width measured at half the peak height (w_h) is:

$$w_h = 2.354\,\tau \quad (2.12)$$

Some workers prefer the latter measurement because it avoids the need to draw tangents. Note, however, that it is a narrower width and may be more difficult to measure precisely.

Measures of column efficiency

A tenet of plate theory is that the solute molecules moving in the gas phase form many successive equilibria with the solute molecules stuck in the liquid phase. The alternative approach taken by rate theory discounts the notion of equilibrium occurring anywhere in the column because the constant movement of the mobile phase will not allow it. Yet the terms **plate number** (previously called the *number of theoretical plates*) and **plate height** have persisted in both theories for evaluating the efficiency of columns.

Plate theory defines the relation between plate number (N), plate height (H), and column length (L) as:

$$N = \frac{L}{H} \quad (2.13)$$

By rearranging Equation 2.13, you can see that plate height is the length of column needed to create one plate:

$$H = \frac{L}{N} \quad (2.14)$$

Plate height

As noted above, standard deviation (σ) is a convenient measure of the width of a solute band in the column as it has the same units as the distance that the band has moved.

Another valid measure of the dispersion of molecules is the variance of the band. The **variance** of a normal distribution is the square of its standard deviation (σ^2).

Many random processes act on the sample molecules from the instant of their injection to the instant of their detection. Statistical theory shows that the variances of these independent and random processes are additive, whereas their standard deviations are not. Therefore, the width of a chromatogram peak reflects the cumulated variance of these dispersive processes.

As a band of molecules migrates along the column it tends to spread; the farther it goes, the wider it gets. On average, its molecules are moving farther and farther from the mean. This continuing dispersion of the molecules increases the variance of the band in direct proportion to the distance it has traveled.

A poor column will facilitate dispersion and allow the variance of the band to increase rapidly with distance traveled, whereas a good column will minimize dispersion and keep the molecules closer together.

To evaluate column performance, we need to know the *rate of dispersion per distance moved* and we use plate height to measure it. You can visualize plate height as the "rate of generating variance" within the column.

This gives plate height (H) an unusual unit of measure, being the variance (σ^2) of the band of molecules divided by the distance it has moved (d):

$$H = \frac{\sigma^2}{d} \quad (2.15)$$

Plate height has length units, typically mm or cm. It's the best measure we have for identifying the causes of dispersion and then prescribing ways of reducing it. Any change in operating conditions that reduces plate height always improves the resolution of peaks on the chromatogram.

Development of peak shape

The molecules in a migrating solute band are normally distributed. At any point in the column, the base width (w_b) of the band measured in length units is:

$$w_b = 4\sigma \quad (2.16)$$

Combining Equations 2.15 and 2.16 gives an equation for the base width after it has moved a distance (d) along the column:

$$w_b = 4\sqrt{H \cdot d} \quad (2.17)$$

Since faster-moving bands travel more distance, the above equation indicates that faster bands will become wider than slower bands. Figure 2.6 illustrates this odd behavior. The gray area to the left shows the spatial distribution of the solute bands within the column at the instant that the air peak reaches the detector (that is, 90 s in this diagram).

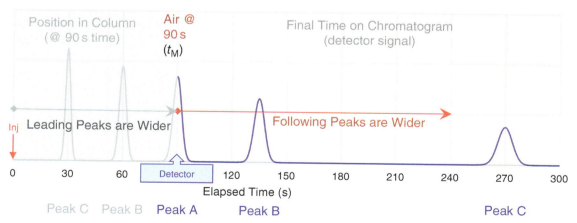

This example shows an air peak (A) and two analyte peaks (B and C). When the air peak reaches the detector, the analyte peaks are still in the column – shown in the gray area to the left of the air peak. At that time, we imagine the three peaks are equally spaced along the column. Then, since Peak B took 90 s to travel two-thirds of the column, it will take another 45 s to travel the remaining one-third, for a retention time of 135 s. Peak C traveled only one-third of the column in 90 s and will need 180 s to reach the detector, for a retention time of 270 s. Thus, the linear spacing in the column turns into logarithmic spacing on the chromatogram.

[This simplified explanation ignores the effect of pressure, which we'll consider in a later chapter.]

Figure 2.6 Spatial and Temporal Separations.

In this simple example, the solute bands within the column are equally spaced. The molecules of Peak C form not only the slowest band, but also the narrowest one. This is strange. Slower peaks being narrow is contrary to our experience with gas chromatographs. We just *know* that the later peaks on our chromatograms are always wider than the early ones.

Yet it's true that fast peaks are wider than slow peaks when *inside the column*, where they are separate in space but not yet separate in time. This reverse-width effect *always* occurs inside a column but is unnoticed when a single column goes straight to a detector.

The reverse-width effect becomes important only when two columns operate in series or when a single column is temperature programmed. When evaluating what is happening inside a column at a certain time, you may need to remember that a slow-moving component peak is still a very compact band of molecules.

Peak shape at elution

Gas chromatography is an elution technique. Analyte molecules don't remain in the stationary phase; they continue to travel until they exit the column and reach the detector, forming peaks on the chromatogram. Of course, all peaks leaving the column have traveled the same distance; that is, the column length (L).

So for eluted peaks, Equation 2.15 becomes:

$$H = \frac{\sigma^2}{L} \quad (2.18)$$

Equation 2.18 defines plate height as the variance per unit length of the column, which is a measure of the dispersion rate of solute molecules within the column. As such, it is constant for the same solute on different lengths of column working under the same conditions.

In practice, it's not possible for all the column operating conditions to be identical. When the column length changes, the carrier gas pressure must also change. But the effect of pressure is usually quite small, so plate height is almost independent of column length.

To get the eluted peak width (w_b), combine Equations 2.16 and 2.18:

$$w_b = 4\sqrt{H \cdot L} \quad (2.19)$$

Since each peak has a constant value of H, the widths of all peaks change in proportion to the square root of column length as per Equation 2.19. This slow growth of peak width explains why longer columns always give better resolution: the separation of the peaks increases linearly with column length, which outpaces the slower increase in peak width.

Plate height is not constant for every component of the injected sample; the physical properties of components differ, leading to different rates of dispersion inside the column. In practice, later-eluting peaks typically have larger plate numbers and smaller plate heights.

Chromatogram peak shape

The right side of Figure 2.6 shows the chromatogram obtained after peaks B and C have exited the column. As expected, later peaks are wider than early peaks.

In passing, it's interesting to observe that the peaks on the chromatogram are no longer equally spaced, which is an effect we first met in Waters (2020, 48–49). The process of transforming the peaks to the time domain converts their linear separation in the column to a logarithmic separation on the chromatogram.

The calculation of column performance parameters from chromatogram measurements of peak width and retention time is straightforward if you pay attention to the units.

> *Since a chromatogram peak is a time distribution, you can't use the length-based plate height (H) to calculate its width. The chromatography literature sometimes misses this important point.*

To avoid confusion, this text introduces a new symbol (H_t) for the plate height expressed in time units, where:

$$H_t = H \cdot \frac{\tau}{\sigma} \quad (2.20)$$

Hence, from Equation 2.10:

$$H_t = H \cdot \frac{t_R}{L} \quad (2.21)$$

Now convert Equation 2.18 to time units by inserting the expressions for σ and H from Equations 2.10 and 2.21, respectively:

$$H_t = \frac{\tau^2}{t_R} \quad (2.22)$$

As expected, this defines H_t as the variance generated per second.

Combining Equations 2.12 and 2.13 now gives the peak base width in seconds:

$$w_b = 4\sqrt{H_t \cdot t_R} \quad (2.23)$$

On a given chromatogram, the plate heights of peaks in *time units* (H_t) tend to remain constant. The peak widths therefore increase in proportion to the square root of their retention times as per Equation 2.23.

Column efficiency

Theories of chromatography use different assumptions about what happens inside a column, yet they all agree on plate number as the best measure of the column's ability to limit the molecular dispersion within a peak. Larger values of plate number indicate less dispersion, so the peaks are relatively narrow compared to their retention time. Consequently, plate number is useful for measuring the overall **column efficiency**.

The original concept of plate number came from distillation theory as the "*number of theoretical plates.*" It's the number of equilibrium steps the peak would have passed through to produce the observed peak width (Martin and Synge 1941, 1359).

If you have already calculated the column plate height (H), get plate number from Equation 2.13.

Alternatively, to estimate plate number (N) from chromatogram values, convert Equation 2.13 to time units by inserting the expression for H from Equation 2.21:

$$N = \frac{t_R}{H_t} \quad (2.24)$$

Now combine Equations 2.22 and 2.24 to obtain the classic definition of plate number:

$$N = \left(\frac{t_R}{\tau}\right)^2 \quad (2.25)$$

From this, substituting τ from Equation 2.11 generates the familiar equation for calculating plate number from chromatogram data:

$$N = 16\left(\frac{t_R}{w_b}\right)^2 \quad (2.26)$$

Some chromatographers prefer an alternative equation derived by substituting τ from Equation 2.12 instead:

$$N = 5.54\left(\frac{t_R}{w_h}\right)^2 \quad (2.27)$$

The latter calculation doesn't need the drawn tangents. It's a simpler procedure, but the measured width is only about 60 % of the base width, so it might be difficult to get an accurate measurement.

Effective plate number

The plate number is not the same for every peak on the chromatogram – it improves when peak retention time increases. This is partly due to the holdup time, which has less effect on the later peaks. Some workers prefer an alternative measure known as the effective plate number, based on the adjusted retention time (t_R'):

$$N_{\text{eff}} = 16\left(\frac{t_R'}{w_b}\right)^2 \quad (2.28)$$

Or alternatively:

$$N_{\text{eff}} = 5.54\left(\frac{t_R'}{w_h}\right)^2 \quad (2.29)$$

Process chromatographers seldom use this alternative measure.

Using the theory

Resolution on longer columns

To evaluate the effect of a longer column on resolution, recall the definition of resolution from Equation 2.8. For adjacent peaks, the resolution (R_s) is the ratio of their separation (S) to their average base width ($\overline{w}_b$):

$$R_s = \frac{S}{\overline{w}_b}$$

At the same temperature and carrier velocity, the retention time of both peaks is directly proportional to the column length. Therefore, their separation (S) is also a linear function of column length.

But on a longer column the width of the peaks does not increase as much as the separation does. From Equation 2.19, the peak width increases in proportion to the square root of column length. It follows that the resolution of adjacent peaks is a function of the square root of column length:

Figure 2.7 shows the effect of increased column length graphically:

➤ The blue line represents the increase of peak width by the square root of column length.

➤ The orange line represents a peak pair with a high separation factor that are easy to resolve. Their separation (S) increases rapidly with column length.

➤ The purple line represents a peak pair that are more difficult to resolve; their separation (S) increases more slowly with column length.

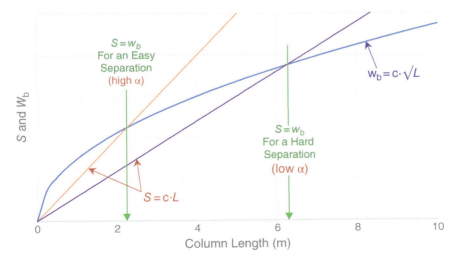

Peak separation (S) increases in direct proportion to column length (orange and purple lines), but peak width (w_b) increases only by the square root of column length (blue curve). Where a straight line crosses the blue curve, peak width is equal to separation giving a resolution of 1.0. To achieve this resolution, the low-alpha separation (purple) needs about 6.3 m of column, whereas the high-alpha separation (orange) needs only 2.2 m of column. In principle, any resolution is achievable given a long enough column.

Figure 2.7 Achieving Resolution.
Source: Giddings, J.C. (1965)/with permission of Taylor & Francis Group.

At the points where the lines intersect:

$$S = \overline{w}_b$$

Therefore, at those intersections the column is just long enough to form a resolution of 1.0. In principle, there must be a column length that will achieve any desired resolution.

In the next chapter, we shall see that the necessary carrier gas pressure poses limits to the length of columns, particularly packed columns. The pressure required for open tubular columns is much lower, so they can be very long and can achieve wonderful resolutions.

The ultimate constraint to column length is the analysis time. A single long column might take hours to resolve a peak pair having a separation factor close to 1.00, yet most process plants need the analysis in just a few minutes. A PGC often solves that problem by using multiple columns. After partial separation of components on one column, the unresolved analytes enter another column having a much larger separation factor.

Estimating plate number

Plate number of a peak is a measure of column efficiency. It comes from the plate theory described in the **SCI-FILE:** *On Plate Theory*. This theory equates plate number to the number of equilibria that would be necessary to form that peak shape on the chromatogram.

Knowing the plate number of a peak is not an academic pursuit, it's a practical necessity. All the practical procedures for optimizing column performance will require you to estimate the plate number from chromatogram measurements. To understand the theory behind those measurements, refer to the SCI-FILE above.

Two measurements – peak position and width – are enough to evaluate the performance of a single column. You won't need an air peak. As an example, let's calculate the plate number (N) for the peak in Figure 2.8 using Equation 2.26:

$$N = 16 \left(\frac{t_R}{w_b} \right)^2$$

Inserting the data from Figure 2.8 gives:

$$N = 16 \left(\frac{360}{60} \right)^2 = 576$$

A plate number of 576 is low and would indicate an inefficient column, since a good column would generate about 2,000 plates per meter. Of course, we drew this peak wide to clearly illustrate the measurements. A real peak would be much narrower: a more typical peak might have one-third of the width and nine times the plate number.

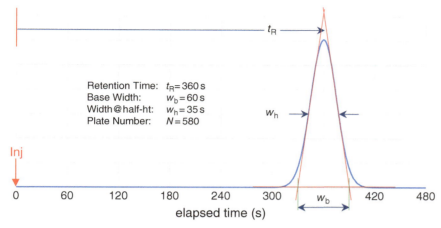

The plate number is a good indication of column performance and needs only two measurements from a single chromatogram peak. Any change of operating conditions that increases plate number will improve the resolution between analyte peaks.

Figure 2.8 Measurements for Plate Number.

Plate number measures the sharpness of a peak and is easy to visualize:

➤ A plate number of 100 means the peak retention time is 2.5 times its base width. That would be a wide peak!

➤ A plate number of 10,000 means the peak retention time is 25 times its base width. That would be a nice sharp peak!

Some chromatographers prefer to calculate plate number using the peak width at half height (w_h) because there's no need to draw tangent lines and it's potentially more accurate. By Equation 2.27:

$$N = 5.54 \left(\frac{t_R}{w_h}\right)^2$$

By inspecting the above equations, you'll see that a higher plate number improves resolution. A higher plate number must be due to an increase in the retention time or a reduction in the peak width, or both. Therefore, a higher plate number will always increase the resolution of adjacent peaks.

Plate number is a good measure of the efficiency of a column. Any change of operating conditions that increases plate number automatically improves the resolution of adjacent peaks.

Plate number is not the same for every peak on a chromatogram. It tends to improve with increased time in the liquid phase, so later peaks typically have larger plate numbers than earlier peaks do. In a similar way, longer columns have larger plate numbers and are more powerful than shorter columns are. In effect, increasing the length of a column is just like adding more plates.

Estimating plate height

Plate height is a fundamental measure of column efficiency because it doesn't vary with column length. To calculate plate height (H), divide the column length (L) by its plate number (N) as per Equation 2.14:

$$H = \frac{L}{N}$$

As calculated above, plate height has units of length, usually millimeters (mm). You can imagine plate height to be the length of column needed to make one equilibrium. This is still a good way to visualize the chromatographic process, although we will shortly discover that modern theorists prefer to downplay the idea of equilibrium occurring in the column.

When calculating plate height, it's convenient to enter the column length in mm, which returns the plate height in mm. An efficient liquid phase column typically has a plate height of about 0.6 ± 0.2 mm. It doesn't take much column to make one equilibrium!

Open tubular columns may have somewhat better (smaller) plate heights than packed columns, but for practical PGC columns the difference is not great. It's often thought that wall-coated columns are inherently more efficient, but that's not their real advantage. They're just longer. Open tubes have low flow resistance and can be very long. That's the main reason they have more plates.

Plate height is a useful parameter as it evaluates the operating efficiency of the separating mechanism itself. Any change in the design or operation of a column that reduces plate height must also increase the plate number and thereby must improve the resolution of adjacent peaks.

> *Plate height is the primary variable used for optimizing column performance. It's a good measure of internal column efficiency and is independent of the column length.*

Plate height can also indicate the condition of a column. A liquid phase packed column exhibiting a plate height of about 0.5 mm is working well and needs no remedial action. Conversely, a column with a plate height of more than 1 mm might not be operating optimally. Later chapters will describe some practical ways to optimize column performance.

A gradual increase in plate height indicates cumulative column damage, perhaps from overheating or from nonvolatile substances in the injected sample.

The complementary theory of chromatography – *rate theory* – also recognizes plate height as the fundamental measure of column efficiency. Rate theory seeks to identify the various causes of peak broadening in columns and includes some more complex notions, so we'll defer it until later.

Knowledge Gained

Practice

- You can measure the holdup time, peak retention time, and peak width on the chromatogram.

- When measuring base width, triangulate the peak and extend the baseline under the peak.

- Alternatively, measure the peak width at half the peak height.

- A chromatogram measurement may be in seconds or millimeters, but always represents a time.

- The air peak doesn't dissolve in the liquid phase, so it travels at the same speed as the carrier gas.

- No peak from the same injection can appear on the chromatogram earlier than the air peak.

- To get through the column, all injected molecules spend the same time moving as the air peak does.

- Additional retention time beyond the air peak time is the time the peak stopped in the liquid phase.

- The air peak divides the chromatogram into two zones: peak time-in-gas and peak time-in-liquid.

- Holdup time is the time each component molecule spends in the gas phase.

- Adjusted retention time is the average time the component molecules stop in the liquid phase.

- Retention time is the sum of time traveling in the gas phase and time stopped in the liquid phase.

- The average carrier gas velocity is the column length divided by the holdup time.

- Retention factor is the adjusted retention time of a solute peak divided by the holdup time.

- Separation factor is the ratio (>1.0) of the adjusted retention times of two peaks on a chromatogram.

- To change the separation factor, use a different liquid phase.

- Separation is the time between the apexes of adjacent peaks on a chromatogram.

- Resolution is the separation of adjacent peaks divided by their average width.

- Under the same operating conditions, separation increases in proportion to the column length.

- Peak width and resolution both increase by the square root of the column length.

- Resolution improves by the square root of column length, but with lower peaks and longer analysis.

- To calculate plate number, use chromatogram measurements of retention time and peak width.

- Any change in operating conditions that increases plate number also increases the resolution of peaks.

- To calculate plate height, divide the column length by its plate number.

- Plate height is a good measure of column efficiency; packed columns should be about 0.6 ± 0.2 mm.

- Wall-coated columns may have lower plate heights and can be much longer, giving even more plates.

Theory

- The times that identical solute molecules arrive in the detector closely follow a normal distribution.

- *A normal distribution is symmetrical, but real chromatogram peaks may not be exactly so.*

- *While in the column, the positions of the individual solute molecules also follow a normal distribution.*

- *The broadening of a peak is the cumulated effect of many random processes occurring in the column.*

- *Standard deviation (σ or τ) is a measure of the width of a "Gaussian" peak (that is, a normal distribution).*

- *Variance is another measure of Gaussian peak width and is equal to σ (or τ) squared.*

- *A Gaussian peak has inflexion points at 60.7 % of its height, where its width is equal to 2σ (or τ).*

- *Tangents from the inflexion points cross the baseline to delineate a peak base width equal to 4σ (or τ).*

- *If measured at half the peak height, the width of a Gaussian peak is equal to 2.354σ (or τ).*

- *Plate theory sees plate height as the length of column equivalent to one equilibrium.*

- *Rate theory sees plate height as the rate of variance accumulation as a peak migrates along the column.*

- *On-column peaks get wider with distance moved; chromatogram peaks get wider with elapsed time.*

- *Peaks that have migrated only a short distance along the column are still a very narrow band.*

- *A linear spacing of peaks inside the column converts to a logarithmic spacing on the chromatogram.*

- *The spatial distribution of peak molecules inside the column has units of millimeters.*

- *The time distribution of peak molecules on a chromatogram has units of seconds.*

- *Plate height is the effective length of column necessary to create one equilibrium.*

- *Plate height (mm) is constant for the same peak on a longer or shorter column, given the same conditions.*

- *Plate height (mm) tends to be smaller for later solute peaks on a chromatogram.*

- *Plate height (s) is about the same for different solute peaks on a chromatogram.*

- *To calculate plate number, use the retention time and width of a peak.*

- *Plate number is a measure of column efficiency, a higher value is always good.*

- *Plate number is the number of equilibria that would be necessary to generate the observed peak.*

- *Plate number tends to be higher when calculated for later peaks on a chromatogram.*

Did you get it?

Self-assessment quiz: SAQ 02

A gas chromatograph uses a single 4.0 m column. On the chromatogram printout, there are two peaks. A pair of intersecting tangents drawn along the sides of each peak extend downwards to cross the baseline. The distance between the injection marker and the intersection of tangents is 370 mm for Peak A and 410 mm for Peak B. The base peak width along the baseline between the two tangent crossovers is 19 mm for Peak A and 20 mm for Peak B.

On Practice

Q1. Estimate the plate number (N) for Peak A.
Q2. Estimate in length units the column plate height (H) for Peak B.
Q3. What is the resolution (R_s) between Peaks A and B?

On Theory

Q4. Estimate in time units the width at half height (w_h) of Peak A.
Q5. Estimate in time units the variance (τ^2) of Peak B.
Q6. Estimate in time units the chromatogram plate height (H_t) for Peak B.
Q7. If working under the same conditions, what column length would give a resolution of 1.5 for Peaks A and B?

Check your SAQ answers with those given in the back of the book.

References

Cited

Ettre L.S. (1993). *Nomenclature for Chromatography: IUPAC Recommendations 1993*. Research Triangle Park, NC: International Union of Pure and Applied Chemistry.

Giddings, J.C. (1965). *Dynamics of Chromatography: Principles and Theory*. New York, NY: Marcel Dekker.

Martin, A.J.P. and Synge, R.L.M. (1941). A new form of chromatogram employing two liquid phases. *Biochemical Journal* **35**, No. (12), 1358–1368.

Waters, T. (2020). *Process Gas Chromatographs: Fundamentals, Design and Implementation*. Chichester, UK: John Wiley & Sons Ltd.

Figures

2.1 Typical Chromatogram Measurements.
2.2 Significance of an Air Peak.
2.3 Measurements for Separation Factor.
2.4 Measurements for Resolution.
2.5 The Gaussian Peak Shape.
2.6 Spatial and Temporal Separations.
2.7 Achieving Resolution.
2.8 Measurements for Plate Number.

Symbols

Symbol	Variable	Unit
d	Distance a band of molecules has moved along the column	mm
H	Plate height of an on-column peak (length units)	mm
H_t	Plate height of an eluted peak (time units)	s

Measures of perfection

k	Retention factor	none
L	Length of column	mm, m
N	Plate number by chromatogram peak	none
N_{eff}	Effective plate number by chromatogram peak	none
R_s	Resolution of adjacent chromatogram peaks	none
S	Separation[4] of adjacent chromatogram peaks	s
t_M	Gas holdup time	s
t_R	Retention time of chromatogram peak	s
t_R'	Adjusted retention time of chromatogram peak	s
$\bar{u}$	Average carrier gas velocity	m/s
w_b	Base width of a solute peak	mm or s
$\bar{w}_b$	Average base width of adjacent solute peaks	mm or s
w_h	Width at half height of a solute peak	mm or s
$\bar{w}_h$	Average width at half height of adjacent solute peaks	mm or s
α	Separation factor of two specified peaks	none
σ	Standard deviation of an on-column peak	mm
τ	Standard deviation of a chromatogram peak	s

Equations

2.1 $\quad t_R = t_M + t_R'$ — Peak retention time as the sum of its time in the gas and liquid phases.

2.2 $\quad \bar{u} = \dfrac{L}{t_M}$ — Average velocity of carrier gas.

2.3 $\quad k = \dfrac{t_R'}{t_M}$ — Retention factor of a specified peak.

2.4 $\quad \alpha = \dfrac{t_R'(B)}{t_R'(A)}$ — Separation factor of adjacent Peaks A and B.

2.5 $\quad S = t_R(B) - t_R(A)$ — Separation of adjacent Peaks A and B.

2.6 $\quad \bar{w}_b = \dfrac{w_b(B) + w_b(A)}{2}$ — Average base width of adjacent peaks.

2.7 $\quad \bar{w}_h = \dfrac{w_h(B) + w_h(A)}{2}$ — Average peak width at half height.

2.8 $\quad R_s = \dfrac{S}{\bar{w}_b}$ — Resolution of adjacent peaks (by base width).

2.9 $\quad R_s = 0.59 \cdot \dfrac{S}{\bar{w}_h}$ — Resolution of adjacent peaks (by half height).

---

[4]Separation is not an officially recognized variable.

2.10	$\dfrac{\sigma}{\tau} = \dfrac{L}{t_R}$	Relationship between peak distribution in distance and time units.
2.11	$w_b = 4\,\tau$	Known base width of a Gaussian peak in time units.
2.12	$w_h = 2.354\,\tau$	Known width-at-half-height of a Gaussian peak in time units.
2.13	$N = \dfrac{L}{H}$	Definition of plate number in length units (frequently used).
2.14	$H = \dfrac{L}{N}$	Definition of plate height in length units (frequently used).
2.15	$H = \dfrac{\sigma^2}{d}$	Plate height defined as variance generated per distance moved.
2.16	$w_b = 4\,\sigma$	Known base width of an on-column Gaussian peak in length units.
2.17	$w_b = 4\sqrt{H \cdot d}$	Base width of an on-column peak versus distance moved.
2.18	$H = \dfrac{\sigma^2}{L}$	Defines column plate height as variance per column length.
2.19	$w_b = 4\sqrt{H \cdot L}$	Peak base width per column length.
2.20	$H_t = H \cdot \dfrac{\tau}{\sigma}$	Definition of plate height in time units.
2.21	$H_t = H \cdot \dfrac{t_R}{L}$	Relation of plate height in time or length units.
2.22	$H_t = \dfrac{\tau^2}{t_R}$	Definition of chromatogram plate height as variance per second.
2.23	$w_b = 4\sqrt{H_t \cdot t_R}$	Base width of a chromatogram peak in time units.
2.24	$N = \dfrac{t_R}{H_t}$	Plate number in time units.
2.25	$N = \left(\dfrac{t_R}{\tau}\right)^2$	Definition of plate number in time units.
2.26	$N = 16\left(\dfrac{t_R}{w_b}\right)^2$	Plate number from chromatogram peak base width (frequently used).
2.27	$N = 5.54\left(\dfrac{t_R}{w_h}\right)^2$	Plate number from chromatogram peak width at half height (frequently used).
2.28	$N_{\text{eff}} = 16\left(\dfrac{t_R'}{w_b}\right)^2$	Effective plate number from chromatogram peak base width.
2.29	$N_{\text{eff}} = 5.54\left(\dfrac{t_R'}{w_h}\right)^2$	Effective plate number from peak width at half height.

New technical terms

When first introduced, these words and phrases were in bold type. You should now know the meaning of these technical terms:

adjusted retention time	**rate theory**
air peak	**resolution**
band	**retention factor**
base width	**retention time**
carrier gas velocity	**selectivity**
column efficiency	**separation**
Gaussian peak	**separation factor**
holdup time	**standard deviation**
mean	**theoretical plates**
normal distribution	**triangulating**
perfect resolution	**unretained peak**
plate height	**variance**
plate number	**width at half height**
plate theory	

For more information, consult the Glossary of terms (pp 379–416).

3

Column technology

"The column is the heart of a process gas chromatograph and if it doesn't perform well, neither will the instrument. Understanding the properties of columns will help you to optimize the measurements and troubleshoot any problems that arise."

Introduction

Among the inestimable contributions to gas chromatography made by Dr Leslie Ettre (1922–2010) is a review article about capillary columns, in which he said:

> *Today, capillary columns **exist**; users don't have to worry how they are made. They simply open a supply house catalog, check the application they are interested in, and order the proper column by its code name and part number (Ettre 2001).*

He made his point well and it remains true today. We buy columns online from a catalog – nobody cares what's inside. You install them in the PGC and they work: what more is there to know?

For routine maintenance, this is just as it should be. You should not need to reinvent the application. If a column fails, you install a new one and restart the instrument; check the carrier flow rates, chromatogram separations, event timing, and calibration; and return the PGC to service. Fast and efficient.

But what if it still doesn't work?

To troubleshoot column failure you'll need to know how columns work. Otherwise, you won't know *why* the column failed – or whether it failed at all. Perhaps the column is not the problem; there are plenty of other reasons for failure. To discover the root cause of failure and prescribe the correct solution, you'll need to know what's inside those coiled tubes.

Process Gas Chromatography: Advanced Design and Troubleshooting, First Edition. Tony Waters.
© 2025 John Wiley & Sons Ltd. Published 2025 by John Wiley & Sons Ltd.

Unless you're a PGC manufacturer, it's unlikely that you will make your own columns. You can go to the catalog for those. Therefore, the limited goal of this chapter is to provide a general understanding of what columns are, how they work, and what affects their performance. We'll leave the column design techniques until the next chapter.

Column types

Two types of column

Industrial gas chromatographs use two different kinds of column: packed columns or capillary columns. Here's the difference:

➢ In a **packed column**, a granular solid containing the stationary phase completely fills the tube, so the carrier gas has to percolate through it. Most packed columns work at constant temperature.

➢ In a **capillary column**, also called an **open-tubular column**,[1] the stationary phase resides on the inside wall of the tube, allowing free passage of carrier gas down the middle. Capillary columns may work isothermally or on a temperature program.

There are many varieties of chromatographic column, but they all fit into one of those two categories.

This chapter will develop a detailed comparison between packed columns and capillary columns. Both types of column have several advantages and disadvantages, but an overall evaluation will often focus on two opposing features:

➢ A capillary column is likely to have a higher plate number than a packed column and thus may provide more resolution between peaks.

➢ A packed column is likely to have a larger **sample capacity** than a capillary column and thus may provide more sensitivity for measuring a low concentration of analyte.

The sample capacity of a column is the maximum amount of an analyte that will produce an undistorted peak shape. We'll look at it in more detail later.

Packed columns

The original PGCs used packed columns made from ¼-inch o.d. (outer diameter) stainless-steel tubing. This quickly changed as theoretical developments predicted narrower columns would be more efficient. By 1980 most PGCs had adopted ⅛-inch o.d. packed columns as standard.

---

[1]For the purposes of this text the adjectives capillary and open-tubular are synonymous.

Today, PGCs often use $\frac{1}{16}$-inch o.d. **micropacked columns** for appropriate applications and they are rapidly supplanting the popular $\frac{1}{8}$-inch o.d. packed columns. Many older PGCs installed in the plants use traditional packed columns, while more recent installations are likely to have the new micropacked columns.

Capillary columns

The invention of the capillary column by Golay (1957, 1958) gave us a glimpse of the future. Golay showed that ultrafast separation was possible by gas chromatography, but clunky 1960s PGC technology wasn't ready for such speed; our sample injectors, valves, detectors, and chart recorders were too big or too slow for rapid analysis. Our equipment is much better now and most PGCs can accept capillary columns.

Figure 3.1 illustrates the three kinds of capillary column used today:

➢ Wall-coated open-tubular (**WCOT**) columns have a thin layer of liquid phase coated or chemically attached to their inner wall.

➢ Porous-layer open-tubular (**PLOT**) columns have a thin layer of adsorbent solid on their inner wall.

➢ Support-coated open-tubular (**SCOT**) columns have a thin layer of liquid-coated support particles on their inner wall.

Capillary columns have low flow resistance, so they can be very long. The practical limit for a packed column is about 12 m, but a capillary column can be over 100 m in length: 30–50 m is common. When operated with temperature programming, such columns can separate hundreds of peaks allowing the analysis of complex and unpredictable mixtures. This ability is highly valued by the analytical laboratory, where most samples are liquid and contain many components. Therefore, most laboratory gas chromatographs today use a single, temperature-programmed capillary column.

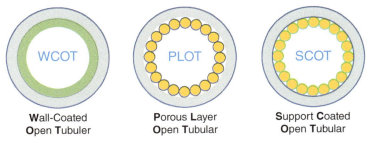

The original capillary columns had steel tubes and gave poor performance for polar analytes that adsorbed onto the metal surface. Desty et al. (1960) were the first to develop glass capillaries. For a long time, these were very popular in laboratory chromatographs but were far too fragile for process use.

Figure 3.1 Three Kinds of Capillary Column.

Table 3.1 Subjective Evaluation of Columns Used in PGCs.

Material[a]	Style	Installed Base	Recent Shipments
Stainless steel >1/8-inch o.d.	Large packed	1 %	0 %
Stainless steel 1/8-inch o.d.	Small packed	61 %	46 %
Stainless steel 1/16-inch o.d.	Micropacked	24 %	33 %
Stainless steel 1/16-inch and less	Open tubular WCOT, PLOT, SCOT	4 %	10 %
Fused silica Most 0.53 mm i.d.	Open tubular WCOT, PLOT, SCOT	10 %	11 %

This anecdotal data illustrates a trend toward micropacked and capillary columns in recently shipped PGCs compared with those already working on site.
[a]Teflon tubing is still used for some highly corrosive samples.
Source: Adapted from Tony Waters.

A process chromatograph isn't like that. Nearly every PGC uses a multiple-column system to analyze a predictable process sample of fairly constant composition. The PGC column system needs to separate a few nominated analytes and it must do it fast. Speed is critical – the industrial process can't wait.

Capillary columns work well in some process applications, but not so well in others. Therefore, we don't expect capillary columns to totally replace packed columns in process chromatographs as they did in the majority of laboratory instruments. For process work, the current trend still favors micropacked columns over open-tubular columns, as evident in Table 3.1.

Mechanism of retention

Mutual affinity

For both packed columns and capillary columns, there are two types of stationary phase, a *solid phase* or a *liquid phase*. The dominant mechanism of retention by a solid stationary phase is **adsorption** – the active surface attracts and holds component molecules. The mechanism for a liquid stationary phase is different; it works mainly by **solvation** – component molecules dissolve in the liquid. Either way, a brief contact with the stationary phase selectively retains component molecules as the carrier gas moves on, causing separation to occur.

The actual mechanisms involved in retention and separation are quite complex and can be difficult to grasp. The generic explanation is that the molecules of different analytes have a different **affinity** for the stationary phase. For some analytes, the mutual attraction is weak, while for others it's strong. The concept of *mutual affinity* is adequate for most purposes.

Put simply, gas chromatography happens because the carrier gas carries injected solute molecules into a column where they touch a stationary phase. That contact briefly retains the molecules. The duration of retention varies for different kinds of molecule, depending on their affinity for the stationary phase, resulting in the separation of one kind of molecule from another kind of molecule.

When discussing the affinity between solutes and liquid phases we often invoke the concept of polarity. A **polar** molecule is a neutral molecule that has an unevenly distributed electronic charge within – and the intensity of that imbalance is what we call polarity. A **nonpolar** molecule is a neutral molecule that has an evenly distributed electronic charge within. The general principle is that polar solutes are more soluble in polar liquid phases, and nonpolar solutes are more soluble in nonpolar liquid phases. There's more to it than that. If you want more, wade into the **SCI-FILE:** *On Mutual Affinity* for a full exposition.

SCI-FILE: *On Mutual Affinity*

Molecular forces

By presuming that gas molecules don't interact with each other, the **kinetic theory** of gases was able to explain Boyle's observation that gas volume and pressure are inversely related to each other. That was a huge jump in understanding, yet we soon found it lacking, as real gases tend to deviate from such ideal behavior.

The behavior of molecules is always a consequence of the forces acting upon them. The observed deviations from Boyle's law imply that real gas molecules actually do affect each other, so forces must be involved. And the same is true of their interaction with a solid or liquid surface.

All intermolecular forces are electronic and occur in three related flavors: dispersion, polar, and ionic.

Dispersion forces

Consider two proximate molecules, neither having an uneven distribution of electrons, as in Figure 3.2a. Due to their uniform electron distribution, no electronic force exists between them. We call them *nonpolar* molecules. For example, all homonuclear diatomic gases and all paraffins (alkanes) are nonpolar.

Yet the electrons in a molecule are in rapid motion. Their random movement inevitably leads to a fluctuating electron density at any chosen location in the molecule. Clearly, an instantaneous high electron density at one location implies a low electron density elsewhere. These high and low electron densities create a fleeting electronic charge differential within the molecule called a **dipole**, illustrated in Figure 3.2b.

Then, if the electrons in the second molecule are free to move, a high electron density in the first molecule will repel them and a low electron density in the first molecule will attract them. Thus, a fluctuating dipole in the first molecule will induce a synchronous dipole of opposite polarity in the second molecule per Figure 3.2c. Opposite charges attract, so the net result is a fluctuating but always attractive force between the two molecules. The aggregate of myriads of such random fluctuating forces forms a small but sustained force of attraction.

The net dispersive force is the weakest force between molecules but is always present. Its strength depends on the contact area of the

44 Column technology

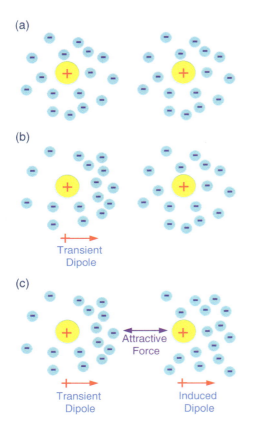

Figure 3.2 Origin of Dispersive Forces. (a) Two adjacent neutral atoms or molecules. (b) Random electron shift causes a transient dipole. (c) Induced dipole forms an attractive force.

molecules, the number of electrons they have, and how tenaciously they hold on to those electrons. The overall ability of the electrons to disperse is the **polarizability** of the molecule. Thus, hydrocarbons containing double or triple bonds are more polarizable than those that don't. Similarly, larger molecules have more electrons and form a stronger mutual attraction.

The universal presence of dispersion forces explains the behavior of all solutes on a nonpolar liquid phase. The *affinity* of a solute for a nonpolar liquid phase depends only on its polarizability. For this reason, larger molecules always elute after smaller ones. And since the boiling points of the solutes are a result of those very same forces, they elute from the column in boiling-point order.

Dispersion forces also explain the retention behavior of paraffins on *any* liquid phase. Paraffin molecules are nonpolar so they elute by molar mass. Again, that's because it's easier to polarize a heavier molecule. As an example, consider the isomers of a normal paraffin. They all have the same molar mass, but the more compact forms are less polarizable and elute before those with longer carbon chains. Therefore, the normal paraffin is always the last one out.

Dispersion forces are often called *London forces* after Fritz London who first identified them (London 1937).

Polar forces

In a molecule containing identical atoms – like nitrogen (N_2), for instance – each atom has the same influence over the electrons present, so the electrons are symmetrically distributed and the molecule has no dipole. But within a neutral molecule having two or more *different* atoms there may be an asymmetric distribution of electrons. Each unique atom in the molecule has its own complement of protons and electrons that unevenly attract or repel the electrons that bond one atom to another atom and in doing so generate a **permanent dipole** within the molecule.

The **electronegativity** of an atom is a calculated property that indicates its ability to attract electrons. In the periodic table of the elements, electronegativity increases by group number (from left-to-right) and decreases by period number (from top-to-bottom). Of the reactive elements, fluorine is the most electronegative element and cesium is the least. When two different atoms form a chemical compound, the more electronegative atom will pull electrons away from the other atom, thus creating a permanent dipole.

In passing, we should also mention that a compound that contains many atoms may exhibit several dipoles due to electron displacement. Each

 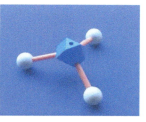

The methane molecule (left) is symmetric and nonpolar, while the ammonia molecule (right) is asymmetric and higly polar.

Figure 3.3 Nonpolar and Polar Molecules.

dipole is a vector acting along the line between two atoms. In an entirely symmetric molecule, the polarity vectors then cancel, resulting in a nonpolar molecule. For example, in the methane molecule [CH_4] depicted in Figure 3.3, the central carbon atom attracts the bonding electrons to create four weak C—H dipoles equally spaced in three dimensions. Since they are symmetrical, the dipole vectors cancel, so the methane molecule is nonpolar.

In contrast, the ammonia [NH_3] molecule also shown in Figure 3.3 is an asymmetric molecule. Ammonia has three N—H dipoles on one side of the nitrogen atom, so their vectors reinforce each other to create a strong permanent dipole. Thus, ammonia is very polar.

A polar molecule will attract another polar molecule – even one of its own kind, as famously occurs between water molecules. The force of attraction is strong, but not as strong as a chemical bond. In these interactions, the electropositive atom is often hydrogen, so the weak bond formed between the two molecules is known as a **hydrogen bond**. By this mechanism, a polar liquid phase will strongly retain a polar solute.

To interact with a nonpolar molecule, a polar molecule must first polarize it and that might not be easy. In general, a polar solute will not be able to polarize a nonpolar liquid phase, limiting their mutual affinity to only the omnipresent dispersion forces. That's why polar solutes tend to elute in boiling point order from a nonpolar column, along with their nonpolar relatives.

A polar column acts differently. The permanent dipole present in the liquid phase clings to the permanent or induced dipole in a polar solute, retaining the solute either strongly or weakly depending on their respective electron potentials. Thus, a polar column often gives an excellent separation of polar or polarizable solute molecules.

Finally, consider a nonpolar solute on a polar liquid phase. Their only mutual affinity comes from the weak dispersion forces. This explains the weak retention of alkanes and other nonpolar solutes on polar columns.

As a general principle, maximum retention occurs between solutes and liquid phases that are chemically alike. Thus, a polar polyethylene glycol column is good for separating alcohols and a nonpolar silicone column is good for paraffins.

Ionic forces

An ion is an atom or molecule that has totally lost or gained an electron, bestowing a full electronic charge and a strong dipole. Ionic interactions are stronger than the dispersive and polar forces discussed herein. When present, all three forces contribute to affinity: dispersive, polar, and ionic.

Ionic liquid phases are a recent development in column technology that provide high stability and good peak shape when separating highly polar solutes. We expect to see more of them in future PGCs. For more about them, see the review by Poole and Poole (2011).

Nomenclature

Dispersion forces are known as **London forces** in honor of the physicist Fritz London (1900–1954).

The interactions between a permanent dipole and an induced dipole are **Debye forces**, after physicist Peter Debye (1884–1956).

The interactions between two permanent dipoles are **Keesom forces**, after physicist Willem Keesom (1876–1956).

A generic term for the above intermolecular forces is **Van der Waals forces**, after Johannes Van der Waals (1837–1923).

Summary

As detailed in the above SCI-FILE, the two major causes of retention, and hence separation, are:

> The effect of dispersion forces: these forces also determine the boiling point of a solute, so nonpolar solutes elute in boiling point order on any liquid phase.

> The *additional effect* of polar forces: these forces occur when a polar solute interacts with a polar liquid phase, and act in addition to the dispersion forces.

Note also that polar solutes can polarize some nonpolar liquid phases adding to their retention time.

While only a simplification, it's helpful to think of the observed retention as the sum of two individual retentions. A nonpolar column retains a peak mostly in proportion to its boiling point, and a polar or polarizable column retains a peak by both its boiling point *and* its polarity.

Packed columns for GSC

When a gas chromatograph uses an adsorbent solid as stationary phase, the full name of the technique is gas-solid chromatography (GSC). In a gas-solid column, the gas is moving and the solid is stationary.

Solid-phase columns

Columns for GSC use an adsorbent granular solid as stationary phase. Typical adsorbent solids include carbon, alumina, silica gel, and porous polymers. These solids have a large surface area that attracts solute molecules, ultimately causing separation. Another useful solid is a **molecular sieve** – a carbon or zeolite packing that works a little differently. It has minute pores that differentiate molecules according to their size and has the useful ability to separate oxygen and nitrogen.

Within the general category of GSC columns, there are two subcategories that exhibit different behaviors: the brittle active solids such as carbon or zeolite molecular sieves, silica gel, and alumina; and the relatively soft porous polymers.

Active solids

The adsorbent active solids are valuable in PGCs for their ability to separate permanent gases like hydrogen, oxygen, and nitrogen, as well as carbon monoxide and methane. They are not suitable for the analysis of more condensable gases as their strong and nonlinear adsorption characteristics can cause severely distorted peaks. Anyway, their analysis times would be long – an anathema to the process industries.

This behavior curtails the use of active solids in process chromatographs. It's true that temperature programming can mitigate this limitation in the laboratory, but it adds unwelcome complexity to a process instrument. Therefore, PGCs tend to use active-solid columns to analyze only the light gases that fail to separate on any other column. Common measurements include:

- Hydrogen or deuterium.
- Helium, neon, argon, and krypton.
- Oxygen, nitrogen, methane, and carbon monoxide.

The less-problematic porous polymers give good separations of carbon dioxide and the light paraffin and olefin gases, and strongly retain water. Consult Chapter 4 for more information.

Operation and maintenance

Solid-phase columns are simple in construction, yet complex in operation. In particular, columns employing the hard, adsorbent solids demand more maintenance attention than those using a liquid phase and are lacking in long-term stability.

The main disadvantage of active solids is their gradual deactivation in use. Depending on the solid used, the retention time of large or polar molecules can be essentially infinite, so these substances gradually accumulate on the solid surface and block access to adsorption sites by other molecules. The squatter molecules may be substances present in the injected sample or in the carrier gas. An effective column design will backflush the harmful components in the sample, but it can't protect the active solid from the carrier gas. Traces of water vapor in the carrier gas will gradually deactivate any active solid column, and carbon dioxide will deactivate a molecular sieve column. Even with stringent precautions, expect gradual deterioration. Over a period of several months, peak retention times will gradually fall with an eventual loss of resolution. To minimize the rate of decline:

- Never leave the PGC without carrier gas flow unless the column system inlets and outlets are all capped to prevent ingress of water or other vapors from headers, vents, or atmosphere. This applies also to shipment from the shelter fabricator.

- The carrier gas must pass through an absorption dryer[2] to remove traces of carbon dioxide, moisture, and other polar substances before it enters the PGC. The maintenance schedule must ensure timely replacement of the dryer.

- The carrier gas purifier should not remove oxygen as traces of oxygen are essential for the wellbeing of some porous polymers.

---
[2] Refer to Waters (2020, 100–101) for more information about carrier gas purifiers.

> A protective column must always precede the active-solid column to retain and divert water and heavies and prevent them from entering the analytical column. This precolumn may contain Sorbitol or a porous polymer such as PoraPak T[3] or HaySep T.[4]

> Routine checking and adjustment may be necessary to compensate for the gradually shortening retention times. The frequency of this maintenance depends on the effectiveness of the carrier gas dryer and the backflush column.

> The analysis method should never reverse the flow in an active-solid column as the packings are friable and don't tolerate pressure pulses well – this includes PLOT columns.

It may be possible to restore an active-solid column by prolonged heating at the manufacturer-specified temperature while continuously purging with a dry inert gas. Don't do this in the analyzer: remove the column and heat it in a laboratory oven. During the activation procedure, a copious amount of steam may emerge from the column. After several hours, turn off the heat and maintain the purge flow while the column cools, then immediately cap the ends to prevent the ingress of moisture. When the column is back in the PGC and in stable operation, adjust Method settings to allow for any change in component retention times.

Porous polymer columns may also need activation, but in a well-designed column system this should be necessary only once, before use, and often done by the column manufacturer. A well-designed column system does not allow any invasive molecules to accumulate over time.

If taking the PGC out of service, don't allow a hot column to cool without carrier gas flow as it will suck in water and oxygen from the ambient air. If it's necessary to stop the carrier gas flow, wait until the column is cold, then immediately seal both ends to prevent the ingress of water vapor and other atmospheric gases. This general *Best Practice* applies to all PGC columns and is particularly relevant to those containing an active solid.

Active-solid packings are friable – particularly graphitized carbon and molecular sieves – so treat such columns gently. Don't drop them on the ground or tightly bend them.

Pros and cons of GSC

In summary, active-solid columns are more troublesome than liquid-phase columns, so PGCs use them only when a liquid phase is not available to achieve the separation. Porous polymers are more versatile and less friable and are popular for the separation of low-molecular-weight gases, such as light hydrocarbons.

---

[3]PoraPak is a trademark of Waters Corporation (no relation to the author).
[4]HaySep is a trademark of Merck KGaA, Darmstadt, Germany.

Packed columns for GLC

When a gas chromatograph uses an involatile liquid as stationary phase, the full name of the technique is gas–liquid chromatography (GLC). In a gas–liquid column, the gas is moving and the liquid is stationary.

GLC columns are by far the most common in use. In a packed column, like those shown in Figure 3.4, an inert granular solid supports a thin coating of the desired liquid phase. The inert **solid support** has a large deactivated surface area which, when coated, holds a very thin and wide layer of liquid. The extensive area of the liquid film and its shallow depth enable a rapid exchange of solute molecules between gas and liquid phases.

Solid support

An ideal support particle would be strong enough to resist abrasion, have a large surface area, and be chemically inert. It should exert no affinity for the analyte molecules. Such perfection is unattainable in practice. We do our best to deactivate the surface but some residual forces remain, which tend to make the peaks asymmetric.

The traditional and still most common solid supports are for self-evident reasons called the pink and white supports. More recently, graphitized carbon has also become common in narrow-bore columns.

The pink and white supports both derive from [5]Celite®, a **diatomaceous earth**. The raw material occurs naturally in rock strata, notably in Europe and North America, and is the fossilized skeletal remains of ancient marine algae. The material is mostly silica and retains the fine structure of the original microscopic organisms, usually as a pair of species-dependent

Figure 3.4 Typical Packed Columns.
Source: Reproduced with permission from Ohio Valley Specialty Company.

---
[5]CELITE is a trademark of Imerys Minerals California, Inc.

half-shells finely perforated with micron-sized holes. Its complex structure gives Celite a large surface area and multiple pathways for gas flow.

The white support (for example, [6]Chromosorb® W) is from diatomaceous earth that has been calcined at 900 °C with sodium carbonate flux. This fuses the particles together and converts the iron oxide into a colorless silicate. The white support has an inert surface that is ideal for polar liquid phases, but the particles are friable and easily fractured during packing.

The pink support (for example, Chromosorb® P) is also from diatomaceous earth. Basically, it's crushed firebrick, made by calcining diatomite at about 1000 °C without the flux. The pink color is due to the metal oxides present. The pink support is harder and can hold larger amounts of liquid phase. Its surface is more active than the white support and is not suitable for polar analytes, but it delivers excellent performance for hydrocarbons and other nonpolar analytes.

The pink support is about twice as dense as the white support. Therefore, if the two supports had an equal weight percent of liquid phase, a column using the pink support would contain twice as much liquid phase as an identical column using the white support. The extra liquid increases peak retention factors and makes the pink support superior for the separation of the lighter hydrocarbons.

For more detailed information about supports, refer to the classic review by Ottenstein (1963).

Several companies produce proprietary solid supports refined from Celite to reduce its surface activity. The Celite is acid washed (AW) to remove traces of metal oxides and reacted with a silicon compound[7] to kill polar hydroxyl groups at the silica surface. Additional treatments, often closely held trade secrets, suppress the surface affinity for targeted polar substances and thereby improve the symmetry of those peaks.

The adsorbent solids occasionally act as solid supports when deactivated by a liquid coating. In addition, some of the porous polymers are effective supports in special applications.

Coating the support

The total quantity of liquid phase in a column is an important parameter. For a packed column, it's specified as the **liquid loading**, that is, the weight percentage of the liquid phase coated on the support. A potential confusion arises because the stated percentage may be relative to the weight of the prepared packing or the weight of support used. According to IUPAC (2014), a column packing made with 20 g of liquid on 80 g of support is a 20 % liquid load, not a 25 % liquid load.

A common procedure for preparing a packing is to dissolve a known weight of the liquid in a suitable solvent and mix the solution with an

---

[6]Chromosorb is a trademark of Imerys Minerals California, Inc.
[7]Common treatments are hexamethyldisilazane (HMDS), dimethylchlorosilane (DMCS), or dimethyldichlorosilane (DMDCS).

appropriate weight of support to form a slurry. Then, to allow the solvent to evaporate naturally or with vacuum assistance.

The liquid phase loading for PGC columns is typically in the range 1–25%. The higher percentages are generally less efficient because the liquid tends to form puddles or bridges between particles which reduce the rate of mass transfer between phases. Even so, the higher loadings may be necessary when injecting large sample volumes to measure low ppm concentrations and are also effective for separating the lighter hydrocarbons. Chapter 5 prescribes a method for optimizing the liquid loading to maximize resolution.

Pros and cons of GLC

Liquid-phase columns are far more versatile than solid-phase columns and PGCs use them in all applications except the small group of GSC applications mentioned above. Their advantages over solid-phase columns are:

- There's a myriad of liquid phases to choose from.
- They are more durable than solid-phase columns and don't need activation (although some conditioning may be necessary).
- They have a higher sample capacity and produce more symmetrical peaks.
- They allow the optimization of retention factor by adjusting percent loading or film thickness.
- They tolerate flow surges well, as in backflush for instance.

The main disadvantage of liquid-phase columns is their limited ability to separate light gases. For instance, a liquid-phase column can't separate hydrogen, methane, carbon or nitrogen oxides, or the gases found in the atmosphere.

Packed column technology

Packed columns have large sample capacity and we can often use them to analyze components that other techniques can't easily separate. In the past, packed columns suffered from unstable liquid phases that produced high rates of column bleed and short column life. In addition, the stainless-steel tubing used for packed columns was not inert, allowing active compounds to interact with its inner surface. These woes are now behind us and modern packed columns are often preferred for PGC.

Since we don't expect you to make your own columns, the information given here is an overview of how the vendors make packed columns, not a full prescription of how to make them yourself.

Tubing

The tubing used for packed columns in PGCs is usually Type 304 or 316 L stainless-steel tubing, and rarely pure nickel or Hastelloy[8] alloy. The tubing must be seamless, not welded, and its inner wall must be super clean. The cleaning procedure uses a sequence of solvents to remove oil and soluble salts from the wall. These may include a dilute mineral acid, pure water, a ketone like acetone, an alcohol like methanol, or a paraffin like n-hexane. It's best to avoid chlorinated solvents because residual traces of chlorine may reach the detector and cause corrosion. Most column makers purchase pre-cleaned tubing from a specialist column supplier.

For analyzing highly reactive analytes, the tubing may have a fused silica lining, but this would be unusual. The inner wall of the tubing is usually **surface-deactivated** by chemical treatments such as **electropolishing** and **silicon coating**. Such passivation procedures are mandatory for accurate low ppm measurements of sticky components like alcohols or hydrogen sulfide.

Generally, PGCs don't use packed columns made from glass, aluminum, copper, or polymer tubing, although you may occasionally encounter a Teflon[9] packed column for highly active analytes.

Older instruments used 3/16-inch or even 1/4-inch o.d. columns, but nearly all packed columns in PGCs now working onsite have an outside diameter no greater than 1/8-inch. Table 3.2 lists some common tube dimensions.

Mesh size

A packed column is a tube tightly packed with a granular material already described. The objective is to reduce the free gas volume to a minimum, while not causing an excessive pressure drop. Ideally, all the individual granules would be the same size, so they form a homogeneous bed when packed inside the tube. Otherwise:

> Larger particles might create open spaces in the bed – called voids – that are devoid of packing. Voids increase peak dispersion and reduce column efficiency.

Table 3.2 Typical Tubing for Packed PGC Columns.

Outside Diameter		Inside Diameter		
(inch)	(mm)	(inch)	(mm)	Comment
1/16	1.59	0.039	1.0	Micropacked
1/8	3.18	0.085	2.2	Packed
3/16	4.76	0.118	3.0	Older analyzers
1/4	6.35	0.180	4.6	Now obsolete

---

[8]Hastelloy is a registered trademark of Haynes International, Inc.
[9]Teflon is a registered trademark of The Chemours Company.

> Smaller particles might clog the gas passages through the bed and cause an unacceptable pressure drop. Smaller particles may also come from rough handling.

> *Column packings are friable and demand gentle handling. Never drop a packed column or bend it aggressively. Otherwise, you may create a pinch point that will change the distribution of pressure along the column and blunt its performance.*

In practice, the granules are not precisely the same size but have a narrow range of diameters. The column maker sieves the packing to remove any larger or smaller particles that would disrupt the uniformity of the bed. The sizing procedure uses two sieves in series. The upper sieve rejects any larger particles and allows the target particles to pass through, while the lower sieve retains the target particles and allows smaller particles to pass. The procedure isn't perfect and about 90 % of the granules in the product are within the mesh sizes of the two sieves.

Unfortunately, the **mesh size** of a laboratory sieve is a somewhat arbitrary dimension based on the number and size of wires in the wire weave. Table 3.3 relates the mesh number to hole size. Most packed PGC columns use a particle size range of 60–80, 80–100, or 100–120 mesh, resulting in medial particle diameters of 214, 163, or 137 µm, respectively.

The choice of particle size is a compromise. Larger particles give less flow resistance while smaller particles enhance column efficiency.

> *You should recall that higher efficiency means narrower peaks and more resolution. A column with the smallest plate height will have the largest plate number and achieve the best resolution of analyte peaks.*

In theory, the lowest attainable plate height is about twice the particle size. Thus, for a column packed with 100–120 mesh particles, the theoretical limit is a plate height $H \approx 0.3$ mm (about 3,300 plates per meter). Real columns

Table 3.3 Selected Mesh Sizes.

Metric Size (mm)	Sieve Number (#)	Hole Size (inch)	Hole Size (µm)
0.420	40	0.0165	420
0.354	45	0.0139	354
0.297	50	0.0117	297
0.250	60	0.0098	250
0.177	80	0.0070	177
0.149	100	0.0059	149
0.125	120	0.0049	125
0.104	150	0.0041	104
0.074	200	0.0029	74

would not realize that utopian value, but it illustrates that smaller particles are inherently good. A practical aspiration for the minimum plate height is three times the average particle size, so 0.5 mm ± 0.1 mm is a fair estimate. That's about 2000 plates per meter. However, we shall see that PGCs often run their columns under suboptimal conditions to achieve the fastest possible analysis time. Don't be surprised to see plate height approaching or exceeding 1.0 mm under those operating conditions. There's more on fast analysis in Chapter 6.

The rub is the pressure drop. Smaller particles require more carrier gas pressure and this limits the column length. In practice, the particles in conventional packed PGC columns are rarely smaller than 100–120 mesh because the column pressure needed for finer particles would be too high for the analyzer. However, a short micropacked column might use smaller particles.

Packing the tube

To ensure effective packing, typical packed columns are seldom more than 6 m long. When a longer column is necessary, the PGC will use two columns in series. Micropacked columns are more difficult to pack and are unlikely to be longer than 2 m.

To make a ⅛-inch column, the column maker installs a sintered stainless-steel frit or glass-wool plug into one end of a straight tube and pours the granular packing into the other end. For short columns, it may be adequate to allow the granular packing to flow into the tubing by gravity, assisted by gentle vibration. Another method is to pressurize the feed hopper or apply suction to the other end of the tube, or both, thereby creating a gas flow that gently drives the packing down the tube. To check for proper filling, the maker records the packing weight and compares it to known standards. When done, they install a frit in the end of the tube to hold the packing in place.

It's usually acceptable to coil columns after packing, although there are some procedures to pack pre-coiled tubing.

To make a micropacked column, use tubing with an internal diameter of at least three times the particle size. The diameter of 100–120 mesh particles is <0.149 mm, so they will even fit inside a 0.53 mm i.d. "megabore" column. One way to ensure tight packing is to vibrate the coiled column tube by immersing it in an ultrasonic bath (Cramers and Rijks 1979).

Instead of sintered frits, micropacked columns may use short lengths of silicon-treated braided wire as end seals, held in place by a crimp in the tubing. To ensure leak-free connections, do not install a ferrule on the crimped area of the tubing.

Conditioning the column

After packing, condition the column by purging it for several hours with helium or nitrogen while heating it to about 20 K above its normal operating temperature. It's best to depower the detector and disconnect it from the column during this purge time. If not for immediate use, continue the purge until the column is cool, then cap both ends to prevent ingress of atmospheric contaminants. Install the conditioned column in the usual way and set the oven temperature and carrier gas flow per the application. Reconnect the detector and observe the baseline. It might take a while before the baseline settles and becomes flat.

Pros and cons of packed columns

Except for a few special applications, packed columns are now quite rare in the laboratory, yet they are still employed by about 80 % of PGCs. Over the years, capillary columns have outperformed packed columns, which have now lost many of their original advantages. Those that tenuously remain are:

- Packed columns enjoy a wider range of liquid phases that may be useful for difficult separations.
- In some applications, they accept a larger sample size enabling the analysis of low parts-per-billion concentrations.
- From a single sample injection, they can separate and measure multiple analytes on widely different measurement ranges.
- They are rugged and durable in the process plant environment and may require less-frequent replacement.
- Their connections are easier to work with.
- They may better tolerate occasional process upsets such as solids, liquids, or polymers present in a gas sample injection.
- They work in older equipment that is too voluminous for capillary columns.
- They are easy to make and less expensive to replace.
- They are familiar and reassuring to process-plant personnel, whose preferences are slow to change.

Newer technology allows dual capillary columns to perform the backflush and heartcut techniques discussed later in this book. This capability nulls some previously touted advantages of packed columns. Now, both kinds of column can achieve these important objectives:

- Multiple column systems can remove all injected components from the columns before injecting another sample. This is a housekeeping task that is essential in a continuous industrial chromatograph.
- Multiple columns often can separate analytes that no single column can.
- When measuring only a few components, multiple columns may achieve a shorter analysis time than is possible on a single column.

Frankly, if a packed column works – use it! Yet, when compared with capillary columns, packed columns have several disadvantages:

➤ They are generally less efficient because their high flow resistance limits column length and plate number.
➤ Their inner surfaces are less inert, so their chromatogram peaks are always a little more asymmetric.
➤ They use a lot more carrier gas.
➤ Their conventional liquid phases may release more vapor or reaction products into the detector, causing additional baseline offset or noise.
➤ They tend to be less reproducible when replaced.
➤ They are less suited to temperature programming.
➤ They suffer large pressure surges during flow reversal.

Micropacked columns

To mitigate the above drawbacks while retaining some of the advantages of the packed column, researchers developed the micropacked column. The tubing most used for micropacked columns in PGCs is 1/16-inch o.d. stainless steel with a 1 mm inside diameter, although 0.95 mm × 0.75 mm tubing is also available. As such, micropacked columns are intermediate in size between regular packed columns and capillary columns. Generally, a micropacked column contains particles closely sized to 10–30 % of the tube bore. Column efficiencies of 3000–5000 plates per meter are attainable (Cramers and Rijks 1979, 134).

Micropacked columns are particularly effective for separating gases as they are more efficient than regular packed columns and contain more liquid than a capillary column, a combination that provides sharp peaks and good retention factors. Applications include ammonia, sulfur gases, and the C_1 to C_6 hydrocarbons.

Micropacked columns are economical in service as they use less carrier gas than a packed column, but they are expensive and commercially available only in lengths of 2 m or less. PGC manufacturers sometimes make longer columns, but they are difficult to pack and less reproducible than regular packed columns.

Most applications use hydrogen carrier gas as any other carrier would incur significant pressure.

A micropacked column can accept a larger sample volume than a capillary column but not as much as a 1/8-inch o.d. packed column, which puts a limit on their ultimate measurement sensitivity. Some applications can overcome this limit by injecting a large sample into a regular precolumn and transferring only the smaller peaks into the micropacked column.

Typical operating conditions

Most PGCs using packed columns or micropacked columns now operate at constant temperature. Subambient operation is possible, but rare. The oven temperature setting for most packed column applications is within the range of 60–150 °C, although a few are much hotter – up to 275 °C. Do not exceed the upper-temperature limit for the stationary phases in use.

Carrier gas flow rates for packed columns vary by application, so always follow the application data. As a guide, typical gas flows are:

- For ≈1 mm i.d. micropacked columns: 5–10 mL/min
  (mostly 1/16-inch o.d.)

- For ≈2 mm i.d. packed columns: 20–40 mL/min
  (mostly 1/8-inch o.d.)

- For ≈4 mm i.d. packed columns: 80 mL/min
  (mostly 1/4-inch o.d. and obsolete)

With helium carrier, a typical micropacked column will need a pressure differential of about:

- For 80–100 mesh packing: 200 kPa/m

- For 100–120 mesh packing: 280 kPa/m

A typical 1/8-inch o.d. packed liquid-phase column will need about:

- For 80–100 mesh packing: 50 kPa/m

- For 100–120 mesh packing: 55 kPa/m

Adsorbent solid-phase columns may need twice these pressures. If using hydrogen carrier, the required pressure is about half of the above values.

Capillary column technology

Capillary tubing

The original capillary columns for gas chromatography used metal or glass tubing. If you are interested in the historical development of such columns, the books by Hinshaw and Ettre (1994) and Jennings (1978) provide an excellent overview.

The early metal and glass capillary columns were capable of much higher resolution than a packed column but were troublesome to make and use. Glass capillaries are brittle and easily broken, and their glossy surface is difficult to coat with an even layer of a liquid phase. Metal capillaries are easier to coat but tend to adsorb polar analytes – as do the metal ions in the walls of a glass tube. Thus, both metal and glass tubes distort the peak shape for low concentrations of a polar analyte.

In 1976, Siemens experimented with glass capillary columns in a Model 200 PGC and found them to be stable; they observed no decline in the resolution of peaks after 16,000 injections of a test sample (Müller and Oreans 1977). Siemens went on to sell many PGCs with glass capillary columns in Europe, establishing the feasibility of capillary column technology in process chromatography (Mahler 2021).

The game changed when Dandeneau and Zerenner (1979) developed a method of making capillary columns from fused silica capillary tubing. They used a machine that melted and extruded a quartz rod, originally intended for making optic fibers for data-transmission lines. Astutely replacing the quartz rod with a quartz *tube*, they got the machine to extrude a long capillary tube instead of a solid optical fiber. To protect each virgin tube from moisture, which can cause cracking, they immediately coated its outer surface with a brownish polyimide sheath. The Hewlett-Packard company received US Patent Number US4293415A for this invention, but they didn't enforce their patent rights and many chromatography supply companies started to produce fused silica columns.

Figure 3.5 pictures fused silica columns in a PGC oven. Fused silica turned out to be an ideal material for capillary column construction: it's highly flexible and almost free of metal ions, providing a more inert surface than the previous soda-lime and borosilicate glasses. In confirmation, Lipsky (1983) thoroughly evaluated fused silica capillary columns and found them

Figure 3.5 Typical PGC Capillary Columns.
Source: Author's Collection.

Fused silica columns in a Siemens Maxum process gas chromatograph. This column system has two parallel analytical trains, each with backflush initiated by valveless column switching.

to be essentially free of the problems noted with soft-glass capillary columns. That was the end-of-the-road for glass capillary columns.

From 1980 onward, chromatographs equipped with a single fused silica column gradually became the preferred laboratory method of analysis. Often running each analysis with a programmed increase of column temperature, these wall-coated open tubular (WCOT) columns now satisfy about 90 % of laboratory applications.

Yet, despite their huge success in the laboratory, fused silica columns have enjoyed only limited acceptance in PGCs. When first used on site, fused silica columns suffered frequent breakages due to process vibration, maintenance clumsiness, or even turbulence in an air-bath oven. They were not popular at all. Nevertheless, their separations were spectacular and many jobsites learned to live with them.

Then, in 1987, Hewlett-Packard requested substantial royalties on the sale of fused silica columns. At that time, Restek was a small chromatography supply company, and the additional royalty cost would have a substantial effect on their business. Of necessity, they set out to find an alternative. They found a way to coat the inside wall of a stainless-steel capillary tube with an exceedingly thin layer of silicon. The inner surface then became similar to a fused-silica tube – but unbreakable and more suitable for a process instrument. PGCs quickly adopted the new product and capillary columns using deactivated stainless-steel tubing are now common in PGCs.

Table 3.4 lists some typical sizes of capillary tubing used for PGC columns and their properties. PGCs often opt for the so-called **megabore tubing**. Its strangely precise 0.53 mm internal diameter came from the laboratory need to insert a standard 0.47 mm syringe needle into the tube for on-column sample injection – not a practice we employ in process analyzers.

Table 3.4 Typical Tubing for PGC Capillary Columns.

Type	Fused Silica	Stainless Steel	Properties
Megabore	Tube i.d.; 0.53 mm Length: 10–105 m	Tube i.d.; 0.53 mm Length: 10–60 m	Capacity: 1–2 mg ΔP: ~0.7 kPa/m $H_{min} = 0.50$ mm
Wide bore	Tube i.d.; 0.32 mm Length: 10–60 m	Tube i.d.; 0.32 mm Length: 10–30 m	Capacity: 110–220 µg ΔP: ~3.4 kPa/m $H_{min} = 0.33$ mm
Narrow bore	Tube i.d.; 0.25 mm Length: 10–100 m	Tube i.d.; 0.25 mm Length: 10–30 m	Capacity: 80–160 µg ΔP: ~4.6 kPa/m $H_{min} = 0.25$ mm

Sample capacity is for each analyte on a liquid phase of similar polarity and may be a lot less on other liquid phases. The approximate differential pressure (ΔP) is for helium carrier. Plate heights (H_{min}) are theoretical minima.
Source: Adapted from de Zeeuw (2015).

Wall-coated open tubular (WCOT) columns

The liquid film coating of capillary tubing is a procedure best left to the professionals. For general information only, there are two ways to do it. The simplest way is to make a solution of the liquid phase in a volatile solvent, and then use dry nitrogen to push a plug of this mixture slowly through the tubing followed by a dry nitrogen purge to evaporate the solvent. It's difficult to control the liquid **film thickness** by this method.

The other way is to fill the capillary with the liquid phase solution, block one end, and then allow the solvent to slowly evaporate from the other end. It's sometimes possible to expedite solvent removal by the judicious application of heat or vacuum. By this method, the film thickness is more precisely known.

The first WCOT capillary columns used the same liquid phases as used in packed columns. These early columns often failed to meet expectations due to a tendency by some liquid phases to form puddles rather than a uniform thin-film wall coating. This imperfection most often occurred with polar liquid phases that failed to wet the ultra-smooth wall of the tube.

Further development focused on the development of liquid phases based on various copolymers of methyl and phenyl siloxanes. Researchers varied the types and percentages of monomers in the copolymer, to make several new liquid phases with very low vapor pressures and a wide range of polarity, as described in Chapter 4.

The next breakthrough was to polymerize the siloxane monomers *in situ* on the column wall, forming a crosslinked polymer coating bonded to the tube surface. The polymer so formed has a very low vapor pressure and is less reactive with trace impurities in the carrier gas, so less column bleed flows into the detector, reducing baseline offset and noise. The bonded phase also occupies the active sites on the tube wall, blocking the adsorption of analyte molecules and reducing peak tailing.

Porous-layer open tubular (PLOT) columns

The inner wall of a PLOT column holds a thin layer of active solid particles – molecular sieve, alumina, or porous polymer. These columns achieve efficient separations but have limited sample capacity and can suffer damage by moisture or other impurities in the carrier gas. The original columns were prone to particle shedding and although later bonding techniques have improved their reliability PGCs seldom use them.

Support-coated open tubular (SCOT)

Due to their increased liquid phase loading, SCOT columns can perform efficient separations of lighter molecules such as the lower hydrocarbons. To make a SCOT column, the manufacturer first coats the inner walls of the capillary tube with a layer of solid support, such as diatomaceous earth, about 30 µm thick. Then they coat the deposited solid support with the chosen liquid stationary phase. PGCs sometimes favor a SCOT column for its greater sample capacity and higher detector sensitivity than a more efficient WCOT column.

Phase ratio

An important property of a WCOT column is the **phase ratio**, the ratio of gas volume to liquid volume in the column.

The phase ratio is a function of the internal diameter of the capillary tubing and the thickness of the liquid film. These two column variables are the equivalent of the liquid loading in a packed column.

The internal diameter of most capillary columns is in the range of 0.1–0.53 mm and their liquid film thickness runs from 0.1 to 18 µm. The narrowest tubes and thinnest films are rare in process work.

By geometry, neglecting the liquid film thickness, the volume of the gas phase (V_G) is proportional to the column length (L) and the square of its internal diameter (d_c):

$$V_G = L \cdot \pi \cdot \left(\frac{d_c}{2}\right)^2 \tag{3.1}$$

Equation 3.1 is accurate enough for thin liquid films. For columns with thicker films, it's best to use a modified equation that allows for the reduction in diameter caused by the liquid film thickness (d_f):

$$V_G = L \cdot \pi \cdot \left(\frac{d_c}{2} - d_f\right)^2 \tag{3.2}$$

The volume of the liquid phase (V_L) is proportional to the tube bore and the liquid film thickness:

$$V_L = L \cdot \pi \cdot d_c \cdot d_f \tag{3.3}$$

Dividing Equation 3.1 by Equation 3.3 gives the effect of column diameter and film thickness on the ratio of gas volume to liquid volume:

$$\frac{V_G}{V_L} = \frac{d_c}{4d_f}$$

This fraction is the phase ratio (β) of the column, an important factor in column design:

$$\beta = \frac{V_G}{V_L} \tag{3.4}$$

$$\beta = \frac{d_c}{4d_f} \tag{3.5}$$

The equations presented here return a β-value uncorrected for pressure in the column, but this is a normal practice when working with capillary columns. The uncorrected value is adequate for comparing one column with another, as similar columns operate at similar pressures. Anyway, most capillary columns operate at low pressure, so any correction factor would be small.

Table 3.5 summarizes the phase ratio of typical columns.

Columns with a low β-value have relatively more liquid in them and this increases the peak retention factors and separations achieved. They are useful for separating gases and light components having low affinity for the stationary phase, and for injecting larger samples for ppm analysis. These are the columns most often used in PGCs.

Columns with a high β-value have higher efficiency and are able to resolve dozens of components. They are useful for separating heavy components that have high affinity for the liquid phase, but they require smaller samples which limits measurement sensitivity.

Table 3.6 illustrates some properties of columns with high and low phase ratio.

Table 3.5 Phase Ratios of WCOT Columns.

d_f (μm) d_c (mm)	0.10	0.15	0.20	0.25	0.50	1.0	3.0	5.0
0.18	450	300	225	180	90	45	15	9
0.20	500	333	250	200	100	50	17	10
0.22	550	367	275	220	110	55	18	11
0.25	625	417	313	250	125	63	21	13
0.32	800	533	400	320	160	80	27	16
0.53	1325	883	663	530	265	133	44	27

Phase ratio (β) calculated by Equation 3.5 for combinations of tube bore (d_c) and film thickness (d_f). Equal-length columns with about the same phase ratio, like the two highlighted values, would give the same analysis time.

Table 3.6 Phase Ratio and Column Performance.

Column Performance	High β-Value	Low β-Value
Peak asymmetry and tailing	Worse	Better
Typical efficiency (plate height)	0.2 mm	0.5 mm
Typical efficiency (plates-per-meter)	5000	2000
Allowable sample volume	<3 nmol	<30 μmol
Resolution of small molecules	Poor	Good
Resolution of large molecules	Good	Excessive
Analysis time	Shorter	Longer

Data for capillary columns compiled from multiple sources.

Two columns of equal length and equal β-value have equal peak retention times and may substitute for each other. Then:

- The narrower thin-film column demands less sample volume and generates sharper peaks. It may also produce less column bleed and less baseline noise.

- The wider thick-film column gives less resolution but improved peak shapes and can accept a larger sample volume to enhance measurement sensitivity.

The phase ratio is a column fabrication variable; it decreases when there is more liquid in the column relative to the open gas space.

While the concept of phase ratio applies to all liquid-phase columns, it's easy to quantify the phase ratio for a capillary column but not so easy for a packed column. Consequently, those using packed columns rarely refer to phase ratio. In packed column parlance, *liquid loading* is the preferred parameter. It's evident that halving the liquid loading is closely equivalent to doubling the β-value and will halve the peak adjusted retention times and separations, and reduce the analysis time. The typical range of phase ratio values for packed columns is from 5 to 35 (Jennings 1978, 9).

Carrier gas flow rate

It turns out that all columns return their maximum efficiency at about the same carrier gas velocity. To maintain the same velocity, it's evident that the carrier flow rate must be proportional to the square of the diameter. Thus, a narrow-bore column will exhibit its full efficiency at a very low flow rate, as noted in Table 3.7. That's not to say that a capillary column *must* run at a low flow rate; in practice, PGCs often run capillary columns fast. In doing so, they sacrifice a part of the enormous separating power of the capillary in favor of a faster analysis.

Jones (2008) provides an interesting insight: the speed of analysis, sample capacity, and analyte resolution form three corners of the performance triangle shown in Figure 3.6. You can't optimize one of these objectives without degrading the other two. In process chromatography, speed is often paramount, so we typically use heavily loaded capillary columns, run them faster

Table 3.7 Optimum Flow Rates for Capillary Columns.

Tube Bore	0.25 mm	0.32 mm	0.53 mm
Nitrogen carrier	0.4 mL/min	0.6 mL/min	0.9 mL/min
Helium carrier	1.4 mL/min	1.8 mL/min	3.0 mL/min
Hydrogen carrier	1.8 mL/min	2.3 mL/min	3.7 mL/min

The listed flow rates are optimum for maximum column efficiency. Most PGC columns operate at higher flow rates to achieve lower analysis times.
Source: Adapted from de Zeeuw (2015).

Optimum speed, capacity, and resolution are mutually exclusive. Improving one automatically limits the other two.

Figure 3.6 The Column Performance Triangle.
Source: Adapted from Agilent Technologies, Inc.

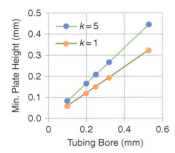

Showing the increased efficiency possible from narrow-bore columns. Real columns achieve about 1.5 times these minima.

Figure 3.7 Theoretical Minimum Plate Height.
Source: Adapted from Equation 3.6.

than optimum, and then inject larger samples. As a result, a PGC may not realize the awesome efficiency that a capillary column is capable of, yet that column is still more efficient than a packed column.

Separating power

It's possible to compute the theoretical minimum plate height (H_{Th}) of a capillary column from its internal diameter (d_c) and the retention factor (k) of a chosen component (Annino and Villalobos 1992, 175):

$$H_{Th} = d_c \cdot \sqrt{\frac{1 + 6k + 11k^2}{12(1 + k)^2}} \qquad (3.6)$$

From this equation, Figure 3.7 displays the theoretical minimum plate height of various column diameters at their optimum flow rate. Real columns never achieve such perfection, so we sometimes express their measured plate number as a percentage of the theoretical maximum, a measure known as the **utilization of theoretical efficiency** (UTE) and a useful indicator of column condition. A good column will typically have a UTE of about 60–70 % when running at its optimum flow rate.

Pros and cons of capillary columns

The advantages of capillary columns are:

➢ Excellent resolution: can separate dozens of peaks.
➢ Fast analysis: quick separation of multiple peaks.
➢ Long life: bonded stationary phases eliminate column bleed.
➢ Economy: carrier gas consumption is minuscule.

The potential disadvantages of capillary columns vary with application. When run at maximum efficiency, they require a **sample splitter** to reduce sample volume, an oftentimes troublesome device. The small sample may also limit the ultimate analyte sensitivity and measurement range. The next section below explores these limitations in detail.

When used with temperature programming, a capillary column may need more frequent replacement. Also, the process instrument becomes more complex and inherently less reliable than an isothermal PGC.

Packed or capillary? A comparison

Acknowledgment

Before starting to examine the advantages and disadvantages of different columns, we need to moderate expectations: this section will give insight into the many variables affecting PGC column technology, but it can't reduce this complex subject to a few simple rules. Column selection is as much art as it is science, and the performance of a PGC is ultimately dependent on the skill of the applications engineer.

For the following discussion, the author acknowledges with gratitude the valuable counsel of our expert contributors listed on pages xxi-xxv, many of whom gave freely from their own unique experiences.

Design of equipment

In the new millennium, PGCs quickly adapted to the peculiar needs of capillary columns. Mostly, the design changes were to reduce the volume of the analytical flow path. The internal volume of a capillary tube is small, so the flow rate is low and any unpurged volumes in the flow path cause rapid dispersal of peak molecules – in other words, the peaks get wider. So you can't install a capillary column in a PGC not designed for it. Small-bore columns require small sample volumes, which may require special injection technology; and the detector volume must be close to zero. Even the interconnecting tubing must be short, with no unpurged cavities for molecules to diffuse into. Chapter 8 examines these extra-column requirements in full detail.

> *You can't install a capillary column in a PGC not designed for it. Capillary columns have stringent requirements for sample size, detector volume, and column-switching techniques.*

Efficiency and resolution

First and foremost, a capillary column has far more separating power than a packed column. It can generate an astounding plate number, typically 50,000–100,000 or more, which is way above the maximum attainable plate number for a packed column, at best 15,000–20,000. This high efficiency allows a capillary column to resolve hundreds of peaks. If that's what you want, this is a *clear win for the capillary column.*

Since resolution increases by the square root of plate number, a capillary column may achieve double the best resolution that a packed column can offer. This ultra-high resolution of analyte peaks is the key performance advantage of capillary columns, making them invaluable for separating complex samples with dozens of analytes. But most process applications are not like that; they commonly measure only one or two analytes. In these applications, resolution is not the main issue, which immediately diminishes the key advantage of the capillary. Then, another performance criterion might swing the decision toward packed columns.

It's useful to know that capillary columns are not especially efficient when measured by plate height. When working at maximum efficiency, a capillary column might achieve a plate height of 0.25 mm, but that's not the way we use them in a process chromatograph. In practice, we mainly use thicker-film columns with a plate height of 0.5–1.0 mm and that's hardly different from the plate height of an efficient packed column. To get the fastest analysis PGC columns usually run faster than optimum, so neither variety of column will operate with minimum plate height – as becomes abundantly clear in the experiment that follows.

Hinshaw and Ettre (1994, 92–96) ran two practical tests to compare a packed column and a thick-layer megabore column, both containing the same liquid phase. Table 3.8 summarizes their results. In the first test, they ran both columns at the same flow rate and injected into each column a test sample containing 10 components. Peak retention times were similar on both columns, giving about the same analysis time; but the capillary column produced narrower and more symmetric peak shapes than those obtained with the packed column, improving resolution.

> *In practice, we find that a 0.53 mm i.d. megabore column with a well-chosen liquid film thickness is often a direct replacement for a 1.0 mm i.d. micropacked column and sometimes for a regular ⅛-inch o.d. packed column.*

A key point to note is that the megabore column in this first test is working under extremely inefficient conditions, much too fast to realize its true capability. Its plate height of 6.28 mm is really terrible, being eight times the plate height of the packed column and delivering only 159 plates-per-meter. The column succeeds only because of its enormous length; 25 m versus 2 m for the packed column.

The second test ran the megabore at close to its optimum flow rate. It then returned a plate height of 1.22 mm and a plate number of 20,540 – almost eight times the plate count of the packed column. This is great, but plate number is not the only goal of column design. The laboratory chemist might be ecstatic to see such a beautiful chromatogram, but the process control engineer would not be happy at all. The analysis time is now six times longer

Table 3.8 Experimental Comparison of Packed and Capillary Columns.

Column Parameters	Packed	Capillary	
Length and bore	2 m × 2.0 mm	25 m × 0.53 mm	
Liquid loading/film thickness	8 %	5.0 μm	
Experimental Data	**Reference**	**First Test**	**Second Test**
Helium flow rate	20 mL/min	20 mL/min	1.4 mL/min
Column temperature	90 °C	90 °C	110 °C[a]
Differential pressure	207 kPa	124 kPa	7 kPa
Retention time (last peak)	172 s	166 s	1029 s
Plate number	2,585	3,978	20,540
Plate height	0.78 mm	6.28 mm	1.22 mm

[a]The second test used a higher temperature to reduce the long retention times.
Source: Adapted from Hinshaw and Ettre (1994, 94).

than it was before. Speed of analysis is often more important to a process measurement than it is to a laboratory test.

In summary, packed and capillary columns have similar plate heights, but capillary columns can achieve awesome resolution because their open-tubular construction offers far less resistance to flow. A long capillary column requires modest carrier gas pressure: for the second test in Table 3.8, the megabore used only 7 kPa for 25 m. This low-pressure gradient allows a capillary column to be extremely long, and a long column has more plates. However, more resolution may not be good if it comes at the price of extra analysis time and severely reduced sample size.

PGC column selection has to balance a number of conflicting variables, including analysis time. Fast analysis puts a strain on the equipment and reduces reliability, yet a fast analysis may be a sales necessity. Purchasing decisions are often influenced by promises made rather than real process requirements. Then, the column designer is forced to prioritize analysis time rather than long-term reliability.

Sample size

Injecting an excessive sample volume into any column will distort the chromatogram peaks and shift their retention times. There are two quite different constraints on sample size:

- The **feed volume** is the total volume of gas entering the column with the sample. It includes the sample volume plus any carrier gas mixed with it and is an indication of the total width of the sample injection profile.

- The **sample capacity** is the maximum amount of a specified analyte that a column can tolerate without excessively distorting its chromatogram peak.

Note that these apparently similar constraints originate from different chromatographic limitations and have opposite effects on the chromatogram peak shape.

Feed volume

The feed volume represents the width of the sample profile as it enters the column. An ideal injection profile is a narrow rectangular pulse, with all the sample molecules clustered together. It's evident, though, that the physical width of even such a perfect injection profile must depend upon both the injected sample volume and the sectional area of the column. Therefore, to maintain the same injection width, a column with half the internal diameter would require one quarter of the sample volume. Thus, narrow capillary columns demand smaller samples to avoid a reduction in efficiency.

In practice, the injection profile is unlikely to be ideal and will probably include some carrier gas, making the feed volume wider than expected.

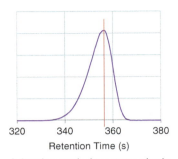

A fronting peak rises more slowly than it falls and may be a symptom of an excessive sample volume.

Figure 3.8 A Fronting Peak.

As an additional complication, an increase in peak width due to a larger feed volume has more effect on the performance of an efficient column than it does on an inefficient column. A narrow, highly efficient capillary column needs an extremely small sample to realize its full separating power. But the automated, reproducible injection of microgram samples is difficult. The PGC can use a sample splitter to reduce sample size and minimize feed volume, but laboratory experience with sample splitters has not been good. Sample splitters cause many of the problems reported with laboratory chromatographs.

A symptom of excessive feed volume, often called **column overload**, is a **fronting peak** on the chromatogram that falls more rapidly than it rises, as in Figure 3.8. This peak shape is due to an increase in solute affinity for the liquid phase at higher concentrations. As the peak rises, more solute dissolves in and becomes part of the liquid phase, which then retains the peak apex more strongly than its base. In extreme cases these peaks can look like right triangles, ramping up slowly and then falling rapidly to the baseline. See Chapter 8 for further details.

Peak fronting also occurs when the vaporization temperature of a liquid sample is too low, leading to a slow injection profile. This is not an issue with sample volume but rather with the speed of injection.

Sample capacity

Sample capacity[10] is the maximum amount of an analyte that a column can accept without excessive peak distortion. The specified sample capacity refers to an individual analyte and not to the whole sample volume. It's not unusual to deliberately overload a column with the major component of the sample when that component is not a target analyte. While excess analyte volume is rarely a problem with packed columns, it can severely distort the shape of an analyte peak on a narrow or lightly loaded capillary column.

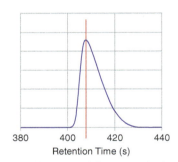

A tailing peak falls more slowly than it rises and may be a symptom of an excessive amount of the analyte.

Figure 3.9 A Tailing Peak.

It's also possible for a large component concentration to generate a flat-top peak. This may be due to saturation of the liquid phase without a change in component affinity, resulting in an on-column band of constant concentration.

A symptom of an excessive analyte amount is a **tailing peak** on the chromatogram that rises more rapidly than it falls, as in Figure 3.9. All normal peaks tail to some extent, so this is more pronounced tailing than normal and affects only the largest peaks on the chromatogram. Generally, this peak shape indicates that the affinity of the analyte for the stationary phase falls off at higher concentrations allowing the peak apex to move faster than its base.

Increasing the sample volume has at first only a small effect on column efficiency, but beyond a certain limit, additional sample volume causes a

---

[10] It's important not to confuse sample capacity with capacity ratio, an obsolete term for retention factor.

progressive increase in peak width and asymmetry. The specified sample capacity of a column permits a minor loss of efficiency – perhaps 20 % reduction in plate number. A minor distortion will not affect the measured area of the analyte peaks, as long as it doesn't spoil their resolution.

Theoretically, it's possible to show that the maximum volume (V_{max}) of a single analyte that a given column can tolerate depends on the retention factor (k) of the peak and the square root of plate number (N):

$$V_{max} \propto \frac{1+k}{\sqrt{N}} \qquad (3.7)$$

We will examine this relationship in Chapter 8. An exact calculation isn't possible as the relevant equation contains an undefined constant (Ettre and Hinshaw 1993, 100–102). Yet Equation 3.7 demonstrates that sample capacity is worse for early peaks in a chromatogram with low retention factors and improves for later peaks with more retention. It also confirms that very efficient columns with enormous plate numbers will not tolerate large samples.

From a practical standpoint, sample capacity is a subjective measure and reported values vary widely. This variation is understandable, considering the several column variables that affect its value. Here are the variables that improve sample capacity by increasing analyte retention factor:

➤ The selected liquid phase: a liquid with higher affinity for the analyte will tolerate a larger sample.

➤ The measurand: analyte peaks with higher affinity for the liquid phase, have increased sample capacity.

➤ The amount of liquid phase in the column: columns with higher percent loading (a lower phase ratio) contain more liquid which dissolves more analyte, thereby increasing retention factor and sample capacity.

➤ The column temperature: a lower column temperature increases analyte solubility in the liquid, thereby increasing retention factor and sample capacity.

And these variables increase sample capacity by reducing plate number:

➤ An increase in plate height of the column: a column with a high rate of carrier gas flow has a larger plate height and will produce less resolution, but it will tolerate a larger volume of each analyte.

➤ Having more liquid in the column (that is, a lower phase ratio): in addition to its effect on retention time, more liquid tends to increase plate height, so wider-bore or thicker-film capillary columns can tolerate larger samples than those with narrower bore or thinner films.

In an attempt to quantify these effects, Hinshaw and Ettre (1994, 109) used a calculation similar to Equation 3.7 to calculate the *relative change*

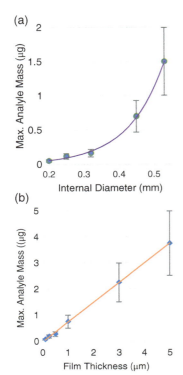

Showing recommended sample capacities for (a) different column diameters, and (b) different liquid film thicknesses. The stated sample capacity is for an analyte of similar polarity as the liquid phase and high retention factor. For a dissimilar analyte, capacity may be drastically less.

Figure 3.10 (a) Sample Capacity and Column Diameter. (b) Sample Capacity and Film Thickness.
Source: Adapted from Agilent Technologies, Inc.

to sample capacity when one column variable changes. According to their calculations:

> *Effect of column diameter*: a 0.53 mm i.d. column with a 1.0 μm liquid film (phase ratio: 53) has 2.5 times the theoretical sample capacity than a 0.32 mm i.d. column with the same film thickness (phase ratio: 32).

> *Effect of film thickness:* a 0.53 mm i.d. column with a 5.0 μm liquid film (phase ratio: 106) has 3.7 times the theoretical sample capacity than the same column with a 1.0 μm liquid film (phase ratio: 53).

These relative values may be useful for comparison, but they assume the column is operating at optimum velocity and the ratios may not be the same at higher flow rates.

Since a theoretical estimate of sample capacity is elusive, these authors decided to directly measure the sample capacity of several capillary columns working at their optimum carrier gas velocity (op. cit., 107–111). Their findings were informative, for instance:

> For a 25 m 0.53 mm column with a film thickness of 5.5 μm and a plate height of 0.472 mm, the empirical sample capacity was 2.5 μg (540 nL of undecane at 130 °C). Phase ratio: 24.

> For a 25 m 0.32 mm column with a film thickness of 0.26 μm and a plate height of 0.217 mm, the empirical sample capacity was 0.115 μg (24 nL of undecane at 110 °C). Phase ratio: 308.

The 0.53 mm thick-film column can tolerate 20 times more analyte than the 0.32 mm thin-film column. The capacity of more efficient columns is even less; their sample capacity drops rapidly with reduced bore size and reduced film thickness.

Larson (2006, 3) calculated the theoretical sample capacity for several equal-length capillary columns of different bores but with film thickness adjusted to give the same phase ratio. Thus, the retention data would be the same on each column. On that basis, he found a fivefold capacity difference between the 0.53 and 0.32 mm columns.

Figure 3.10 displays the effect of these two variables – bore size and film thickness – on the maximum mass of analyte recommended by a column manufacturer. These data are the maximum loading for an analyte with strong affinity for the liquid. For analytes with weak affinity, the sample capacity is much reduced. Contrarily, the capacity factor increases when a column operates inefficiently to yield a lower plate number, possibly because it has a thick liquid film or a high carrier gas flow rate. Such inefficient operating conditions are quite normal in process work as optimal flow rates can be painfully slow.

Finally, a note about units. Column vendors specify sample capacity in nanograms, the preferred unit of the laboratory. For process work, we need to think in volumetric terms. Here's a comparison:

> 1 μg of propane liquid is about 2.0 nL. That's a 1 % peak from a 0.2 μL liquid injection.

➢ 1 µg of propane gas is about 0.63 µL at 60 °C. That's a 1,000 ppm peak from a 0.63 mL gas injection.

Measurement sensitivity

Since the capacity factor is a constraint on the maximum concentration of an analyte, it has no effect on the minimum detectable concentration of that analyte. More relevant to ultimate sensitivity is the effect of feed volume. Larger sample volumes increase the width of the sample injection pulse, which makes the peaks a little wider and reduces the efficiency of the column. Clearly, the detrimental effect of a small width increase is much more pronounced in a high-efficiency column than it is in a low-efficiency column. This looks like a *big win for a packed column.*

Again, that would be true if comparing two columns at their maximum efficiencies. In process chromatography, though, we often separate only a few analytes and it's rare to need the maximum efficiency achievable by narrow-bore, thin-film capillary columns. We tend to use less efficient capillary columns and run them fast. Such columns can still provide plenty of resolution and a small loss of column efficiency due to a larger sample volume is not a big concern. It's not unusual to deliberately overload the first column when measuring low concentrations.

When making comparisons, it's always important to distinguish between performance variables quoted at maximum efficiency and the real-world conditions experienced by the columns. In process chromatography, it's rare for a column to run at maximum efficiency because it's too slow. Low-efficiency capillary columns have two additional advantages over packed columns:

➢ They produce better peak shapes: their thick liquid films block adsorption sites in the column walls leading to taller, symmetrical peaks.

➢ They produce less baseline noise: their more refined or bonded liquid phases reduce the bleed of foreign molecules into the detector.

In summary, capillary columns often compensate for reduced sample size by better chromatographic performance. However, when trying to resolve two peaks with similar retention factors, the demand for maximum efficiency will certainly limit the sample size and the achievable measurement sensitivity. That would be the time to look for a different liquid phase.

Analyte rangeability

Here, we consider the maximum difference in measurement range for two analytes on the same column. The sample size would be set to provide sufficient detector sensitivity for the analyte with the lower concentration. If that analyte requires a very low measurement range, the sample size would be large. Then, the analyte with the higher concentration might exceed the sample capacity of the column for that analyte. This is unlikely to occur with a packed column.

There are too many variables to prescribe any mathematical rules, but we can quickly see some general principles. The sensitivity of the detector determines the sample size, so the focus is on the larger peak. How large can it be? To maximize the sample capacity for this peak:

> A larger retention factor increases sample capacity, so choose a liquid phase having a strong retention for the larger peak. It's best for the trace peaks to come out first.

> A lower plate number increases sample capacity, so choose a liquid phase having a high separation factor to reduce the plate number needed for resolution.

Since small-bore capillary columns with thin liquid films have high plate numbers and lower sample capacity, those are the columns most likely to limit rangeability. Luckily, a PGC rarely needs such enormous plate numbers, so we tend to use the wider-bore capillaries with thicker films, and we run them fast. Such conditions increase plate height, reduce plate number, and usually provide adequate rangeability. Even so, you should be wary of applications that attempt to use capillary columns to measure percent and ppm ranges from the same sample injection. A packed column might be a better choice.

Column bleed

When carrier gas leaves a column and enters a detector it always contains traces of foreign molecules picked up during its passage through the last column. If the detector is sensitive to these molecules, they will elevate the chromatogram baseline and exacerbate detector noise. The presence of these alien molecules is known as **column bleed**. Although all columns bleed, it's the bleed created by the column feeding the detector that is of primary concern; that column may permanently retain or smooth the bleed from prior columns.

The column bleed may comprise molecules from various sources. Always present are molecules of the liquid phase. The carrier gas emerging from a liquid-phase column inevitably contains traces of that liquid phase due to its saturated vapor pressure at the oven temperature. To minimize this component of bleed, we choose liquid phases of high molar mass and low vapor pressure and operate them at the lowest workable temperature. Anyone familiar with a programmed-temperature chromatogram would attest to the increase in baseline seen as the temperature rises. That's why most liquid phases now in use are long-chain polymers having very low vapor pressures.

In isothermal applications, the liquid-phase bleed is constant, and the PGC automatically zeroes the detector signal to compensate for any offset.

Yet it's still necessary to minimize the bleed to avoid baseline upsets due to column switching or minor temperature variations. Capillary columns have the advantage here as they can use crosslinked polymers that bond to the active site in the column wall, practically eliminating bleed by vapor pressure.

Another consequence of liquid phase vapor pressure is the gradual loss of liquid phase from the column. Some packed columns in older PGCs used volatile liquid phases and suffered a progressive loss of resolution and limited column life. This is no longer a problem due to advances in column technology, but it's still worth noting that PGC columns should always operate far below the maximum temperature specified for their liquid phase.

A second cause of column bleed occurs with a new column and may be due to residual traces of the solvent used for its preparation, random molecules absorbed from the air, contaminated tube fittings, or fingerprints. These impurities elute as multiple broad peaks on an elevated baseline that exponentially approaches its final level. With a sensitive detector, it may take many hours for the baseline to become flat enough to use, and several days to become truly flat and level.

A third cause of column bleed is from chemical reactions occurring inside the column. These reactions are probably between the liquid phase and traces of oxygen or moisture that are always present in the carrier gas. In addition, thermal cracking of polymeric liquid phases may occur at higher temperatures. It's found that capillary columns with thick liquid films have more bleed than those with thin films, presumably due to more reaction products.

Detector signal noise seems to be correlated to the amount of liquid in the column, so reaction products may be a major contributor to random baseline noise. Signal noise may also correlate to the flow rate of carrier gas entering the detector. On both counts, this is a clear win for capillary columns. Those are unproven conjectures, but the empirical observation is that capillary columns exhibit less detector signal noise than packed columns, allowing the detector to measure smaller analyte peaks. This increased detector sensitivity often compensates for the smaller sample capacity of the capillary.

Finally, we should note that column bleed can also damage the detector. We don't use halogenated liquid phases because the products of combustion would corrode a flame detector. Of liquid phases used today, only silicon derivatives might harm the detector. Any bleed from a polysiloxane column would form silica deposits in a flame detector. However, most capillary columns use crosslinked and bonded polymers that have very low vapor pressure and the carrier gas flow rate is much lower than an equivalent packed column. These features severely limit the amount of silicon entering the detector.

Table 3.9 summarizes the advantages and disadvantages of the various packed and wall-coated columns.

Table 3.9 Comparison of PGC Column Types.

Column Type	Advantages	Disadvantages
Packed i. d. > 1 mm.	Wide variety of available stationary phases. Larger sample capacity gives good measurement linearity. High sensitivity for permanent gases. Tolerant of older hardware. More durable at jobsite. Least expensive option.	The high pressure drop limits column length and plate number. Limited resolution. Consumes more carrier gas. Liquid phase may be unstable. High column bleed. Low maximum temperature.
Micropacked i. d. ≤ 1 mm	Fast analysis for a few peaks. More efficient; can resolve many peaks. Uses less carrier gas than packed column.	Higher pressure drop limits length and plate number. Lower sample capacity may limit analyte sensitivity. Columns are expensive.
Megabore 0.53 mm i.d. With thick liquid film ($d_f > 1$ µm)	Practically unlimited column length. Can quickly resolve several analyte peaks: fast analysis. More inert, less peak tailing. High sample capacity for good analyte sensitivity. Can directly replace a micropacked column. Low carrier gas consumption.	Risk of blockage. Fused silica type is fragile. Not suitable for analytes with high affinity for liquid phase. Peaks may tail. More column bleed than thin films. Limited selection of liquid phases.
Megabore 0.53 mm i.d. With thin liquid film ($d_f \leq 1$ µm)	Practically unlimited column length. More efficient, can resolve dozens of peaks. Bonded phases produce less column bleed. resulting in less baseline offset and noise. Low carrier gas consumption. Metal capillaries are durable.	Risk of blockage. Low sample capacity may limit analyte sensitivity. Can't measure multiple analytes with large disparity of measurement range. Not suitable for analytes with low affinity for liquid phase. Fused silica type is fragile. Needs compatible hardware.
Wide bore 0.32 mm	Resolves scores of peaks. Lower column bleed. Less baseline noise. Less often used in PGC.	More risk of blockage. Lower sample capacity may limit analyte sensitivity. Difficult to ensure small sample injection repeatability. Tolerates less measurement range disparity. May need a temperature program.
Narrow bore ≤ 0.25 mm	Resolves hundreds of peaks.	Not supported by most PGCs. Needs sample splitter.

Knowledge Gained

About Theory

- *In gas chromatography, the stationary phase may be a solid (GSC) or a liquid (GLC).*
- *In a GLC column, separation is due to solvation of molecules into a stationary liquid.*
- *In a GSC column, separation is due to adsorption of molecules onto a solid surface.*
- *Separation is due to sample molecules having differing affinity for the stationary phase.*

- *Three electronic forces contribute to the overall affinity: dispersion, dipole, and ionic forces.*
- *Random electron movement causes multiple transient forces between molecules.*
- *This electronic dispersion force is ubiquitous between molecules but is weak.*
- *Larger molecules experience larger dispersion forces causing longer retention times.*
- *Boiling points also depend on dispersion forces, so they correlate with retention times.*
- *Retention on a nonpolar column is by dispersion forces and homologs elute in boiling point order.*
- *Excessive feed volume will overload the column and produce fronting chromatogram peaks.*
- *Excessive analyte volume may exceed sample capacity and produce a tailing analyte peak.*
- *Retention on a polar column is additional and is caused by an attraction between dipoles.*
- *A dipole is an electronic charge due to the displacement of molecular electrons.*
- *A molecule is permanently polarized if it has an asymmetric structure with dissimilar atoms.*
- *A neutral molecule can be polarized by the dipole of an adjacent polar molecule.*
- *Ionic phases offer the highest polarity and may become popular in the future.*
- *A good rule-of-thumb is to use a liquid phase that is chemically similar to the analytes.*
- *The phase ratio (β) used in column design is the volume ratio of gas and liquid in the column.*
- *Same β-value columns give the same separation but the thinner film gives better resolution.*
- *Column design is always a compromise between speed, capacity, and resolution.*
- *To get a faster analysis, PGCs often run faster than the optimum carrier gas flow rate.*
- *The sample capacity for an analyte is higher for lower plate numbers and for later peaks.*

About packed columns

- *A packed column is full of particles that contain the stationary phase and are all about the same size.*
- *Most PGCs now use ⅛-inch o.d. packed columns or ¹⁄₁₆-inch o.d. micropacked columns.*
- *Before packing, clean the column tubing with a sequence of polar and nonpolar solvents.*
- *A GSC packing may be solid particles of activated charcoal, silica gel, or molecular sieves.*
- *A GSC packing may be solid granules of a polymer material such as crosslinked polystyrene.*
- *A GLC packing may be inert support particles coated with an involatile liquid phase.*
- *The carrier gas must be free of water vapor, particularly for active solid columns.*
- *It may be possible to reactivate an active solid column by heating it with a dry nitrogen purge.*
- *In theory, the best plate height is twice the particle size (≈ 0.3 mm) but is about 0.5 mm in practice.*
- *Micropacked columns can achieve plate heights of 0.2–0.3 mm and are now popular in PGCs.*
- *Packed columns have a high flow resistance so they have limited length and limited plate number.*

- *Packed columns have a larger sample capacity, which may result in more detector sensitivity.*
- *They may generate more column bleed and detector noise that limits their overall sensitivity.*
- *They have a wide dynamic range for measuring multiple analytes at different concentrations*

About capillary columns

- *A capillary column is an open tube with stationary phase coated on or bonded to its inner walls.*
- *To use capillary columns, a PGC must have low internal volumes and no unswept flow paths.*
- *Most laboratory chromatographs use capillary columns made from fused silica tubing.*
- *But PGCs prefer silicon-coated stainless-steel tubing, as it's more durable than fused silica.*
- *Open tubular columns are porous layer (PLOT), support coated (SCOT), or wall coated (WCOT).*
- *PLOT columns have a thin layer of adsorbent solid on their inner wall.*
- *SCOT columns have a thin layer of liquid-coated support particles on their inner wall.*
- *WCOT columns have a thin layer of liquid phase on their inner wall, 0.1–10 µm thick.*
- *The liquid phase is often a copolymer of methyl and phenyl siloxanes that may bond to the tube wall.*
- *PGCs can't use the most efficient columns that use <0.25 mm i.d. tubing and ≈0.1 µm liquid film.*
- *Such columns get plate heights of ≈0.2 mm and high plate numbers but need very small samples.*

- *Most PGCs use packed columns, although capillary columns are popular for certain applications.*
- *Most PGCs use packed columns isothermally to separate only a few analyte peaks.*
- *Most PGCs use the 0.53 mm i.d. or 0.32 mm i.d. columns with relatively thick liquid films.*
- *As used in PGCs, capillary and packed columns achieve similar plate heights.*
- *Capillary columns have low flow resistance so they can be long, generating very high plate numbers.*
- *They need smaller sample volumes than packed columns, which may decrease detector sensitivity.*
- *To achieve fast injection, a sample splitter may be necessary and these can be troublesome.*
- *Capillary columns may generate less column bleed and detector noise, enhancing their overall sensitivity.*
- *They have limited ability to measure high and low concentrations on the same chromatogram.*
- *Most laboratory GCs use capillary columns with a temperature program to separate many analytes.*
- *As yet, capillary columns are less popular than packed columns in process gas chromatographs.*

Did you get it?

Self-assessment quiz: SAQ 03

Q1. True or false?

Generally speaking, when compared with capillary columns, packed columns:
- **A.** Can be much longer.
- **B.** Have smaller internal diameters.
- **C.** Have lower plate heights and are less efficient.
- **D.** Have higher plate numbers.
- **E.** Can accept larger sample size.
- **F.** Are more likely to employ a temperature program.

Q2. True or false?

Generally speaking, when compared with packed columns, capillary columns:
- **A.** Can be much longer.
- **B.** Have smaller internal diameters.
- **C.** Have lower plate heights and are less efficient.
- **D.** Have higher plate numbers.
- **E.** Can accept larger sample size.
- **F.** Are more likely to employ a temperature program.

Q3. True or false?

When compared with regular packed columns, micropacked columns generally:
- **A.** Can be much longer.
- **B.** Have smaller internal diameters.
- **C.** Have lower plate heights and are less efficient.
- **D.** Have lower plate heights.
- **E.** Can accept larger sample size.
- **F.** Are more likely to employ a temperature program.

Q4. Consider a sample mixture of three components: isobutane, 1,3-butadiene (a polarizable molecule), and n-butane, having boiling points of −11.7, −4.4, and −0.5 °C, respectively. Predict their order of elution from (a) a nonpolar column and (b) a polar column.

Q5. Is it best to use a column having a high phase ratio (β) to separate sample components of low molar mass? Explain.

Q6. Nominate and explain at least two reasons for running a column much faster than its optimum carrier velocity.

Q7. Calculate the theoretical maximum plate number for a peak having a retention time of five times the holdup time on a capillary column that is 30-m long and has 0.53 mm bore.

Check your SAQ answers with those given in the back of the book.

References

Further reading

This chapter is about the columns used in *PGC*s. For a more general overview of column technology in analytical chemistry and research, refer to:

➤ An up-to-date review article: *Evolution and Evaluation of GC Columns* by Mametov et al. (2019).

- A comprehensive encyclopedia article: Basic Overview on Gas Chromatography Columns by Rahman et al. (2015).
- A short book by two expert professors: Columns for Gas Chromatography by Barry and Grob (2007). Excellent.
- A fascinating personal reminiscence: Evolution of Capillary Columns for Gas Chromatography by Ettre (2001).
- A delightfully clear and non-theoretical handbook: Introduction to Open-Tubular Column Gas Chromatography by Hinshaw and Ettre (1994). Highly recommended.

Cited

Annino, R. and Villalobos, R. (1992). *Process Gas Chromatography: Fundamentals and Applications*. Research Triangle Park, NC: Instrument Society of America.

Barry, E.F. and Grob, R.L. (2007). *Columns for Gas Chromatography: Performance and Selection*. Hoboken, NJ: John Wiley & Sons, Inc.

Cramers, C.A. and Rijks, J.A. (1979). Micropacked columns in gas chromatography: An evaluation. In: *Advances in Chromatography*, Volume **17** (eds. J.C. Giddings, E. Grushka, J. Cazes, and P.R. Brown), 101–162. New York, NY: Marcel Dekker, Inc.

Dandeneau, R.D. and Zerenner, E.H. (1979). An investigation of glasses for capillary chromatography. *Journal of High Resolution Chromatography & Chromatography Communications* **2**, 351–356. https://doi.org/10.1002/jhrc.1240020617

de Zeeuw, J. (2015). *The impact of GC parameters on the separation, Parts 1–6*. A series of technical articles published by Restek Corporation, Middelburg, Netherlands. Accessed 2020/10/27 at: www.restek.com/pdfs/Impact-of-GC-Parameters_Part1.pdf et seq.

Ettre, L.S. (2001). *Evolution of capillary columns for gas chromatography*. LCGC North America, Vol. 19, No. 1 (January). Accessed 2020-10-25 at: https://www.chromatographyonline.com/view/evolution-capillary-columns-gas-chromatography.

Ettre, L.S. and Hinshaw, J.V. (1993). *Basic Relationships of Gas Chromatography*. Cleveland, OH: Advanstar Communications, Inc.

Golay, M.J.E. (1957). Theory and practice of gas-liquid partition chromatography with coated capillaries. *Gas Chromatography; Proceedings of the 1957 ISA Analysis Instrumentation Division Symposium at Lansing, MI* (eds. V.J Coates, H.J. Noebels, and I.S Fagerson), 1–13. New York, NY: Academic Press.

Golay, M.J.E. (1958). Theory of chromatography in open and coated columns with round and rectangular cross-sections. In *Gas Chromatography – Amsterdam 1958* (ed. D.H. Desty), 36–55 and 62–68. London, UK: Butterworths Scientific Publications Ltd.

Hinshaw, J.V. and Ettre, L.S. (1994). *Introduction to Open-Tubular Column Gas Chromatography*. Cleveland, OH: Advanstar Communications, Inc.

IUPAC (2014). *Compendium of Chemical Terminology*, Second Edition. Compiled by A.D. McNaught and A. Wilkinson (last revised February 2014). Oxford, UK: Blackwell Scientific Publications. Accessed 2020/04/17 at: https://goldbook.iupac.org/terms/view/D01814.

Jennings, W. (1978). *Gas Chromatography with Glass Capillary Columns*. New York, NY: Academic Press Inc.

Jones, S. (2008). *Secrets of GC Column Dimensions*. PowerPoint presentation dated May 20, 2008. Accessed 2020/05/18 at: www.agilent.com/cs/library/slidepresentation/Public/Secrets_GC_Column_Dimensions.pdf.

Larson, P. (2006). Column technology in gas chromatography. In *Encyclopedia of Analytical Chemistry*. Chichester, UK: John Wiley & Sons, Ltd. https://doi.org/10.1002/9780470027318.a5502.

Lipsky, S.R. (1983). Flexible soft glass capillary columns vs. flexible fused silica capillary columns for gas chromatography – a critique. *Journal of High Resolution Chromatography* **6**, 359–365. https://doi.org/10.1002/jhrc240060703.

London, F. (1937). The general theory of molecular forces. *Transactions of the Faraday Society* **33**, 8–26.

Mahler, H. (2021) Personal communication, 2021/02/21.

Mametov, R., Ratiu, I.A., Monedeiro, F., Ligor, T., and Buszewski, B. (2019). Evolution and evaluation of GC columns. *Critical Reviews in Analytical Chemistry*. https://doi.org/10.1080/10408347.2019.1699013.

Müller, F. and Oreans, M. (1977). Experience and problems with capillary glass columns in process chromatography. *Chromatographia*, **10**, No. 8 (August), 473–477.

Ottenstein, D.M. (1963). Column support materials for use in gas chromatography. *Journal of Chromatographic Science*, **1**, No. 4 (April), 11–23. https://doi.org/10.1093/chromsci/1.4.11.

Poole, C.F. and Poole, S.K. (2011). Ionic liquid stationary phases in gas chromatography. *Journal of Separation Science* **34**, 888–900. Wiley-VCH Verlag GmbH & Co. KGaA, Weinheim.

Rahman, M.M., El-Aty, A.M.A., Choi, J.H., Shin, H.C., Shin, S.C., and Shim, J.H. (2015). Basic overview on gas chromatography columns. In: *Analytical Separation Science*, First Edition (eds. V.P. Estévez, A.M. Stalcup, J.L. Anderson, and A. Berthod), 823–834. Weinheim, Germany: Wiley-VCH Verlag GmbH & Co. KGaA.

Tables

3.1 Subjective Evaluation of Columns Used in PGCs.
3.2 Typical Tubing for Packed PGC Columns.
3.3 Selected Mesh Sizes.
3.4 Typical Tubing for PGC Capillary Columns.
3.5 Phase Ratios of WCOT Columns.
3.6 Phase Ratio and Column Performance.
3.7 Optimum Flow Rates for Capillary Columns.
3.8 Experimental Comparison of Packed and Capillary Columns.
3.9 Comparison of PGC Column Types.

Figures

3.1 Three Kinds of Capillary Column.
3.2 Origin of Dispersive Forces.
   (a) Two adjacent neutral atoms or molecules. (b) Random electron shift causes a transient dipole. (c) Induced dipole forms an attractive force.

Column technology

3.3 Nonpolar and Polar Molecules.
3.4 Typical Packed Columns.
3.5 Typical PGC Capillary Columns.
3.6 The Column Performance Triangle.
3.7 Theoretical Minimum Plate Height.
3.8 A Fronting Peak.
3.9 A Tailing Peak.
3.10a Sample Capacity and Column Diameter.
3.10b Sample Capacity and Film Thickness.

Symbols

Symbol	Variable	Unit
d_c	Internal diameter of column tubing	mm
d_f	Film thickness of liquid phase in column	µm
H_{Th}	Theoretical minimum plate height	mm
L	Column length	mm
k	Retention factor of component peak	none
N	Plate number of column	none
V_G	Gas volume of column	mL
V_L	Liquid volume of column	mL
V_{max}	Maximum volume of analyte injected	mL
β	Phase ratio	none

Equations

3.1 $V_G = L \cdot \pi \cdot \left(\dfrac{d_c}{2}\right)^2$ Volume of the column gas phase (approximate).

3.2 $V_G = L \cdot \pi \cdot \left(\dfrac{d_c}{2} - d_f\right)^2$ Volume of the column gas phase (exact).

3.3 $V_L = L \cdot \pi \cdot d_c \cdot d_f$ Volume of the column liquid phase.

3.4 $\beta = \dfrac{V_G}{V_L}$ Phase ratio.

3.5 $\beta = \dfrac{d_c}{4 d_f}$ Calculation of phase ratio for a capillary column.

3.6 $H_{Th} = d_c \cdot \sqrt{\dfrac{1 + 6k + 11k^2}{12(1+k)^2}}$ Theoretical minimum plate height of a capillary column.

3.7 $V_{max} \propto \dfrac{1+k}{\sqrt{N}}$ Theoretical maximum analyte injection volume.

References

New technical terms

When first introduced, these words and phrases were in bold type. You should now know the meaning of these practical terms.

capillary column	**molecular sieve**
column bleed	**packed column**
column overload	**open-tubular column**
diatomaceous earth	**PLOT column**
electropolishing	**sample splitter**
film thickness	**SCOT column**
fronting peak	**silicon coating**
ionic liquid phase	**solid support**
liquid loading	**tailing peak**
megabore tubing	**WCOT column**
mesh size	
micropacked column	

You should also know the meaning of these theoretical terms.

adsorption	**nonpolar column**
affinity	**permanent dipole**
Debye forces	**phase ratio**
dipole	**polar**
dispersive forces	**polar column**
electronegativity	**polarization**
feed volume	**polarizability**
hydrogen bond	**sample capacity**
induced dipole	**solvation**
Keesom forces	**surface deactivation**
kinetic theory	**UTE**
London forces	**Van der Waals forces**
nonpolar	

Consult the Glossary at the end of the book for more information.

4

The stationary phase

"Most PGC columns use liquid stationary phases. There are vastly more liquid phases available than there are solid phases, so our primary focus is on liquid phase selection and use. However, solid phases do separations that no other column can do, so they are important too."

Liquid phases

Since it's unlikely that you'll make your own columns, this text discusses liquid phases in general terms with typical examples. For a more complete listing of liquid phases and their properties, refer to the resources in the *Further Reading* section at the end of the chapter. Be aware, though, that manufacturers are continually developing new liquid phases, so consult the sales literature of a specialist GC supplier for the most recent data.

The need for variety

Packed columns employ a wider variety of liquid phases than capillary columns do. The reason is partly historical – early researchers tried just about every available liquid as a stationary phase! But, a more pressing issue is the limited efficiency of the rugged columns preferred in most PGCs. We shall see that low-efficiency columns need a high separation factor between peaks, sometimes called **selectivity**. And high separation factors need a wide range of liquid phases.

In contrast, a typical laboratory gas chromatograph can achieve most of the separations it needs using just a few capillary columns operating at high efficiency. When a column has a high plate number, the separation factor is less important and there's no need for a large selection of liquid phases.

Process Gas Chromatography: Advanced Design and Troubleshooting, First Edition. Tony Waters.
© 2025 John Wiley & Sons Ltd. Published 2025 by John Wiley & Sons Ltd.

It's certainly true that future PGCs will use more capillary columns, but many of those columns will not operate at high efficiency. To achieve fast analysis, large sample capacity, and adequate detector sensitivity, it's likely that the capillary columns in future PGCs will continue to exhibit modest plate numbers. And again, that will demand liquid phases with high selectivity between analyte peaks. A wide choice of stationary phase will always be a high priority for a PGC.

Table 4.1 Vendor Codes for Columns and Liquid Phases.

Code	Manufacturer
ArD	ASDevices
AT	Alltech
BP	Trajan (SGE)
CP	Agilent Varian
DB	Agilent J&W
DC	Dow Corning
EC	Alltech
HP	Agilent HP
MXT	Restek
OPTIMA	Macherey-Nagel
OV	Ohio Valley
QF	Dow Corning
Rtx	Restek
SE	Agilent (J&W)
SPB	Supelco
TG	Thermo Scientific
TR	Thermo Scientific
VB	Valco Instruments
VF	Agilent Varian
ZB	Phenomenex
007	Quadrex

Column identification

In the beginning, PGC manufacturers always specified the columns used, including full details of the stationary phase, liquid loading, mesh size, and solid support. Before long, third-party suppliers entered the replacement column business, so to protect their spare parts revenue the PGC vendors resorted to tagging their columns with arcane part numbers instead of giving us detailed and useful information. Then, as the competition grew, aftermarket suppliers were themselves obliged to label their columns and even their liquid phases with cryptic code names. Now it's hard to make any sense of their product catalogs.

Most suppliers label their proprietary liquid phases by an alpha-numeric code, often two alpha characters followed by a number. There are over 1,000 branded code names for GC liquid phases. Nobody knows them all. Table 4.1 lists the alpha codes for the main suppliers.

With so many brand names it's confusing to search for a suitable liquid phase. The largest suppliers recommend the type of analysis each phase is good for – like "aromatics" – but that's not helpful if you need a specific separation. They also provide an equivalence table showing for each of their phases which competitive products they consider essentially equivalent. That might be useful if you are replacing a known column, but there can be subtle differences between column manufacturers that can spoil critical-pair resolution or analyte peak symmetry.

Liquid phase polarity

We loosely categorize liquid phases as nonpolar, semipolar, and strongly polar. This polarity scale alludes to the *dipole moment* of the electronic charges occurring within the liquid phase molecules and their ability or inability to form *hydrogen bonds*, concepts we explored in Chapter 3.

In reality, polarity is an imperfect way to categorize a liquid phase because the affinity of a solute for a stationary phase depends on a combination of electronic forces that varies with the solute:

> *Polarity is an imperfect measure of liquid phase behavior. It's like knowing the strength of a magnet: it cannot tell you the force of attraction until you know what it's attracting.*

Table 4.2 Typical Polarity of Solutes.

Solute Type	Typical Examples	Typical Stationary Phases
Nonpolar Solutes		
Permanent gases	Hydrogen, oxygen, nitrogen, methane, carbon monoxide	Molecular Sieves, n-Octane on Res-Sil C, HayeSep, Carbosieves, Carboxen
	Argon in oxygen	ArDSieve
	Carbon oxides, noble gases	Porous polymers
Saturated hydrocarbons with single C—C bonds	n-Paraffins, isoparaffins, cycloalkanes	Porous polymers Dimethyl silicones, low-phenyl silicones (XX-5), SE-30, SPB-Octyl
Symmetric sulfides and halides	Carbon disulfide, symmetric mercaptans, halocarbons	Dimethyl silicones, low-phenyl silicones
Polarizable Solutes		
Hydrocarbons with double or triple C—C bonds	Alkenes, alkynes, aromatics, carbon disulfide	Fluorosilicones, cyanosilicones, polyols
	Olefin cis and trans isomers	High-cyano cyanosilicones (for example, OV-275, SP-2340)
Weakly Polar Solutes		
Hydrocarbons containing an ether, carbonyl, or nitrile group *These compounds are also proton acceptors and form hydrogen bonds with a proton-donor liquid phase*	Ethers	Polyesters, FFAP, TCP, polyols
	Aldehydes and ketones	Fluorosilicones, dialkyl phthalates, FFAP, polyols
	Esters	Fluorosilicones
	Nitrile, nitro, tertiary amines, or sulfur compounds	Fluorosilicones
Moderately Polar Solutes		
Hydrocarbons containing a hydroxy or amine group and asymmetric alkyl halides *These compounds are also proton donors and form strong hydrogen bonds with a proton-acceptor liquid phase*	Alcohols, phenols, organic acids, halocarbons, freons	Cyanosilicones, polyols, dialkyl phthalates
	Ammonia, amines	Porous polymers, not polyols
	Hydrogen cyanide	Polyesters
Highly Polar Solutes		
Compounds containing multifunctional groups, and some compounds of sulfur, phosphorus, or halogens	Hydroxyl acids, glycols, thiols	Porous polymers, dicyano and phenylcyano silicones, TCEP
Ionizable substances	Water, acids	Ionic liquid phases

See Tables 4.4, 4.5, 4.9, and 4.10 for an interpretation of stationary phase names.
Source: Adapted from Supelco (2015a).

Therefore, **solute polarity** is just as important as liquid phase polarity. As a general guide, Table 4.2 indicates the relative polarity of some typical solutes. The table is imprecise because the observed polarity of a solute also depends on the liquid phase it's contacting.

Liquid phase polarity is still a useful guide for column selection. Liquid phase suppliers sometimes have their own scale of polarity, often based on squalane as 0-polarity and TCEP *(1,2,3-tris(2-cyanoethoxy) propane)* as 100-polarity. Other polarity scales use the silicone equivalents of those traditional phases instead, as discussed herein.

Solute affinity

Recall that, for separation to occur, the sample components must dissolve in the liquid phase. The chemist's adage that "like dissolves like" is helpful here. Simplistically, we should choose a liquid phase having a molecular structure similar to the analytes we wish to separate.

> *Since "like dissolves like," it's often good to choose a liquid phase that is chemically similar to the analytes we wish to measure.*

When evaluating the behavior of different solutes on a liquid phase, recall from Chapter 3 that four electronic forces are potentially at work inside the column, three of them polar:

➤ Dispersive *London Forces* proportional to the size of the interacting solute and solvent molecules.

➤ Polar *Debye Forces* due to the interaction of a permanent dipole with a dipole induced in another molecule.

➤ Polar *Keesom Forces* due to the presence of permanent dipoles in both of the interacting molecules.

➤ Polar hydrogen bonding, a strong attraction typically between a hydrogen atom and an electronegative atom like oxygen or sulfur.

The *affinity* of a solute for a given liquid phase is the cumulative effect of all four of these forces, so the size of the interacting molecules is as important as their polarity. Let's examine how this affects the retention time of a solute.

A fully nonpolar liquid phase has no dipole moment, cannot be polarized, and has no hydrogen-bonding capability; so it cannot exert a polar force on a solute molecule regardless of solute polarity. On such a column, the adjusted retention time of a peak depends solely on the net dispersion force between solvent and solute. It turns out that the vapor pressure of a substance also depends on dispersion forces, so the adjusted retention time of a solute on a nonpolar column is a function of its vapor pressure. In other words, the peaks come out in boiling point order.

Recall that dispersion forces are always present, so all columns produce an underlying separation in boiling point order. On top of that, a polar liquid phase exerts an additional affinity for polar solutes, retaining them longer than their nonpolar siblings. Polar columns can therefore separate substances having similar boiling points.

But even a highly polar liquid phase can't exert a dipole attraction for a fully nonpolar solute because that solute has zero dipole moment and

Table 4.3 Strength of Solvent–Solute Interaction.

Solvent \ Solute	Nonpolar	Polar
	Peak order on chromatogram	
Nonpolar	By boiling point	By boiling point (BP)
Polar	By boiling point	By BP plus polarity

cannot share hydrogen bonds. Common experience tells us that a polar solvent (think water) won't dissolve much of a nonpolar solute (think oil). Thus, nonpolar solutes elute rapidly from a polar column in boiling point order.

Table 4.3 summarizes these interactions. To get a polar attraction, solute and solvent must both have some degree of polarity. Luckily, many polar liquids are available, so we have more opportunities for separating polar solutes than nonpolar solutes.

Polar or Nonpolar?

When there's a choice, we prefer nonpolar phases. However, for polar solutes like alcohols and ketones, a polar liquid phase is often needed to get good separations and avoid distorted peaks. The general rule is to use a liquid phase similar to the analytes you wish to measure.

Nonpolar liquid phases tend to be more stable and resistant to oxygen or moisture in the carrier gas or sample. They give better coating efficiencies and sharper peaks than polar phases and create less **column bleed**. And, of course, their separations dutifully follow the solute boiling points: it's all very bland, and reliably so.

> *In contrast, polar phases are the spices that raise column design to a work of art. The skilled practitioner chooses from a delightful array of delicate flavors to separate and measure just about anything.*

PGC's often use two or more columns to get a fast analysis and it's not unusual for one column to be nonpolar and the other to be polar. For instance, a common PGC column system uses a nonpolar first column to hold back **heavies** (high-boiling components) until the lighter components have entered the following polar column. The first column is then backflushed to remove the heavies. Since the heavies never enter it, the second column can strongly retain the polar components without fear of them colliding with any heavies peaks. Another application might do the opposite: a polar first column removes the highly polar components so the nonpolar second column can separate the paraffins.

For more about backflush systems, refer to Chapter 10.

Silicone liquid phases

In any discussion about liquid phases, long chemical names are inevitable. Unless you are a chemist, such names can be daunting at times, but try not to let them intrude in your understanding of how liquid phases work. Actually, the mechanism of retention is entirely electronic!

For reference, the Glossary of Terms at the end of the book has a short explanation of each chemical name used herein.

Methyl silicones

The most-used liquid phases nowadays are pure silicone oils or viscous gums. Although there are dozens of varieties, all silicones are **siloxane** polymers. These polysiloxanes are ideal liquid phases as they exhibit low vapor pressure and high structural stability. In a PGC, a packed or capillary column with a silicone liquid phase will operate reliably over a wide range of column temperatures with low levels of column bleed.

In a laboratory gas chromatograph operating on a temperature program, the column temperature can sometimes get high enough to fracture their polymer chains (Blomberg 2001). The volatile fragments then released are visible as an elevated baseline on the chromatogram.

The basic silicone, chemically speaking, is poly[dimethylsiloxane] and is by far the most used liquid phase. As seen in Figure 4.1, the backbone of this polymer comprises a large number (n) of alternate oxygen (O) and silicon (Si) atoms. In addition, each silicon atom bonds to two methyl groups ($\cdot CH_3$).

For economy of space, we can represent that structure as:

$$(CH_3)_3Si \cdot [O \cdot Si(CH_3)_2]_n \cdot O \cdot Si(CH_3)_3$$

Poly[dimethylsiloxane] is an almost nonpolar liquid phase. Compared with squalane, the classic zero-polarity liquid phase, it's slightly polar but has

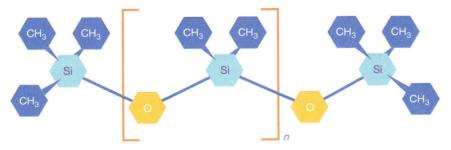

The oxygen–silicon chain structure of the polysiloxanes gives them low vapor pressure and high stability – ideal for process gas chromatography. Shown is the structure of poly[(dimethyl)siloxane]. Note that each silicon atom in the chain bonds to two methyl groups.

Figure 4.1 Structure of Poly[dimethylsiloxane].

lower vapor pressure and higher stability, both desirable features for process gas chromatography.

This silicone oil is also the base from which we build all the other silicones. By replacing some of its methyl groups with other functional groups, we can tune the silicone for low, medium, or high polarity. This is the secret of its success.

Phenyl silicones

There are two ways to install replacement groups in the polymer; either change the structure of the dimethylsiloxane monomer or make a copolymer from a mixture of two different monomers.

The first method is simply to replace one or both of the methyl groups in the dimethylsiloxane monomer with another group. If we replace one of the methyl groups ($\cdot CH_3$) with a phenyl group ($\cdot C_6H_5$), the polymer becomes poly[methylphenylsiloxane], a more polar silicone:

$$(CH_3)_3 Si \cdot [O \cdot Si(CH_3)(C_6H_5)]_n \cdot O \cdot Si(CH_3)_3$$

This polymer owes its high polarity to the large number of phenyl groups in its structure – one on each silicon atom. Note that the polar phenyl groups have replaced exactly 50 % of the nonpolar methyl groups along the polysiloxane chain, so the polymer is predicably symmetric. But that sets a limit to the flexibility of this technique because each attached group inevitably becomes 50 % of the total attachments.

The second method is more flexible. To make a bespoke phenyl-silicone of any desired polarity, we adjust the phenyl content of the polymer by starting from a known mixture of two monomers. This is how it works: a copolymer of $(100 - n)$ % dimethylsiloxane and n % diphenylsiloxane has those same percentages of dimethyl and diphenyl nodes randomly scattered along its backbone:

$$(CH_3)_3 Si \cdot [O \cdot Si(CH_3)_2]_{(100-n)\%} \cdot [O \cdot Si(C_6H_5)_2]_{n\%} \cdot O \cdot Si(CH_3)_3$$

The polarity of the copolymer depends on the percentage of phenyl in its structure. Multiple varieties are available with 5–75 % of phenyl, and their polarity increases accordingly. The trade-off is that the efficiency of a silicone column falls off as its polarity increases; that is, columns with higher polarity tend to have a reduced plate number.

The copolymer made by this method is not quite the same as the polymer made by the first method because each silicon atom now carries two identical groups, rather than one of each. For instance, in the 50 % phenyl copolymer half of the silicon atoms have two methyl groups and the other half have two phenyl groups. However, when we test the two liquids we find very little difference in performance. It's the total amount of phenyl that dictates polarity, not the location of those phenyl groups.

For their competitive products, manufacturers know that their customers are looking for a specific phenyl percentage, so they mark their products accordingly. But they market their unique formulations without stating the copolymer percentages, or even as "proprietary" with no information at all. It's safe to say there are many different and unknown percentages in use.

The presence of phenyl groups in the silicone increases its affinity for both nonpolar and polar solutes. The increase in nonpolar retention is due to stronger dispersion forces between the large phenyl groups and solute molecules. In addition, the phenyl groups are highly polarizable and will selectively retain olefinic and aromatic analytes, as well as polar solutes containing oxygen, nitrogen, or sulfur atoms. These strong interactions allow us to adjust the selectivity of the liquid phase to suit the desired analytes simply by changing its phenyl content.

More possibilities

To get an even larger range of polarities and some subtle differences in specific separations, polysiloxanes can contain novel functional groups. These new functional groups can substitute for one or both of the methyl groups in the original dimethylsiloxane monomer. The modified monomer can then form a polymer alone, or as a copolymer with a known percentage of another siloxane, original or modified.

The key to understanding this bewildering array of silicone liquid phases is to realize that there are four points of attachment in the copolymer, represented as A, B, C, and D in the structure:

$$(CH_3)_3Si \cdot \left[O \cdot Si{-A \atop -B}\right]_{(100-n)\%} \cdot \left[O \cdot Si{-C \atop -D}\right]_{n\%} \cdot O \cdot Si(CH_3)_3$$

The manufacturer can select each of the groups A–D from a wide range of potential attachments and then decide the percentage (n) of each monomer in the copolymer product. That's enormous flexibility. Yet many of the proprietary concoctions offered for sale are effectively equivalent to each other. Their composition remains secret, hidden behind a cryptic product code. Table 4.4 gives several examples of these modified silicones and illustrates their wide polarity range.

Nonpolar silicones

Dimethyl silicone is considered to be nonpolar, but in practice it's about 5 % polar when compared with squalane, a branched-chain C_{30} paraffin. You can see this in Table 4.4. Replacing some of its methyl groups with longer paraffinic groups, such as n-octyl ($\cdot C_8H_{17}$) makes the phase even less polar. The 50 % n-octyl polymer closely mimics the performance of squalane, while capable of operating at a much higher temperature. Some manufacturers peg it as zero polarity on their own polarity scale.

The stationary phase

Table 4.4 Popular Polysiloxane Liquid Phases.

Siloxane Monomer	Co-Monomer	Phenyl	Polar	Typical Liquid Phases
Methyl and Phenyl Polysiloxanes				
100 % dimethyl	none	0 %	5	XX-1, CP-Sil 5CB, SE-30, SP-2100, DC200, OV-101
95 % dimethyl	5 % diphenyl	5 %	8	XX-5, CP-Sil 8CB, SE-52, SE-54, PE-2, 007-2
94.5 % dimethyl	5.5 % diphenyl	5.5 %	10	OV-73
80 % dimethyl	20 % methylphenyl	10 %	10	OV-3, PE-3
60 % dimethyl	40 % methylphenyl	20 %	14	XX-20, OV-7, 007-7
80 % dimethyl	20 % diphenyl	20 %	15	XX-20, VOCOL
67 % dimethyl	33 % diphenyl	33 %	19	OV-61
30 % dimethyl	70 % methylphenyl	35 %	19	XX-11, DC-710
65 % dimethyl	35 % diphenyl	35 %	19	XX-35
100 % methylphenyl	none	50 %	21	XX-17
50 % dimethyl	50 % diphenyl	50 %	21	XX-50, SP-2250, CP-Sil 24
70 % methylphenyl	30 % diphenyl	65 %	26	OV-22, XX-65
35 % dimethyl	65 % diphenyl	65 %	26	XX-65
50 % methylphenyl	50 % diphenyl	75 %	28	OV-25
Fluoro and Cyano Polysiloxanes				
methyl(cyanopropyl)	dimethyl	none	11	OV-105, ZB-50
6 % phenyl(cyanopropyl)	94 % dimethyl	3 %	13	XX-624, XX-1301
14 % phenyl(cyanopropyl)	86 % dimethyl	7 %	19	XX-1701, BP-10, CP-Sil 19CB
100 % methyl(trifluoropropyl)	none	none	37	XX-210, SP-2401, QF-1, AT-201, OV-202, OV-215
50 % methyl(cyanopropyl)	50 % methylphenyl	25 %	58	XX-225, CP-Sil 43CB, SP-2300
80 % bis(cyanopropyl)	20 % phenyl(cyanopropyl)	10 %	67	SP-2330, Silar 3C
90 % bis(cyanopropyl)	10 % phenyl(cyanopropyl)	5 %	70	SP-2380
100 % bis(cyanopropyl)	none	none	88	XX-88, XX-2560, BPX-90, Silar 10CP
100 % dicyanoallyl	none	none	101	OV-275

Data is typical and not exclusive. The polarity scale is a linear interpolation between squalane as 0 and TCEP as 100 (see Table 4.8). Capillary columns with bonded and crosslinked phases may exhibit small polarity variations. The notation "XX" refers to the brand codes of multiple suppliers as listed in Table 4.1. These suppliers often sell capillary columns with the same suffix to indicate the incorporated liquid phase. Column brands with the additional suffix "MS" (not shown) use a low-bleed bonded and crosslinked liquid phase suitable for use with a mass spectrometer detector. At this time, only one commercial PGC has an MS detector (Wasson).
Sources: Compiled from Agilent (2021), Barry and Grob (2007), Restek (2021b), Rotzsche (1991), and Supelco (1999, 2013).

Fluoro silicones

Replacing one or more of the methyl groups in the siloxane monomers with a fluorinated alkyl group yields a moderately polar silicone oil with unique selectivity: it retains carbonyl groups more than hydroxyl groups (Yancy 1994b).

A common replacement is the trifluoropropyl (·C$_2$H$_4$CF$_3$) group, yielding poly[methyl(3,3,3-trifluoropropyl)siloxane], a silicone with 50 % trifluoro and 50 % methyl content. This is the formulation of the classic QF-1 liquid phase.

The trifluoro group imparts a dipole moment to the liquid phase that can polarize and attract olefins and aromatics. More importantly, the electronegative fluorine atoms have a strong affinity for solutes having free electron pairs, notably those containing oxygen or nitrogen atoms. Thus, they strongly retain nitriles (·CN) and carbonyl compounds (:CO) such as aldehydes and ketones.

Cyano silicones

Replacing some of the silicone methyl groups with cyanoalkyl groups like cyanopropyl (·C$_3$H$_6$CN) increases the polarity and selectivity of the phase for certain solutes. Poly[methyl(cyanopropyl)siloxane] is the basic polymer that contains a fixed 50 % of cyano groups. As with other substitutions, though, a copolymer of dimethyl and bis(cyanopropyl) siloxanes can have any desired percentage of cyano nodes, allowing manufacturers to tune the product for optimal selectivity.

Each cyano or "nitrile" group (·CN) is distant from the polymer backbone, which bestows a strong dipole moment for attracting polar solutes and polarizable molecules like olefins and aromatics. Some of these phases are adept at separating *cis*- and *trans*-olefin isomers. In a key distinction from the fluoro silicones, the cyano nitrogen atom can attract only electron acceptor molecules, giving the phase a strong affinity for alcohols, phenols, and nitro compounds. The cyano-silicones are the most polar of the siloxane polymers and can even supplant TCEP as #100 on the polarity scale. They can also withstand much higher column temperatures than TCEP.

However, these highly polar phases have very low retention for nonpolar solutes like the normal paraffins, so their high McReynolds[1] constants may be somewhat deceptive.

Multifunctional silicones

Finally, there are endless possibilities for replacing the four copolymer methyl groups with two or more mixed functional groups to yield a unique silicone oil having the ability to perform a desired peak separation. Here's an example; can you draw its structure?

poly[methyl(cyanopropyl)methylphenylsiloxane].

The profusion of possible copolymers explains the diversity of the silicone liquid phases, yet also conveys a warning:

> *Proprietary liquid phases said to be equivalent are unlikely to be identical and may not give the same performance.*

Even two liquid phases with the same polymeric formula might not meet your expectations as the polymer chain length and crosslinking may not

---

[1] See the **SCI-FILE:** On Retention Data, below.

be the same. To be sure, it's wise to install only the spare columns supplied by the original PGC manufacturer.

Bonded phases

The copolymer production technique led to an important advancement in capillary column technology: the **bonded phase** column. By reacting the monomer inside the column itself, it's possible to form a thin polymer film chemically attached to the inside walls of the capillary tube. The polymer layer can also be crosslinked by adding a little methyl vinyl siloxane into the monomer mix. During the conditioning process, the vinyl double bond opens to build bridges between the long polysiloxane strands. These techniques reduce the vapor pressure of the liquid phase and increase its resistance to high temperatures. For long-term operation in a PGC, the bonded and crosslinked liquid phases are more durable than wall-coated columns. Their liquid phase deteriorates slowly, if at all, so it generates far less of the classic carrier gas contamination known as column bleed, resulting in reduced detector signal noise and enhanced sensitivity.

In the past, a drawback of using a silicone packed column with a flame detector was the silica deposits forming on the flame jet due to oxidation of the column bleed. With a bonded-phase capillary, though, the low bleed and low flow rate pretty much eliminate this problem. You can even rinse a bonded-phase capillary column with a solvent without damaging its liquid film. Consequently, when using capillary columns in a PGC, we prefer to use bonded phases whenever possible. However, many of the highly polar liquid phases are not available in bonded form.

A precaution

Silicone oil columns are very stable and should perform reliably for a very long time. Nevertheless, the silicone may react with traces of oxygen or moisture in the carrier gas (particularly at high column temperature), which eventually degrades its performance. As a precaution, install oxygen and moisture absorbers in the carrier gas line.

Prevent any strong acids or bases in the sample from reaching a silicone column as they may react with the liquid phase and reduce column life (Yancey 1985a). However, it's often possible to cut off and discard the first half-meter of a damaged capillary column with little detriment to its performance.

Non-silicone liquid phases

Dialkyl esters

Organic esters were among the first liquid phases used in PGCs. They exhibit a wide range of polarities and have useful selectivities, but they also have relatively low temperature limits and are unsuitable for capillary columns. Consequently, their usage has declined.

You'll find these traditional non-polymer PGC liquid phases only in packed columns. They are pure chemical compounds derived from the alpha–omega dicarboxylic acids; those with acid groups at both ends of their carbon chain. Replacing both of the acidic hydrogen atoms with identical alkyl groups (R) forms the dialkyl ester. Typical attachments are dibutyl, di(2-ethylhexyl), dioctyl, dinonyl, di(iso-decyl), and didecyl; all nonpolar. Another kind of attachment includes methoxy or ethoxy groups that are more polar.

Common esters are:

➢ The dialkyl succinates: $RO_2C \cdot (CH_2)_2 \cdot CO_2R$

These diesters are generally too volatile for use in industrial PGCs.

➢ The dialkyl adipates: $RO_2C \cdot (CH_2)_4 \cdot CO_2R$

For example, di(isodecyl) adipate (DIDA).

➢ The dialkyl sebacates: $RO_2C \cdot (CH_2)_8 \cdot CO_2R$

For example, di(2-ethylhexyl) sebacate (DEHS).

➢ The dialkyl phthalates: $RO_2C \cdot C_6H_4 \cdot CO_2R$

For example, dinonyl phthalate (DNP).

Table 4.5 provides further examples. Having different dialkyl groups tends to change the physical properties of the ester, like its vapor pressure and viscosity, but has little effect on phase selectivity. However, the optional methoxy and ethoxy groups do provide additional polarity.

The moderate polarity of the dialkyl adipates and sebacates derives from the double-bonded oxygen atoms in their structure. The electron pairs on these atoms provide targets for solutes like alcohols, capable of hydrogen bonding.

The phthalates enjoy the additional polarizability of the central benzene ring structure ($\cdot C_6H_4 \cdot$), which bestows good selectivity for olefins and aromatics.

PGCs today rarely use the dialkyl esters. While they exhibit interesting separation factors for special applications, they lack long-term reliability. Their high volatility and low stability can cause more detector noise than an equivalent silicone oil. In fact, the silicone-phase column may have an upper temperature limit as much as 200 °C higher than the dialkyl ester column it's replacing.

Polyesters

These polymer liquid phases have a wider range of polarities and higher temperature limits than the dialkyl esters. An example is [poly] diethylene glycol succinate (DEGS), popular for its high polarity. Note that the names of these polyesters usually omit the "poly" designator, potentially causing confusion. Like the organic esters, the core structure of a polyester is from its parent dicarboxylic acid, from whom it inherits its name: succinate, adipate,

Table 4.5 Popular Non-silicone Liquid Phases.

Chemical Name	Code	Polar	Min °C	Max °C	Typical Liquid Phases
Moderately-Polar Ester Phases					
Di(2-ethylhexyl) sebacate	DEHS	16		130	Octoil S, PE-B
Bis(2-methoxyethyl) adipate	BMEA	17			
Diisodecyl phthalate	DIDP	18			
Dinonyl phthalate	DNP	19	20	130	Narcoil 40
Bis(2-ethoxyethyl) sebacate	BEES	30			
Bis(2-butoxyethyl) phthalate	BBEP	30			
Tricresyl phosphate[a]	TCP	34	20	110	
High Polarity Polyol, Polyester, and Cyano Phases					
Poly[propylene glycol]	PPG	20		100	Ucon LB-550X
Poly[propylene glycol]	PPG	38		160	Ucon 50-HB-2000
Poly[ethylene glycol] PEG-20M	PEG	55	70	250	XX-Wax, Carbowax 20M, BP-20
Free fatty acid phase (modified PEG)	FFAP	61	60	240	OV-351, XX-FFAP, SP-1000, BP-21
Poly[butane-1,4-diol succinate]	BDS	64	50	200	
Poly[ethylene glycol adipate]	EGA	65	100	190	Reoflex 400[b]
Poly[diethylene glycol adipate]	DEGA	66	50	190	
Poly[ethylene glycol] PEG-1540	PEG	67	50	175	Carbowax 1540[c]
Poly[diethylene glycol succinate]	DEGS	82	20	200	PE-P, Polyester A
Poly[ethylene glycol succinate]	EGS	90	50	180	
1,2,3-Tris(2-cyanoethoxy) propane	TCEP	100	20	170	Fractonitril III[d]
Extreme-Polarity Ionic Liquid Phases					
Ionic liquid phases:		>100			SLB-IL100 (#indicates polarity)

[a]CAUTION: Tricresyl phosphate may contain a mixture of isomers and the ortho isomer is highly toxic.
[b]Reoflex is a trademark of Obschestvo s ogranichennoy otvetstvennostyu "EKOPOL."
[c]Carbowax is a trademark of Dow, formerly Union Carbide, Inc.
[d]Fractonitril is a trademark of Merke, E.
Data is typical and not exclusive. The polarity scale is a linear interpolation between squalane as 0 and TCEP as 100 (see Table 4.8). The notation "XX" refers to the brand codes of multiple suppliers as listed in Table 4.1.
Sources: Data compiled from Anderson and Armstrong (2005), Barry and Grob (2007), Poole and Poole (2011), Restek (2021a), Rotzsche (1991), and Supelco (1999, 2013).

sebacate, or phthalate. Each unique polyester is a copolymer of the chosen acid and a diol. For instance:

➤ Poly[ethylene glycol succinate] is the copolymer of ethylene glycol ($HO \cdot C_2H_4 \cdot OH$) and succinic acid ($HOOC \cdot C_2H_4 \cdot COOH$), with the structure:
$$HO[\cdot C_2H_4O \cdot CO \cdot C_2H_4 \cdot COO\cdot]_n H$$

➤ Poly[diethylene glycol succinate] (DEGS) has two glycol units for each succinic acid unit:
$$HO[\cdot (C_2H_4O)_2 \cdot CO \cdot C_2H_4 \cdot COO\cdot]_n H$$

Like the alkyl esters, the polyesters were popular for packed columns in older PGCs and are now becoming obsolete, in favor of cyano-substituted silicone oils with similar selectivity and superior long-term stability.

Waxy polymers

Poly[ethylene glycol], also known as PEG or by its trade name Carbowax®, is always a popular choice of stationary phase. Chemically, it's a polyol (or polyether), having a polymer chain containing many oxygen atoms:

$$H \cdot [O \cdot C_2H_4]_n \cdot OH$$

The numeral suffix, as in PEG-1540, indicates the average molar mass of the polymer. Since the repeating unit has a molar mass of 44 da, one can estimate the polymer chain length. For Carbowax 20M, it's about 450 repeating units. This waxy polymer has become the favorite PEG liquid. It performs well and is less volatile than the lower-mass products.

PEGs are particularly good at separating polar solutes and remain popular even for capillary columns, where they invariably go by the name "WAX." Their strong selectivity is due to their oxygen atoms forming hydrogen bonds with solutes containing OH groups. They strongly retain alcohols and phenols.

Conversely, the PEG phases are terrible for separating paraffins. As with other highly polar phases, all nonpolar solutes elute quickly from a PEG column with little separation achieved.

The polyethylene glycols are very susceptible to damage by reaction with oxygen or moisture in the carrier gas at high oven temperatures. Install suitable absorbers in carrier gas lines and run the column at the lowest feasible temperature.

Modified PEGs are available for acidic samples. The reaction of Carbowax 20M with 2-nitroterephthalic acid forms Carbowax 20M-TPA. This phase is suitable for acidic samples and is the basis of the popular free fatty acid phase (FFAP), among others. These are general-purpose phases and are especially good for the analysis of acidic organic compounds such as phenols and carboxylic acids. They are not for basic compounds like amines nor for aldehydes. Other formulations are available for those.

Nitrile ethers

These highly polar liquid phases are pure chemical compounds containing multiple cyanoethoxy groups ($\cdot OC_2H_4CN$). The most popular of these is 1,2,3-tris(2-cyanoethoxy)propane (TCEP), which is available in packed or capillary columns. TCEP is traditionally the most polar liquid phase and acts as #100 on the 0–100 polarity scale.

Column temperature effects

Liquid phase temperature rating

Most PGC columns operate at moderate temperature, only occasionally above 180 °C. Even so, it's advisable for the liquid phase rating to be as high as possible, as that is an indication of its probable long-term stability. The temperature ratings of chromatographic-quality silicone phases are generally above 300 °C and they are ideal for process chromatographs.

Temperature-programmed chromatographs often need to reach higher temperatures, so temperature rating then becomes an important factor in the choice of liquid phase.

For most solutes, column operating temperature has only a small effect on separation factors, although there are some exceptions.

Liquid phase viscosity

A polymer isn't a defined substance. It can have any number of repeating monomer units. The range is enormous: a dimethyl silicone, for instance, can have as few as 80 units (liquid) or as many as 35,000 units (gum), resulting in a molar mass anywhere between 6,000 and 2,500,000 and a viscosity as low as 10 mPa·s or as high as 10 kPa·s (Rotzsche 1991, 207). Thus, commercial silicones said to be similar may differ in molar mass and viscosity. Luckily, these differences have little effect on the selectivity and polarity of the phase, but they may affect column efficiency.

In a later chapter, we'll see how a high viscosity liquid phase reduces the mobility of solute molecules and thereby increases the plate height of a column, reducing its plate number. From that perspective, we prefer low-viscosity phases. Paradoxically, however, it's not easy to smoothly coat a capillary surface with a thin film of low-viscosity liquid, particularly a polar one. It's likely to cling together and form isolated globules. A high-viscosity phase is more likely to completely cover the capillary walls or solid support for maximum column efficiency. Capillary column suppliers seek to mitigate this difficulty by improving the passivation of the contact surfaces and by bonding and crosslinking the polymer film.

Choosing a liquid phase

The practical method

The practical approach to column selection in process chromatography is rarely the scientific or mathematical endeavor you might imagine it to be. Indeed, a prominent professor of chromatographic science once found it appropriate to write:

> *The bottom line is that stationary phase selection is often largely guesswork (Miller* 2005, 107).

More recently, three well-known professors offered this homely advice: *Ask someone who knows* (McNair et al. 2019, 69). You might get the impression that these experts don't expect the well-developed theories of column performance to be much help in selecting a stationary phase. Sadly, it's true.

The difficulty comes down to the vast number of potential solute-solvent pairs and the many unquantifiable interactions between them. It's not possible to know them all. So PGC column designers work from their own experience, with some help from the performance data published by the manufacturers of chromatographic stationary phases.

For all their diversity, many liquid phases give similar separations, so a PGC applications team tends to limit their columns to a few stationary phases they know well. A long time ago, Hawkes et al. (1975) concluded that six liquid phases would be enough to perform most analyses. This was their recommended toolkit:

➢ A polydimethylsiloxane like OV-101, SE-30, or SP2100.
➢ A 50 % phenyl silicone like OV-17 or SP2250.
➢ A polyethylene glycol like Carbowax 20M.
➢ Diethylene glycol succinate (DEGS).
➢ A cyanopropyl silicone like Silar 10C or SP-2340.
➢ A trifluoropropyl silicone like OV-210 or SP-2401.

A few years later, Yancey (1986) published a comparable list but with no mention of DEGS. Today, most PGC column designers work with a similar set of liquid phases, although there have been some developments since 1975:

➢ For capillary columns, silicone phases with a wide range of polarities have replaced most of the traditional non-silicone columns. A notable exception is poly[ethylene glycol], now universally dubbed "wax".
➢ For packed columns, the rush to silicones is more like a dawdle. It's not that the advantages of silicones don't apply; it's more about tradition and habit. Of course, the nonpolar dimethyl silicone has been a favorite since the very first PGCs. For more polar separations, though, tradition favors the alkyl esters, polyesters, and polyols.

It's understandable that PGC application chemists tend to favor the few phases they know well; they would need to invest in some experimental time before adding new liquid phases to their armory. In a busy factory, research time is always in short supply – it's quicker to copy columns from a previous project that worked well. Yet, if you are a column designer, perhaps now would be a good time to consider a wider choice of liquid phases? A new approach might improve the long-term reliability of your product.

In addition to choosing the stationary phase, the column designer must decide whether to use packed or capillary columns. Again, tradition plays a role, but other motivators might factor into this decision: capillary columns are more expensive than traditional packed or micropacked columns and they take more space in the oven – a larger oven might be required. At the time of writing, micropacked columns are becoming the strong favorite for process applications.

Using retention data

Notwithstanding the influence of tradition, some science remains. For an unfamiliar separation, an applications chemist or troubleshooter needs a way to evaluate many liquid phases and select the best one for the job. We have seen that liquid phase polarity is an imperfect guide to column performance because *polarity* has multiple dimensions. Struggling with this complexity, investigators sought to understand (and then predict) the nuanced affinity of liquid phases for different solutes.

They almost succeeded.

In the early years of chromatography, a host of researchers published the retention times of hundreds of solutes on a plethora of liquid phases. Unfortunately, this surfeit of data was not all that useful because retention times are totally dependent on column design and operating conditions. We needed a standard data format.

The **SCI-FILE:** *On Retention Data* that follows describes the standard data reporting formats and liquid phase classifications that arose, flourished, then faded away. The SCI-FILE is optional reading as these data formats are no longer in general use, although prevalent in the scientific literature. Yet we encourage you to study these techniques of column classification, as they reveal a lot about how columns work and will help you understand the multidimensionality of *polarity*.

SCI-FILE: *On Retention Data*

Retention index

To reduce the uncertainty in published retention data, Kováts (1958) proposed the **retention index**. This data reporting system is based on two properties of columns that you already know:

- The *separation factor* between two solutes on a given liquid phase is largely independent of column structure and operating conditions.
- In *isothermal* separations, the adjusted retention times of **homologs** such as the normal paraffins increase exponentially with carbon number.

Kováts realized he could express the retention time of any solute relative to the retention times of two nearby straight-chain paraffins on the same column. Just like a separation factor, this relative value is the ratio of two peak retention times and tends to be independent of the column operating conditions. It's a true property of the liquid phase.

Kováts used the known linear relation between the number of carbon atoms (n) in the normal paraffins and the logarithm of their adjusted retention times. To avoid having decimals, he assigned a retention index (I) of $100n$ to each paraffin. For example, n-pentane (C_5) has a defined retention index of 500 on any column under any operating conditions.

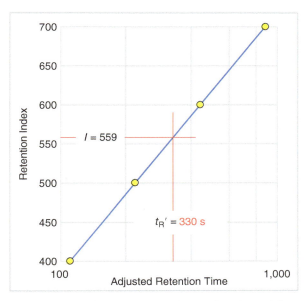

The blue line plots the adjusted retention time of four paraffins, C₄ – C₇, on a logarithmic scale. The red lines indicate that the retention index of a component peak halfway between C5 and C6 is 559.

Figure 4.2 Principle of Retention Index.

Figure 4.2 demonstrates a graphical way to estimate the retention index of a solute. It's clear that a peak eluting between *n*-pentane and *n*-hexane must have a retention index between 500 and 600.

To directly calculate the retention index of an analyte, you would need to know, under identical column conditions, the adjusted retention times (t_R') of the analyte peak, the preceding *n*-paraffin peak, and the following *n*-paraffin peak. There are techniques for using more distant paraffins, but the assumption of linearity is best reserved for those with consecutive carbon numbers.

Then, use Equation 4.1 to get the analyte's retention index (I_i), a dimensionless number:

$$I_i = 100\left[n + \frac{\log_{10}(t_R')_i - \log_{10}(t_R')_n}{\log_{10}(t_R')_{n+1} - \log_{10}(t_R')_n}\right] \quad (4.1)$$

Wherein the subscripts indicate:

- *i* is the measured peak.
- *n* is the immediately preceding *n*-paraffin.
- *n* + 1 is the immediately following *n*-paraffin.

Due to its logarithmic nature, 100 retention index units represents a fixed multiplier of adjusted retention time – about double for most homologs: therefore, *n*-hexane ($I = 600$) has about twice the adjusted retention time of *n*-pentane ($I = 500$) and four times the adjusted retention time of *n*-butane ($I = 400$). While there are exceptions, this predictable retention behavior is true for many homologous series. An additional methylene group in a solute's molecular structure adds about 100 to its retention index and doubles its adjusted retention time on that column.

The retention index of a solute immediately tells you where its peak will appear on the chromatogram. For instance, if $I = 559$, the analyte peak is between the *n*-pentane and *n*-hexane peaks on any column using that liquid phase. What's more, due to the logarithmic scale, any peak with an index ending in 59 is midway between two paraffin peaks on the chromatogram.

The Kováts index is simpler for columns operating on a temperature program. A uniform temperature ramp creates a linear relation between the adjusted retention time of an *n*-paraffin and its carbon number. Equation 4.1 then reduces to a simple function of retention times (t_R):

$$I_i = 100\left[n + \frac{(t_R)_i - (t_R)_n}{(t_R)_{n+1} - (t_R)_n}\right] \quad (4.2)$$

McReynolds constants

Retention index was a key improvement for reporting experimental data as it expresses the retention time of a solute peak relative to the retention time of paraffin peaks, thereby eliminating the effect of several column variables like column length, liquid loading, and carrier flow. Yet it's limited to a specified chemical compound on a specified liquid phase and doesn't provide a useful guide to column polarity.

The problem is that "polarity" is a vague concept that tries to capture the affinity of a liquid phase along a single dimension, whereas in reality affinity is due to many different forces between the molecules. From that realization, researchers sought to discover an adequate number of polarity dimensions.

Table 4.6 McReynolds Probes.

#	Probe	Reveals liquid affinity for:
X'	Benzene	Aromatics, olefins
Y'	1-Butanol	Alcohols, nitriles, acids, and halogen compounds
Z'	2-Pentanone	Aldehydes, ketones, ethers, esters, and epoxides
U'	Nitropropane	Nitro and nitrile compounds
S'	Pyridine	Heterocyclic compounds

Each of the probe samples purports to test the liquid phase for its affinity for different structural features in the solute molecules. These affinities are related to the different intermolecular forces involved in polarity.

Rohrschneider (1966) proposed that the retention index of five standard solutes – known as **probes** – would reveal the affinity of a liquid phase to specific chemical structures in the sample molecules. Later, McReynolds (1970) modified those standard probes and suggested five more. However, it turned out that the last three of the new probes were ineffective predictors, leaving seven at most. For most purposes, though, the first five are adequate, and these are the ones usually specified. Table 4.6 lists these probes and the affinities they purport to reveal.

To classify a liquid phase, one measures the retention index of each probe at 120 °C and then subtracts its retention index on squalane under the same conditions. The difference is the **McReynolds Constant** (MRC) for that probe on that liquid phase. We deem squalane to be completely nonpolar, so the derived MRC is solely due to the polarity of the phase *for that probe*.

Suppliers often use the average of five McReynolds constants to indicate the overall polarity of a liquid phase. Mostly, they adopt a linear scale that has squalane at zero polarity and TCEP at 100, although they may use the nearest paraffins instead.

Several authors studied the effect of column temperature on the MRC. Ashes and Haken (1973) found the MRC of nonpolar silicones are almost independent of temperature, while the MRC of polar silicones tended to increase with temperature. In a more detailed study, Santiuste and Takács (1999) discovered negative and positive variations in each MRC value for eight liquid phases, but the effects were not highly significant. So the effect of temperature is not of concern when using the MRC values to choose a suitable liquid phase.

Unfortunately, the maximum operating temperature for squalane ($C_{30}H_{62}$) is about 120 °C. To overcome this constraint, several investigators studied Apolane-87 ($C_{87}H_{176}$) as a potential zero-polarity liquid phase for classification schemes (Riedo et al. 1976; Matisová et al. 1988; Poole et al. 1989). This synthetic paraffin has a higher temperature limit and is available in pure form. Not much became of this idea because the new fused-silica capillary columns were so efficient that liquid phase selection tools quickly became irrelevant.

Yet PGC columns sometimes need higher selectivity than those used in the laboratory and an applications chemist may need to consider a wider range of liquid phases. McReynolds constants might still be useful for that more-complex task.

Using McReynolds

McReynolds constants for liquid phases are quoted in suppliers' literature and remain a useful way to explore the five dimensions of polarity; they're a good place to start when looking for a suitable column. Table 4.8 lists MRC for an illustrative set of phases, but the vendors publish far more.

One use for the McReynolds constants is to confirm that certain liquid phases are equivalent. Consider the five silicones listed in Table 4.7. Their MRCs show them to be nonpolar phases, all with essentially identical selectivity.

Table 4.7 Confirming Equivalent Liquid Phases.

Phase	X'	Y'	Z'	U'	S'
SE-30	15	53	44	64	41
OV-1	16	55	44	65	42
OV-101	17	57	45	67	43
SP-2100	17	57	45	67	43
DC-200	16	57	45	68	44

The McReynolds constants show that these five dimethyl silicone phases are practically identical in selectivity and will give much the same separations. However, the phases may differ in viscosity and temperature rating.
Source: Adapted from Ohio Valley Specialty Company.

The McReynolds constants contain more information than is immediately apparent. Recall that the science behind the retention index system is the logarithmic increase of solubility – and hence adjusted retention time – of the n-paraffins. Each additional methylene group ($\cdot \text{CH}_2 \cdot$) in a paraffin increases its retention index by 100, and its adjusted retention time about doubles. This predictable behavior also occurs in other chemical series that differ only by methylene count. When you know the retention index of one member of the series, you can instantly predict the index of another member.

As an example, consider the tricresyl phosphate data in Table 4.8. For 1-butanol, the MRC value is: $Y = 321$. From NIST (2018), the retention index for 1-butanol on squalane is 644. Therefore, the retention index for 1-butanol on TCP is:

$$I_{1-\text{C}_4\text{OH}}^{\text{TCP}} = 644 + 321 = 965$$

That index value indicates that the 1-butanol ($\text{C}_4\text{H}_9\text{OH}$) will elute between n-nonane and n-decane as shown in Figure 4.3. In addition, we know that alcohols follow the methylene rule: each methylene group contributes 100 units to the retention index. We thereby deduce that:

- 1-Propanol ($\text{C}_3\text{H}_7\text{OH}$) has an index value of about 865 and elutes between n-octane and n-nonane with about half the adjusted retention time of the 1-butanol.

Table 4.8 Selectivity of Common Liquid Phases.

Liquid Phase	X'	Y'	Z'	U'	S'	Mean	Polarity	
Squalane (reference phase)	0	0	0	0	0	0	0	Nonpolar
50:50 Poly[methyl (n-octyl) siloxane]	3	14	11	12	11	10	1	
Apolane-87	21	10	3	12	25	14	2	
Poly[dimethyl siloxane]	16	55	44	65	42	44	5	
20:80 Poly[diphenyl and dimethyl siloxanes]	69	113	111	171	128	118	14	Slightly polar
Di(2-ethylhexyl) sebacate – DEHS	72	168	108	180	125	131	16	
Dinonyl phthalate	83	183	147	231	159	135	16	
35:65 Poly[methylphenyl and dimethyl siloxanes]	102	142	145	219	178	157	19	
50:50 Poly[diphenyl and dimethyl siloxanes]	125	175	183	268	220	194	23	Moderately polar
50:50 Poly[diphenyl and methylphenyl siloxanes]	160	188	191	283	253	215	26	
Tricresyl phosphate – TCP	176	321	250	374	299	284	34	
Poly[trifluoropropyl methyl siloxane]	146	238	358	468	310	304	37	
Poly[neopentyl glycol adipate] – NPGA	234	425	312	402	438	362	46	
Carbowax® 20M[a]	322	536	368	572	510	462	56	
Poly[diethylene glycol adipate] – DEGA	378	603	460	665	658	553	66	
Carbowax 1540 (PEG1540)	371	639	453	666	641	554	67	Highly polar
Poly[diethylene glycol succinate] – DEGS	496	746	590	837	835	701	84	
Poly[bis(3-cyanopropyl) siloxane]	520	757	659	942	800	736	88	
1,2,3-Tris(2-cyanoethoxy) propane – TCEP	594	857	759	1,031	917	832	100	
Poly[dicyanoallyl siloxane]	629	872	763	1,106	849	844	101	

[a]Carbowax is a registered trademark of Dow Chemical.
McReynolds probes: X' = benzene, Y' = 1-butanol, Z' = 2-pentanone, U' = nitropropane, and S' = pyridine.
The overall polarity of a liquid phase is a linear interpolation of its average McReynolds constant, with squalane = 0 and TECP = 100. Data is typical, not exclusive, and compiled from manufacturer's published data sheets. Reported McReynolds constants vary by about ±4 index units, but such small variations usually have an insignificant effect on separation.
Sources: Ohio Valley (2013), Phenomenex (2020), Restek (2021a, 2021b), and Supelco (1997).

The stationary phase

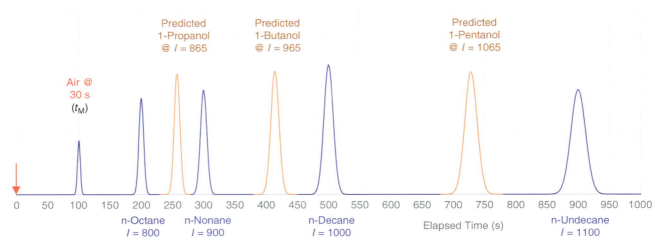

Predicted peak positions for alcohols on a tricresyl phosphate column. See text for explanation.

Figure 4.3 Predicting Retention by McReynolds Constants.

- 1-Pentanol ($C_5H_{11}OH$) has a retention index of about 1065 and elutes after n-decane with about twice the adjusted retention time of the 1-butanol.

Figure 4.3 shows the predicted positions of the alcohol peaks on the TCP chromatogram.

System constants

From an academic perspective, McReynolds constants give an imprecise insight into the interactions between the liquid phase and solute. An updated approach after Abraham et al. (1999) attempts to model the mechanisms involved in the solvation process. As it happens, this solvation parameter model also results in five liquid phase descriptors known as **system constants**.

In addition, each solute has five moderating factors – one for each system constant – to calibrate the ability of that solute to participate in the specified intermolecular interaction. The goal is to develop procedures to calculate these solute factors from the known structure of a solute molecule.

In theory, the retention factor of any solute on any liquid phase obtains from the sum of the modified system constants (Poole 2012, 140):

$$\log k = c + eE + sS + aA + bB + lL \qquad (4.3)$$

Wherein, lowercase letters represent the fixed system constants of the liquid phase and uppercase letters are the structural modifiers of the solute.

At present, column designers don't find the solvation parameters model useful in their daily work of selecting liquid phases for PGC columns.

Making a choice

Are McReynolds constants useful?

The SCI-FILE explains the origin and use of the McReynolds constants to categorize liquid phases. We recommend that you learn how they work. They certainly give insight into the retention behavior of a column and *almost succeed* in predicting its behavior with different analytes. Yet most practical chromatographers don't use them and some suppliers no longer specify them for their products.

It seems there are three reasons for this neglect:

> Most gas chromatographs perform standard analytical methods with specified columns; their owners have no incentive to experiment with different liquid phases. Vendors now sell columns for standard applications without even specifying what the column contains.

> Most laboratory applications now use capillary columns, so the choice of liquid phase is not a critical issue. A few highly efficient capillary columns can separate almost any mixture of analytes.

> The McReynolds constants don't provide an intuitive way to select a liquid phase. There's no direct connection between McReynolds values and the separation of specific peaks; it takes knowledge and experience to interpret their meaning. It's often easier to try a few columns and see what works.

Nevertheless, PGC applications engineers need to design columns for the fastest possible analysis in a wide variety of applications. On occasion, they may need some help to find a suitable liquid phase. Perhaps then, McReynolds will be useful. Table 4.8 gives examples of the McReynolds constants for a few popular liquid phases. For those not listed, visit the vendor's website or consult the reference works listed in Recommended Reading at the end of this chapter.

Adsorption columns

Distribution in GLC and GSC

Unlike the nonvolatile liquid used in gas-liquid chromatography (GLC), the stationary phase in gas-solid chromatography (GSC) is an **adsorbent solid**.

The theories of chromatographic peak formation for solid-phase columns and liquid-phase columns are similar. In their simple forms, both theories presume that a sequence of instantaneous equilibria occurs as an analyte migrates through the column. In all these equilibria, the ratio of analyte concentration in the stationary phase $[A]_S$ to analyte concentration in the gas phase $[A]_M$ is constant and is the **distribution constant** (K_c), where:

$$K_c = \frac{[A]_S}{[A]_M} \qquad (4.4)$$

Given this core assumption, the **theoretical plate theory** introduced in Chapter 2 predicts that chromatogram peaks will follow the symmetric Gaussian shape. In GLC, our best peaks really do follow this expectation of symmetry, albeit slightly distorted due to the imperfections discussed in

Chapter 8. This empirical evidence supports the idea that the distribution constant for a solute between a liquid phase and a gas phase is indeed constant, at least for small sample volumes.

We find no such assurance in GSC. With solid adsorbents, peak asymmetry is the rule, rather than the exception. Unlike solvation in a liquid, adsorption on a solid is often a *nonlinear* process. There are two forms of nonlinearity that can affect adsorption. In both cases, the distribution of adsorbate molecules between the solid phase and the gas phase is not constant but varies with the amount of analyte involved.

The most common type of nonlinearity is a reduction in the distribution constant as the analyte concentration increases: fewer molecules stick. Theorists conject that a finite number of adsorption sites exist at the surface and as these become filled, gas molecules have fewer opportunities to adhere.

A less common type of nonlinearity occurs when the first few molecules adsorbed on a surface facilitate the adsorption of more molecules. This results in an increase of the distribution constant as the concentration of adsorbate increases: more molecules stick.

Adsorption isotherms

The adsorption process is temperature dependent; less gas adsorbs at a higher temperature. An isothermal graphic that relates the amount of an analyte adsorbed to the amount of that analyte remaining in the gas phase is known as an **adsorption isotherm**. Any deviation from linearity is clearly visible in the isotherm.

Figure 4.4 illustrates the nonlinear adsorption behavior predicted by the **Langmuir isotherm**, a simple model of gas-solid equilibrium that predicts a lower rate of adsorption at high solute concentrations (Langmuir 1918). The curves in this figure are for illustrative purposes and don't pertain to a real gas-solid system. The two curves are the same isotherm plotted at different sensitivity. The green curve expands the lower end of the blue curve by 25 times and reveals that the quantity of gas adsorbed can be quite linear when the solid surface is not close to saturation.

The solid surface is more likely to approach saturation when an analyte is at a high concentration or when its molecules have a strong affinity for that surface. Thus, the chromatogram peaks for low-concentration analytes and low-affinity gases are often quite symmetrical, but the peaks for strongly-held substances may be severely distorted.

Nonideal peak shape

To understand the effect of inconstant adsorption on peak shape, recall that the migration rate of peak molecules through a column depends on the percentage of solute molecules retained by the stationary phase. If this percentage isn't constant, neither is the rate of migration.

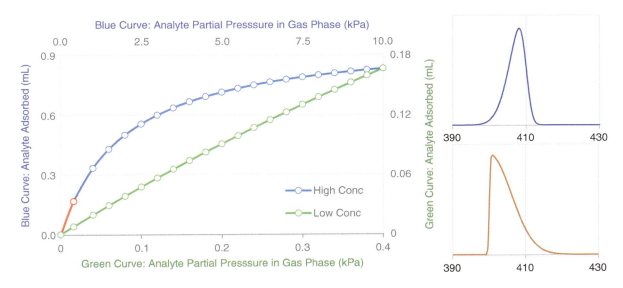

The same isotherm plotted at low gas concentrations (green and red curves) and at 25 times higher gas concentration (blue curve).

The amount of gas adsorbed onto a solid surface may be closely linear with concentration at low gas concentrations (green and red curves) but significantly nonlinear at higher concentrations (blue curve). This nonlinearity distorts the peak shape. Commonly, a lower percentage of the gas adsorbs when the gas concentration is high (blue curve), so the center of the peak moves faster than its base and causes a tailing chromatogram peak (blue peak). Rarely, a higher percentage of the gas adsorbs when the gas concentration is high (isotherm not shown), so the center of the peak moves slower than its base and causes a fronting chromatogram peak (orange peak).

Figure 4.4 Effect of Adsorption Isotherm on Peak Shape.

A chromatogram peak represents a varying analyte concentration. So when fewer molecules adsorb at higher concentrations, the top of the peak migrates faster than its base, forming a **tailing peak** on the chromatogram like the blue peak in Figure 4.4.

Less commonly, the percentage of the analyte molecules adsorbed rises as their concentration increases, driving the adsorption isotherm to curl upwards instead of drooping downwards like the blue curve does in Figure 4.4. The extra adsorption at higher concentrations causes the top of the peak to migrate slower than its base thereby forming a **fronting peak** on the chromatogram as shown by the orange peak in Figure 4.4.

Note that an irregular peak shape due to the asymmetric distribution of analyte molecules doesn't affect the detector sensitivity, and all the molecules are counted. Therefore, the area of the peak is unaffected by its asymmetry and is still valid. However, peak height measurement is inappropriate with asymmetric peaks.

Activation

Active-solid adsorbents strongly retain water, polar solutes, or oil mist in the sample – an irreversible process at low-column temperature. The adsorbed molecules then block access to the adsorption sites by the analyte molecules. You can slow the deactivation process by the meticulous removal of moisture from the carrier gas and by using a protective column arrangement, but you can never eliminate it. The separation always deteriorates over time, eventually requiring column replacement or reactivation.

You can reactivate an adsorption column by heating it for several hours at a specified temperature, typically (300 °C ± 50 °C), with carrier gas flow.

Molecular sieves

These traditional column packings are based on natural minerals known as zeolites, although the pure forms needed for industrial applications are manufactured products. Their crystalline structure incorporates cage-like formations about the same size as gas molecules. These formations retain some gas molecules and allow others to pass by, yielding a separation by molecular size. There are hundreds of known structures, but only two are useful as stationary phases:

> Molecular sieve 5A has a pore size of 5 Å (0.5 nm) and is a calcium aluminum silicate.

> Molecular sieve 13X has a pore size of 13 Å (1.3 nm) and is a sodium aluminum silicate.

These columns are fragile and need special care. The carrier gas must be devoid of water vapor and carbon dioxide as these gases are permanently adsorbed, spoiling the separation of analytes. The adsorbent particles are friable and don't tolerate shock, including shock from column switching. Never backflush a molecular sieve column.

PGCs use the zeolite molecular sieves only for the permanent gas separations listed in Table 4.9, notably the separation of nitrogen from oxygen. Packed columns can't separate argon from oxygen at typical PGC column temperatures.

It's now possible to coat the wall of a metal capillary column with a layer of a molecular sieve to form a rugged porous-layer open-tubular (PLOT) column. A "thick-film" (50 μm) PLOT column can separate all permanent and noble gases – even argon from oxygen (de Zeeuw 2000).

Carbon

The two forms of carbon used as a stationary phase are carbon molecular sieves and graphitized carbon. A wide array of materials are available that differ in particle form, pore size and shape, and surface treatment. For detailed technical information refer to the informative review by Supelco (2015b).

Carbon molecular sieves are rigid spherical beads made by the controlled pyrolysis of a pure polymer. Unlike the zeolites, their mechanism of separation relies on molecular velocity rather than molecular size. Their nonpolar surface chemistry enables them to separate polar molecules without excessive adsorption. For example, hydrogen sulfide, formaldehyde, and water, as well as the inorganic gases and C_1–C_3 hydrocarbons (Poole 2019).

Graphitized carbon is granular and friable. When coated on the wall of a capillary tube, it forms a PLOT column. The manufacturer can treat the carbon particles with potassium hydroxide or picric acid for separating basic or

Table 4.9 Typical Active Solid Stationary Phases.

Type	Product Name[a]	Typical Applications
Molecular sieves	Type 5A Type 13X zeolites	Separation of permanent gases: hydrogen, oxygen + argon, nitrogen, methane, and carbon monoxide
	ArDSieve®	Separation of permanent gases including argon and oxygen
	Carbosieve®, Carboxen®, or Spherocarb®	Permanent gases, carbon oxides, light hydrocarbons, dinitrogen oxide, sulfur dioxide, and hydrogen sulfide in the presence of water vapor
Graphited carbon	Carbopack®, CarboBlack®	Versatile when coated with a liquid phase
Porous silica	Silica gel	Light hydrocarbons (rarely used in PGC)
	Unibeads® C	Permanent gases, carbon monoxide, and carbon dioxide
	Unibeads® S	Carbon dioxide; C_1–C_4 paraffins, olefins, and acetylenes
Other adsorbents	Salted alumina	Can separate all of the C_1–C_4 hydrocarbons
	Bentone® 34	Separates ethylbenzene and the three xylene isomers

[a] ArDSieve is a trademark of ASDevices; Bentone is a trademark of Elementis Specialties, Inc.; CarboBlack is a trademark of Restek Corporation; Carboxen, Carbopack, Carbosieve, and Spherocarb are trademarks of Sigma-Aldrich Inc.; Chromosorb is a trademark of Manville Corporation; Unibeads is a trademark of Mitsubishi Chemical Company.
For a highly detailed wall chart on porous polymers and other solid stationary phases, refer to Camsco (2009).
The temperature limits for these stationary phases are way above PGC operating temperatures, mostly 400 °C.
Sources: Data compiled from Barry and Grob (2007), Rotzsche (1991), Hepp and Klee (1987), and Supelco (2003, 2015a, 2015b).

acidic compounds, respectively. Another option is to coat the particles with a non-silicone liquid phase, which gives the carbon a unique selectivity, even for polar analytes like alcohols in water. As stationary phases, graphitized carbon comes in two varieties: a product with low surface area that can hold up to 1 % of a liquid phase, and another with high surface area that can hold up to a 10 %. Favorite liquid phases are Carbowax 20M or TCEP.

Silica gel or porous silica

Silica gel was the first column packing used in gas chromatography. When activated by heating to remove adsorbed water, silica gel can separate some of the permanent gases and light hydrocarbons. PGC columns rarely use activated silica gel nowadays because it inexorably adsorbs water and gradually deactivates, spoiling the separation.

For packed columns, porous silica beads have a range of characteristics that depend on their pore size. For example, Unibeads 1S, 2S, and 3S are spherical silica beads with pore sizes 2.5, 7, and 10 nm, respectively, and three size ranges: 30–60, 60–80, and 80–100 mesh. All are good for operation up to 200 °C.

Unipak S beads can separate methane, ethane, ethylene, and acetylene. When using a temperature program, they can also separate many of the C_3–C_5 hydrocarbon gases. A very popular column for light hydrocarbon separation is n-octane bonded on silica (Burger 2013).

Silica can also coat the wall of a PLOT column giving excellent selectivity and rapid analysis of the C1–C3 hydrocarbons or halocarbons. Sharp peaks obtain also for sulfur compounds such as hydrogen sulfide, sulfur dioxide, and carbonyl sulfide (de Zeeuw 2000).

Alumina

Activated alumina was a common packing in early chromatographs but was unreliable because of its propensity to absorb water and other polar molecules, thereby deactivating itself. It's also friable. A modern approach is to deactivate the alumina with salts like potassium chloride or sodium sulfate. Alumina PLOT columns are excellent for laboratory petroleum analysis as they can separate all of the C_1–C_4 hydrocarbons (about 20 compounds). Even so, they are generally too unstable for autonomous online operation and rarely used by PGC's.

Porous polymers

Porous polymers are the synthetic polymer beads first introduced to gas chromatography by Hollis (1966). Here are the main commercial products:

- Chromosorb column packings offer a choice of eight different polymers identified by a suffix numeral from 101 to 108. The numeral is nominal and doesn't indicate polarity. They are available in three mesh sizes 60–80, 80–100, and 100–120. As noted in Table 4.10, several of the original varieties are now obsolete; discontinued by the manufacturer.

- HayeSep column packings come in three mesh sizes 60–80, 80–100, and 100–120, and in 10 varieties designated in increasing polarity as D, Q, P, S, R, C, A, B, N, and T. Hayes claims that HayeSep and Porapak products with identical letter grades are interchangeable, although the match may not be perfect.

- Porapak column packings are available in three mesh sizes 50–80, 80–100, and 100–120, and in 6 varieties designated in increasing polarity as P, Q, R, S, N, and T (Waters 2008). There are also silanized versions of P and Q designated as PS and QS.

Even when their polymer formulations are nominally identical, products from different suppliers may differ in selectivity. Table 4.10 identifies some approximate equivalents.

Typical PGC applications for porous polymers are process gases or volatile liquids: permanent gases; light hydrocarbons; carbon, nitrogen, or sulfur oxides; halocarbons; nitrogen or sulfur compounds; and water. They can uniquely produce a symmetric peak for water and you can locate it anywhere on the chromatogram by selecting an appropriate polymer.

The porous polymers are intermediate in affinity, typically stronger than a liquid phase but weaker than an active solid adsorbent. Chemically, they mostly comprise crosslinked styrene-divinylbenzene polymer chains. The more polar products incorporate various additives into the polymer chain structure to increase polarity.

Hepp and Klee (1987) studied the polarity of Chromosorb and Porapak polymers using three of the McReynolds probes. They classified polymer polarity along three dimensions: proton donating, proton accepting, and dipole attraction. Their results indicate that Chromosorb 104 is extremely

Table 4.10 Typical Porous Polymer Stationary Phases.

Chromosorb	HayeSep	Porapak	Max T °C	Typical Applications
				Note: The products grouped together have similar but not identical selectivity. The Chromosorb brand products listed in parens are no longer available from the manufacturer, but column supply companies may have limited stocks. We included them here for reference only.
Lower polarity				
	D		290	Light hydrocarbons, carbon oxides, acetylene, hydrogen sulfide, methanol, ethylene oxide, and impurities in water
106	Q	Q	275	Alkane–alkene–alkyne separation, organics in water (water elutes quickly), hydrogen sulfide, alcohols, nitrogen oxides, and freons
101	P	P	275 250	Permanent gases, light hydrocarbons, ammonia, water, alcohols, aldehydes, ketones, and chloroform (no amines)
	S	S	250	Alcohol isomers, carbonyls, and halogens (no nitro compounds)
(102) (105)			250 200	Alkane–alkene–alkyne separation and organics in aqueous solution
Higher polarity				
	R	R	250	Alcohols, chlorine, hydrogen chloride, ethers, esters, nitriles, nitroalkanes, and water
107			225	Medium polarity analytes, sulfur gases, and formaldehyde
(104)	C		250	Hydrogen cyanide, nitriles, nitroalkanes, sulfur dioxide, hydrogen sulfide, vinyl chloride, carbon dioxide, and water
	A		165	Permanent gases, argon–oxygen, carbon monoxide, nitrogen oxide, hydrogen sulfide, and water
(103)	B		190	Alkali gases: ammonia, light amines, hydrazine, phosphine, water, alcohols, aldehydes, and ketones (no glycols or nitro compounds)
	N	N	165	Ammonia, carbon dioxide, and water; ethane, ethylene, and acetylene; hydrogen sulfide and carbonyl sulfide
	T	T	165	Light hydrocarbons, ammonia, alcohols, and aldehydes. Strongly retains water
108			225	Polar analytes, oxygenates, and water holdup.

[a]Chromosorb is a trademark of Imerys Minerals California, Inc.; HayeSep is a trademark of Hayes Separations Inc.; Porapak is a trademark of Waters Corporation.
Arranged approximately in ascending order of polarity according to Hayes (2020). For a highly detailed wall chart on porous polymers and other solid stationary phases, refer to Camsco (2009).
Sources: Data compiled from Barry and Grob (2007), Hayes (2020), Hepp and Klee (1987), Ohio Valley (2013), Rotzsche (1991), Supelco (2019), and Waters (2008).

polar, an evaluation which differs from that of Hayes (2020) but is now moot; the manufacturer has discontinued Chromosorb 104. Yet it serves to remind us that polarity is an elusive concept, highly dependent upon the variables measured and the column operating conditions. In addition, all porous polymers are susceptible to small batch variations and ageing during use.

We can now make porous polymers by *in situ* polymerization, depositing a layer of crosslinked polymer on the inside tube wall of a passivated metal capillary tube. These PLOT columns can separate most volatile substances and are durable enough for process applications.

Knowledge Gained

Theory

- Adjusted retention times of the n-paraffins increase in exponential proportion to their carbon numbers.

- Kováts assigned a retention index (I) to n-paraffins equal to 100 times their carbon numbers.

- I is the carbon number (×100) of a hypothetical n-paraffin with the same retention as the solute.

- I indicates the logarithmic location of a solute peak relative to the two-bracketing n-paraffin peaks.

- Adding a methylene group to any compound increases its retention index by about 100.

- The retention index with a constantly increasing column temperature is linear, not logarithmic.

- Rohrschneider said that five probe solutes would reveal multiple dimensions of liquid phase polarity.

- McReynolds measured the retention index of 10 probes on the tested liquid phase and on squalane.

- Each probe intends to represent a different kind of electronic interaction with the tested liquid phase.

- McReynolds deducted the probe retention indices on squalane from those on the tested liquid phase.

- McReynolds constants (MRC) are the observed retention index differences for each probe solute.

- It's now thought that five probes are enough to adequately characterize a liquid phase.

- MRC values can confirm that two liquid phases will produce essentially equivalent separations.

- MRC values can predict the position of certain peaks (but not all) on the chromatogram.

- The weakness of MRC is that all probe solutes have multiple interactions with the liquid phase.

- A newer theory uses System Constants to isolate individual interactions, but interest has waned.

Liquid phases

- Most PGCs use low-efficiency packed or capillary columns and need good separation factors.

- Therefore, process GCs employ many more liquid phases than laboratory GCs do.

- PGC manufacturers no longer label their columns with full specifications – they use codes instead.

- Nominally equivalent columns from two vendors might not perform identically.

- If changing a PGC application, consider a new column system from the original PGC vendor.

- The concepts of affinity and polarity are vague and subject to subjective interpretation.

- Polarity is a unique and complex mix of forces that depend upon the solute, as well as the solvent.

- For experienced designers, liquid phase polarity is still a useful guide for column selection.

- A useful starting point is to choose a liquid phase that resembles the analytes: "like dissolves like."

- *A nonpolar column separates all analytes by vapor pressure and peaks are in boiling point order.*

- *A polar column also separates in boiling point order, then adds retention from polar interactions.*

- *There exists a profusion of polar liquid phases giving the column designer many options.*

- *We prefer nonpolar liquid phases for their stability, predictability, efficiency, and ease of coating.*

- *But polar liquid phases create many unique separations that nonpolar liquids cannot do.*

- *The best way to choose a liquid phase may be to ask someone who knows.*

- *Most PGC column designers select from about one dozen stationary phases they know well.*

- *The standard toolbox will include several methyl, phenyl, fluoro, and cyano silicones and a PEG wax.*

- *Packed columns are popular for PGCs and are less expensive than wall-coated metal capillaries.*

Silicone phases

- *The many silicones are all polysiloxanes having low vapor pressure and high stability.*

- *The polymer chain length may vary, so silicone oils and gums may have widely different viscosities.*

- *The basic silicone, polydimethylsiloxane, is almost nonpolar and the most popular of all liquid phases.*

- *The dimethyl silicone has a long polymer chain of alternate silicon and oxygen atoms.*

- *The polymer chain has two methyl groups attached to each silicon atom.*

- *In other silicone phases phenyl, fluoro, or cyano groups replace some of the methyl groups.*

- *A polymer from a single monomer can have 50 % or 100 % of methyl groups replaced by new groups.*

- *A copolymer from multiple monomers can have any chosen percentage of many substitute groups.*

- *Replacing the methyl groups with longer paraffinic attachments makes a silicone even more nonpolar.*

- *Adding phenyl groups increases the retention of both nonpolar and polar solutes.*

- *Phenyl silicones have an increased affinity for olefins, aromatics, and polar solutes.*

- *Fluoro silicones have enhanced polarity, and a unique affinity for carbonyl and nitrile compounds.*

- *Cyano silicones have the highest polarity, and an affinity for alcohols, phenols, and nitro compounds.*

- *The formulation of many silicones is propriety and may include multiple functional groups.*

- *By polymerizing the siloxane monomers in situ, the liquid phase chemically attaches to the tube wall.*

- *Bonded liquid phases have reduced column bleed, allowing higher detector sensitivity.*

- *A small addition of vinyl siloxane will cross-link the polymer chains for additional phase stability.*

- *To avoid damage to a silicone column, remove all oxygen and moisture from the carrier gas.*

Non-silicone phases

- *Older PGCs often used dialkyl esters like dinonyl phthalate but these are now becoming obsolete.*

- *With medium polarity, the diesters can separate olefins, acetylenes, alcohols, and aromatics.*
- *But diester phases have high volatility, low-temperature limits, and low long-term stability.*
- *Polyesters like DEGS have low volatility and high polarity and remain popular in packed columns.*
- *Polyesters are losing market to cyanosilicones with similar selectivity and higher temperature ratings.*
- *Polyols (sometimes called polyethers) remain popular for their high polarity and selectivity.*
- *Polyethylene glycol (reimagined as "wax") is the favorite polar phase in packed or capillary columns*
- *The name of a polyethylene glycol includes its molar mass as a suffix; like Carbowax 1500.*
- *PEG columns are highly polar and give excellent separations of polar solutes, notably alcohols.*
- *Hydrocarbons and other nonpolar analytes are not well separated on highly polar columns like PEG.*
- *Remove moisture and oxygen from the carrier gas to avoid damaging a PEG column.*
- *Modified polyethylene glycol phases are available for acidic or basic samples.*
- *Nitrile ethers like TCEP have long been the liquid of choice to separate polar analytes.*

Solid phases

- *The distribution constant for a gas–solid equilibrium is unlikely to be constant.*
- *Often, the proportion of molecules adsorbed by a solid decreases as their concentration increases.*
- *A decrease in the percent adsorbed at higher concentration can create a severely fronting peak.*
- *Rarely, the proportion of molecules adsorbed by a solid increases as their concentration increases.*
- *An increase in the percent adsorbed at higher concentration can create a severely tailing peak.*
- *An isotherm is a plot of amount adsorbed versus analyte concentration at constant temperature.*
- *An isotherm is linear only if the distribution constant is really constant.*
- *Solid adsorbents lose their retention ability due to adsorbing water and other polar substances.*
- *Zeolite molecular sieves will separate permanent gases including oxygen and nitrogen.*
- *A PLOT molecular sieve column can separate all noble and permanent gases.*
- *Water doesn't deactivate carbon molecular sieves and they can separate many gases and vapors.*
- *Graphitized carbon PLOT columns can be acid or alkali treated or coated with a non-silicone liquid.*
- *Porous silica PLOT or packed columns separate light hydrocarbons and other volatiles.*
- *A salted alumina PLOT column can separate all of the C_1–C_4 hydrocarbons.*
- *Porous polymers can separate a wide range of compounds and several polarities are available.*
- *Polymerizing the styrene in situ forms a thin porous-polymer coating on the PLOT column wall.*

Did you get it?

Self-assessment quiz: SAQ 04

Q1. What kind of forces cause the affinity that a liquid phase has for a solute? Select one answer:
  A. Chemical forces
  B. Electronic forces
  C. Magnetic forces
  D. Kinetic forces
  E. Gravitational forces

Q2. What properties of an analyte can we use to predict its retention on a nonpolar column? Select all the correct answers:
  A. The polarity of the analyte molecule.
  B. The size of the analyte molecule.
  C. The vapor pressure of the analyte at column temperature.
  D. The boiling point of the analyte.
  E. The analyte retention index on that column.
  F. The McReynolds constants of that column.

Q3. What kind of column would you use to separate some paraffins having the same formula; that is, isomers of each other?
  A. A poly[dimethylsiloxane] column like OV-1.
  B. A poly[ethylene glycol] column like Carbowax 20M.

Q4. There are two silicone liquid phases with 50 % phenyl groups along their polymer backbone (that is, discounting the ends). In what way are their molecular structures different? The two liquid phases are:
  A. A polymer of methyl phenyl siloxane.
  B. A 50:50 copolymer of dimethyl siloxane and diphenyl siloxane.

Q5. Researchers using beeswax as a stationary phase found it's McReynolds constants are: 43, 110, 61, 88, and 122. What is the polarity of beeswax on the 0–100 scale used in Table 4.8?

Q6. Given that an isothermal column has a holdup time of 50 seconds, and a retention time of 350 seconds for *n*-pentane, estimate the retention time of a component that has a retention index of 550.

Q7. When using a solid adsorbent column, what shape are the peaks when:
  A. The adsorption isotherm is linear?
  B. The adsorption isotherm curves downward for increasing adsorbate concentrations?
  C. The adsorption isotherm curves upward for increasing adsorbate concentrations?

Check your SAQ answers with those given in the back of the book.

References

Further reading

For technical information about liquid phase selection and performance, refer to:

- The detailed update on silicone liquid phases and their properties; *Silicone Stationary Phases for Gas Chromatography* by Lars Blomberg (2001).

- The lucid encyclopedia article: *Liquid Phases for Gas Chromatography* by Nicholas Snow (2012).

- The comprehensive textbook: *Columns for Gas Chromatography: Performance and Selection* by Barry and Grob (2007).
- The cyclopedic reference work: *Stationary Phases in Gas Chromatography* by Harald Rotzsche (1991).
- The now-dated but still relevant early textbook: *Stationary Phases in Gas Chromatography* by Baiulescu and Ilie (1975).

There's also a series of articles by Yancey (1985–1994) that may be helpful. Joel Yancey was a petroleum chemist at BP America, so his experience is relevant to gas chromatography in the oil and chemical industries.

For online selection tools and practical information, visit the suppliers' websites and also download their helpful literature, for example:

- *Gas Chromatography Supplies Catalog 62* from Ohio Valley (2013).
- *GC Column Selection: Tips and Guidelines and Separation Solutions Guide* from Phenomenex (2017, 2020).
- *GC Column Selection Guide from Supelco* (2013).
- *Rxi GC Columns and Guide to GC Column Selection and Optimizing Separations* from Restek (2021a).

Unfortunately, the more recent editions of supplier guides are becoming less informative, following a trend in the industry to promote proprietary columns for specific applications.

Cited

Abraham, M.H., Poole, C.F., and Poole, C.K. (1999). Classification of stationary phases and other materials by gas chromatography. *Journal of Chromatography A* 842, 79–114.

Agilent (2021). *Standard Polysiloxane GC Columns*. Santa Clara, CA: Agilent Technologies. Accessed 2021-01-11 at: www.agilent.com. Inc.

Anderson, J.L. and Armstrong, D.W. (2005). Immobilized ionic liquids as high-selectivity/high-temperature/high-stability gas chromatography stationary phases. *Analytical Chemistry* 77, 6453–6462.

Ashes, J.R. and Haken, J.K. (1973). The effect of temperature on the retention behavior and polarity of several polysiloxane stationary phases. *Journal of Chromatography A* 84, 231–239.

Baiulescu, G.E. and Ilie, V.A. (1975). *Stationary Phases for Gas Chromatography*. Oxford, UK: Pergamon Press.

Barry, E.F. and Grob, R.L. (2007). *Columns for Gas Chromatography: Performance and Selection*. Hoboken, NJ: John Wiley & Sons, Inc.

Blomberg, L.G. (2001). Silicone Stationary Phases for Gas Chromatography. In: *LC•GC Europe*, February 2001.

Burger, B. (2013). *Res-Sil C Bonded GC Packings for Analyses of Light Hydrocarbons*. Bellefonte, PA: Restek Corporation. Product literature accessed 2023/12/21 at: www.restek.com/en/technical-literature-library/articles/res-sil-C-bonded-GC-packings-for-analyses-of-light-hydrocarbons.

Camsco (2009). *Camsco Sorbent Selection Chart*. Houston, TX: Camsco, Inc.

Hawkes, S., Grossman, D., Hartkopf, A., Isenhour, T., Leary, J., and Parcher, J. (1975). Preferred stationary liquids for gas chromatography. *Journal of Chromatographic Science* 13, No. 3, 115–117.

Hayes (2020). *Specifications — HayeSep® porous polymers*. Accessed on 2021/03/07 at: www.vici.com/hayesep/polyspec.php.

Hepp, M.A. and Klee, M.S. (1987). Characterization of porous polymers by polar strength and selectivity. *Journal of Chromatography A* 404, 145–154.

Hollis, O.L. (1966). Separation of gaseous mixtures using porous polyaromatic polymer beads. *Analytical Chemistry* 38, No. 2, 309–316.

Kovats, E. (1958). Gas-chromatographische Charakterisierung organischer Verbindungen. Teil 1: Retentionsindices aliphatischer Halogenide, Alkohole, Aldehyde und Ketone. *Helvetica Chimica Acta* 41, 1915. https://doi.org/10.1002/hlca.19580410703.

Langmuir, E. (1918). The adsorption of gases on plane surfaces of glass, mica and platinum. *Journal of the American Chemical Society* 40, No. 9, 1361–1403. https://doi.org/10.1021/ja02242a004

Matisová, E., Moravcová, A., Krupčík, J., Čellár, P., and Leclercq, P.A. (1988). Problems with the reproducibility of retention data on capillary columns with hydrocarbon C_{87} as the stationary phase. *Journal of Chromatography A* 454, 65–71.

McNair, H.M., Miller, J.M., and Snow, N.H. (2019). *Basic Gas Chromatography*, Third Edition. Hoboken, NJ: John Wiley & Sons, Inc.

McReynolds, W.O. (1970). Characterization of some liquid phases. *Journal of Chromatographic Science* 8, 685–691.

Miller, J.M. (2005). *Chromatography: Concepts and Contrasts*, Second ed., Hoboken, NJ: John Wiley & Sons, Inc.

NIST (2018). *Chemistry WebBook: NIST Standard Reference Database Number 69*. Accessed 2021/02/09 at: https://webbook.nist.gov.

Ohio Valley (2013). *Gas Chromatography Supplies Catalog 62*. Marietta, OH: Ohio Valley Specialty Company.

Phenomenex (2017). *GC Column selection: Tips and Guidelines*. Product brochure: BR44870117_I. Torrance, CA: Phenomenex, Inc.

Phenomenex (2020). *LC & GC Separation Solutions Guide – Chemical Industry*. Product Brochure: BR67480319_W. Torrance, CA: Phenomenex, Inc.

Poole, C.F. (2012). *Gas Chromatography*. Oxford, UK: Elsevier.

Poole, C.F. (2019). Gas chromatography: column technology. In: *Encyclopedia of Analytical Science*, Third Edition, 118–134. Amsterdam, Netherlands: Elsevier B.V.

Poole, C.F., Pomaville, R.M., and Dean, T.A. (1989). Proposed substitution of apolane-87 for squalane as a nonpolar reference phase in gas chromatography. *Analytica Chimica Acta* 225, 193–203.

Poole, C.F. and Poole, S.K. (2011). Ionic liquid stationary phases in gas chromatography. *Journal of Separation Science* 34, 888–900. Weinheim: Wiley-VCH Verlag GmbH & Co. KGaA.

Restek (2021a). *Guide to GC Column Selection and Optimizing Separations*. Product brochure: Lit. Cat.# GNAR1724A-UNV. Bellefonte, PA: Restek Corporation.

Restek (2021b). *GC Column Cross-Reference: Columns by Phase*. Product Brochure. Bellefonte, PA: Restek Corporation.

Riedo, F., Fritz, D., Tarján, G., Kováts, E.sz. (1976). A tailor-made C87 hydrocarbon as a possible non-polar standard stationary phase for gas chromatography. *Journal of Chromatography A* 126, No. 3, 63–83.

Rohrschneider, L.J. (1966). Eine methode zur chrakterisierung von gaschromatographischen trennflüssigkeiten (A method of characterization of gas chromatographic stationary phases). *Journal of Chromatography A* 22, 6–22.

Rotzsche, H. (1991). Stationary phases in gas chromatography. *Journal of Chromatography Library* 48. Amsterdam, Netherlands: Elsevier Science Publishers B.V. ISBN: 9780444987334.

Santiuste, J.M. and Takács, J.M. (1999). Temperature dependence study of several polarity scales used in gas–liquid chromatography stationary phase characterization. *Journal of Chromatographic Science* 37, No. 4, 113–120. https://doi.org/10.1093/chromsci/37.4.113.

Snow, N.H. (2012). Liquid phases for gas chromatography (updated). In: *Encyclopedia of Analytical Chemistry*. Chichester, UK: John Wiley & Sons, Ltd.

Supelco (1997). *The Retention Index System in Gas Chromatography: McReynolds Constants*. Bulletin 880 T194880. Bellefonte, PA: Supelco division of Sigma-Aldrich Company LLC.

Supelco (1999). *Packed Column Application Guide*. Bulletin 890A T195890A. Bellefonte, PA: Supelco division of Sigma-Aldrich Company LLC.

Supelco (2003). *Carboxen GC PLOT Capillary Columns*. Product brochure T403146B. St. Louis, MO: Sigma-Aldrich Co. LLC.

Supelco (2013). *GC Column Selection Guide: Achieve Optimal Method Performance*. Product Brochure KCX 11873/T407133C 1103. St. Louis, MO: Sigma-Aldrich Co. LLC.

Supelco (2015a). *Separation of Hydrocarbons by Packed Column GC*. Bulletin 743L T100743L. St. Louis, MO: Sigma-Aldrich Co. LLC.

Supelco (2015b). *Specialty Carbon Adsorbents*. Product bulletin MQR 82950/T410081. St. Louis, MO: Sigma-Aldrich Co. LLC.

Supelco (2019). *GC Column Selection Guide*. Product catalog KCX11873/T407133C. St. Louis, MO: Sigma-Aldrich Company LLC.

Waters (2008). *Porapak Gas Chromatography Column Packing Materials*. Care and use manual: WAT027255 Rev 4. Milford, MA: Waters Corporation.

Yancey, J.A. (1985a). Liquid phases used in packed gas chromatographic columns. Part I. Polysiloxane liquid phases. *Journal of Chromatographic Science* 23, No. 4, 161–167. https://doi.org/10.1093/chromsci/23.4.161.

Yancey, J.A. (1985b). Liquid phases used in packed gas chromatographic columns. Part II. Use of liquid phases which are not polysiloxanes. *Journal of Chromatographic Science* 23, No. 8, 370–377. https://doi.org/10.1093/chromsci/23.8.370.

Yancey, J.A. (1986). Liquid phases used in packed gas chromatographic columns. Part III. McReynolds constants, preferred liquid phases, and general precautions. *Journal of Chromatographic Science* 24, No. 3, 117–124. https://doi.org/10.1093/chromsci/24.3.117.

Yancey, J.A. (1994a). Review of liquid phases in gas chromatography, Part I: intermolecular forces. *Journal of Chromatographic Science* 32, No. 8, 349–357. https://doi.org/10.1093/chromsci/32.8.349.

Yancey, J.A. (1994b). Review of liquid phases in gas chromatography, part II: applications. *Journal of Chromatographic Science* 32, No. 9, 403–413. https://doi.org/10.1093/chromsci/32.9.403.

de Zeeuw, J. (2000). Gas–solid gas chromatography. In: *Encyclopedia of Separation Science* (ed. I.D. Wilson), 481–489. Amsterdam, Netherlands: Academic Press. doi: 10.1016/B0-12-226770-2/00151-4.

Tables

4.1 Vendor Codes for Columns and Liquid Phases.
4.2 Typical Polarity of Solutes.
4.3 Strength of Solvent–Solute Interaction.
4.4 Popular Polysiloxane Liquid Phases.
4.5 Popular Non-Silicone Liquid Phases.
4.6 McReynolds Probes.
4.7 Confirming Equivalent Liquid Phases.
4.8 Selectivity of Common Liquid Phases.
4.9 Typical Active Solid Stationary Phases.
4.10 Typical Porous Polymer Stationary Phases.

Figures

4.1 Structure of Poly[dimethylsiloxane].
4.2 Principle of Retention Index.
4.3 Predicting Retention by McReynolds Constants.
4.4 Effect of Adsorption Isotherm on Peak Shape.

Symbols

Symbol	Variable	Unit
$[A]_S$	Concentration of solute "A" in the stationary phase	mol/mL
$[A]_M$	Concentration of solute "A" in the mobile phase	mol/mL
e	(Also s, a, b, l)	System constants of liquid phase.
E	(Also S, A, B, L)	System constants of solute molecule.
I	Retention index	None
k	Retention factor of peak	None
K_c	Distribution constant	None
n	Number of atoms or functional groups in a molecule	None
S'	Fifth McReynolds probe: pyridine	
t_R	Retention time of peak	s
t_R'	Adjusted retention time of peak	s
U'	Fourth McReynolds probe: nitropropane	
X'	First McReynolds probe: benzene	
Y'	Second McReynolds probe: 1-butanol	
Z'	Third McReynolds probe: 2-pentanone	

Equations

4.1 $\quad I_i = 100\left[n + \dfrac{\log_{10}(t_R')_i - \log_{10}(t_R')_n}{\log_{10}(t_R')_{n+1} - \log_{10}(t_R')_n}\right]$ Isothermal retention index.

4.2 $\quad I_i = 100\left[n + \dfrac{(t_R)_i - (t_R)_n}{(t_R)_{n+1} - (t_R)_n}\right]$ Retention index under a linear temperature ramp.

4.3 $\quad k = c + eE + sS + aA + bB + lL$ System constants.

4.4 $\quad K_c = \dfrac{[A]_S}{[A]_M}$ Distribution constant.

New technical terms

When first introduced, these words and phrases were in bold type. You should now know the meaning of these technical terms.

adsorbent solid	**McReynolds constants**
adsorption isotherm	**probe**
bonded phase	**retention index**
distribution constant	**siloxane**
homolog	**solute polarity**
Langmuir isotherm	**system constants**

This chapter also introduced many chemical names. Refer to the Glossary for an explanation of these terms.

5

PGC column design

"Column designers have the awesome responsibility of cajoling a gas chromatograph to do exactly what the user needs it to do. They act like computer programmers, using their own arcane language to breathe purpose into a lifeless machine, thereby transforming it into an autonomous analytical device. Here, we attempt to penetrate some of their secrets."

The design objective

Most process gas chromatographs rely on several columns working in concert to achieve the desired analysis. The overall **column system** must achieve an adequate resolution between each analyte peak and all other components of the sample. But an *individual column* has a more limited and focused role.[1] It might, for example, only be responsible for holding back heavies so they can be backflushed.[2] Or perhaps for separating just two peaks after another column has removed all other components and flushed them to vent.

*Each column should reliably perform its designated role so the column system as a whole delivers **sufficient resolution** of each desired analyte from all other components in the sample and does so in **minimum time**.*

For measurement, the resolution doesn't need to be perfect, but it must be good enough to achieve accurate results. A process chromatograph has some clever ways to compensate for a not-quite-perfect resolution, but we should be careful not to expect too much of the data processing routines. It's also a

---

[1] Clearly, having columns in series complicates the design process. This chapter has the limited goal of understanding the design and operation of a single column. Chapter 9 considers the additional complexity of columns in series.
[2] Backflush is a technique for removing heavies by reversing the flow in the first column. Refer to Chapter 10 for more information.

Process Gas Chromatography: Advanced Design and Troubleshooting, First Edition. Tony Waters.
© 2025 John Wiley & Sons Ltd. Published 2025 by John Wiley & Sons Ltd.

mistake for the method to rely on critical settings that might be beyond the skills of the maintenance personnel.

In other words, always start with good chromatography.

Notice the additional requirement of minimum analysis time. Even a busy laboratory can benefit from more analyses per hour. For a process gas chromatograph, it's critical: if the analysis time is too long, the analyzer might be completely useless for its intended purpose.

Design strategy

The column designer has to ensure that each column performs a specific task, efficiently and reliably. The designer must therefore come up with a peak separation strategy, and each application might require a different approach. There are many questions to answer and decisions to make. For instance:

- Should the analyte peaks initially stay together, so heavies can be backflushed?
- In what order will the analyte peaks occur on the chromatogram?
- What column will move the unmeasured peaks out of the way?
- Will the separation of certain analytes require a second or third stationary phase?
- What is the best column design for samples with both high and low analyte concentrations?
- What is the effect of operating a column in series with another column?
- Is it quicker to analyze duplicate samples in parallel column trains?
- ... and many more.

What's more, designers have to realize that their initial plan might fail and they must be flexible enough to change direction when it's not working out. We will study all these issues in later chapters of the book. For now, it's sufficient to know that each column has a specific job to do and the application engineer must ensure that it does it expeditiously and reliably.

It's always about resolution

For each column in a process chromatograph, there is always a **critical pair** of adjacent peaks that it must resolve. This mandate might arise from the necessity of separating a peak for measurement, or because two or more peaks must pass into a different column for further separation.

> *Therefore, the prime objective of a single column is to generate adequate resolution between two peaks or two groups of peaks, while not detracting from the separation of other peaks.*

When a column is feeding a measurement detector, the resolution of the analyte peaks must be enough to precisely measure their concentrations. In contrast, when a column feeds another column, it might need to create an even greater resolution – enough for switching designated peaks into a second column while diverting others to a different fate.

It turns out that only three column design parameters affect the resolution of adjacent peaks and they might not be the ones you would expect. The plate theory of chromatography identifies the variables involved and reveals a useful relationship between them known as the **resolution equation**. This useful equation employs only the simple chromatogram measurements we made in Chapter 2, peak retention times and widths.

Before discussing the three parameters and how to use them in column design, the following **SCI-FILE:** *On Resolution* explains how the resolution equation follows from first principles. It's not difficult to understand, so if you want to follow along, delve right in. Else, you can skip the theory and continue with the practical discussion that follows. Either way, you will need to recognize the three column-design parameters and comprehend their combined effect on resolution.

SCI-FILE: *On Resolution*

Theory

The complete theory of chromatographic separation is complex and is the subject of many full-length books. An attempt to cover it here would be futile. Yet, from the simple equations you already know, we can discover the column factors that determine resolution. Then, we can calculate the length of column necessary to obtain the desired analysis.

The resolution equation

Chapter 2 defines the resolution of adjacent peaks as a function of their retention times (t_R) and their average base width. Consider the resolution (R_s) of two adjacent peaks with Peak A eluting before Peak B. Since the peaks are close, it's reasonable to assume their base widths (w_b) are approximately equal. Therefore, from Equation 2.8, we get:

$$R_s = \frac{t_{RB} - t_{RA}}{w_b} \quad (5.1)$$

Chapter 2 also introduces *plate number* as the prime measure of column efficiency, since it relates to the peaks being well retained and narrow. From Equation 2.26, the observed plate number (N) is a function of peak retention time (t_R) and peak base width (w_b):

$$N = 16\left(\frac{t_R}{w_b}\right)^2$$

Hence:

$$w_b = 4 \cdot \frac{t_R}{\sqrt{N}} \quad (5.2)$$

To discover the plate number (N_r) required to achieve a desired resolution (R_s), substitute for w_b in Equation 5.1:

$$R_s = \sqrt{N_r} \cdot \frac{t_{RB} - t_{RA}}{4\, t_R}$$

$$\sqrt{N_r} = 4 R_s \cdot \frac{t_R}{t_{RB} - t_{RA}}$$

Now use Equation 2.1 to convert the retention times to adjusted retention times (t_R'):

$$t_R = t_M + t_R'$$

$$\sqrt{N_r} = 4R_s \cdot \frac{t_M + t_R'}{t_{RB}' - t_{RA}'} \quad (5.3)$$

Equation 5.3 has two solutions, one for Peak A and one for Peak B. The solutions are similar but not identical and both appear in the literature, potentially causing confusion. Peak B has the longer retention time so it yields the more conservative estimate of required plate number.

Updating Equation 5.3 for Peak B yields:

$$\sqrt{N_r} = 4R_s \cdot \frac{t_M + t_{RB}'}{t_{RB}' - t_{RA}'}$$

Divide by t_{RB}':

$$\sqrt{N_r} = 4R_s \cdot \frac{t_M/t_{RB}' + 1}{1 - t_{RA}'/t_{RB}'} \quad (5.4)$$

Now we're getting close. The two ratios in the above equation are inverted forms of the *retention factor* (k) and the *separation factor* (α) defined in Chapter 2:

$$k = \frac{t_R'}{t_M}$$

$$\alpha = \frac{t_{RB}'}{t_{RA}'}$$

Substitute (k) and (α) into Equation 5.4 and rearrange:

$$\sqrt{N_r} = 4R_s \cdot \frac{1/k_B + 1}{1 - 1/\alpha}$$

$$\sqrt{N_r} = 4R_s \cdot \frac{\alpha}{\alpha - 1} \cdot \frac{1 + k_B}{k_B}$$

Finally, the required plate number for Peak B is:

$$N_r = 16R_s^2 \cdot \left(\frac{\alpha}{\alpha - 1}\right)^2 \cdot \left(\frac{1 + k_B}{k_B}\right)^2 \quad (5.5)$$

Rearranging Equation 5.5 forms an equation for calculating the resolution obtainable from a column:

$$R_s = \frac{\alpha - 1}{\alpha} \cdot \frac{k_B}{1 + k_B} \cdot \frac{\sqrt{N}}{4} \quad (5.6)$$

This well-known *resolution equation* reveals that the best resolution requires optimum values for separation factor (α), retention factor (k_B), and plate number (N). If one of these values is too low, an adequate resolution will not be possible. It's good to see that all three of these values come from simple chromatogram measurements.

Purnell (1960 and 1962, 114) was the first to derive this useful equation. Several authors have commended its use in PGC, including Annino and Villalobos (1992, 349). We discuss its implication for column design in the main text that follows.

Endnote

Other authors, including Knox (1961) and Scott (1998, 59), derive a variant of Equation 5.5 that calculates the required plate number based on the retention time of Peak A. Since you may see this equation in the literature, we simply state it here:

$$N_r = 16R_s^2 \cdot \left(\frac{1}{\alpha - 1}\right)^2 \cdot \left(\frac{1 + k_A}{k_A}\right)^2 \quad (5.7)$$

If you are up for a challenge, see if you can derive this equation for yourself. It's a good way to learn.

From Equation 5.3, it's clear that a calculation for Peak A must return a lower value for required plate number than a calculation for Peak B. Thus, Equation 5.7 will return a lower value for the required plate number. For closely adjacent peaks, the difference is small.

Achieving resolution

The resolution equation

If you skipped the **SCI-FILE:** *On Resolution*, you should know that Equation 5.6 therein clearly identifies the three factors that affect the resolution (R_s) of two chromatogram peaks. Here it is again:

$$R_s = \frac{\alpha - 1}{\alpha} \cdot \frac{k_B}{1 + k_B} \cdot \frac{\sqrt{N}}{4}$$

This is the resolution equation. It looks rather complex at first, but for two adjacent peaks on a chromatogram, it simplifies to:

$$R_s = a \cdot b \cdot c$$

Each of the parameters (a), (b), and (c) is a function of just one column variable, so it's easy to assess the effect of each variable on the obtained resolution. This book devotes a chapter to each variable:

> Parameter (a) is a function of the *separation factor* (α) between the two peaks of interest. Getting an adequate **α-value** is the main focus of Chapter 4.

> Parameter (b) is a function of the *retention factor* (k_B) of the second peak. Optimizing the **k-value** is the highlight of this chapter.

> Parameter (c) is a function of the *plate number* (N) of the second peak. We'll see how to optimize the column **N-value** in Chapter 6.

Chapter 2 provides procedures for calculating these values from simple chromatogram measurements.

In practice, experienced PGC column designers rarely calculate column variables explicitly. Instead, they rely on intuition gained from their long experience with columns. They instinctively understand that getting more theoretical plates is not the only route to resolution. Now, just by looking at Equation 5.6, those of us with less experience can also see that a high plate number is not the only goal. Getting an adequate separation factor and retention factor is also important. To obtain the desired resolution in minimum time requires an optimum combination of all three variables.

Interaction between factors

We can use the resolution equation to examine the relative effect of the separation and retention factors. Figure 5.1 is a graphical plot of Equation 5.6 for various values of separation factor (α) and retention factor (k_B). It shows the plate number necessary for a resolution of $R_s = 1.0$ between two peaks.

Before getting into detail, recall that a resolution of 1.0 is *barely adequate* to measure equal peaks. To get a baseline separation of those equal peaks, we'll need a resolution of $R_s = 1.5$. Equation 5.6 also shows that resolution is proportional to the square root of plate number, so a baseline separation

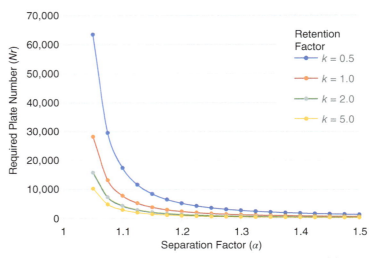

Figure 5.1 Plate Number for Resolution.

Showing plate number required (N_r) versus separation factor (α) for a minimal resolution of $R_s = 1.0$ at different values of retention factor (k). The required column length in mm is roughly equal to the plate numbers shown. This is because a baseline resolution needs about twice these plate numbers and an efficient column has a plate height of about 0.5 mm.

will require 2.25 times the plate numbers shown in Figure 5.1. To resolve a pair of unequal peaks, you might need $R_s = 2.0$ and four times the indicated plate numbers!

Figure 5.1 shows how the separation and retention factors interact. It's evident that a resolution of $R_s = 1.0$ is easy to achieve when $\alpha = 1.5$ or greater. A small plate number will do it. It gets more difficult when the separation factor is lower. For instance, for two peaks with a separation factor $\alpha = 1.1$, we see that:

➤ The column needs 3,000 plates when $k = 5$ (yellow line).
➤ The column needs 18,000 plates when $k = 0.5$ (blue line).

Packed columns can easily deliver such efficiencies. But when the liquid phase yields an even lower separation factor, a packed column may be inadequate for the task. Consider the plate number you would need for two peaks with a separation factor $\alpha = 1.05$:

➤ The column needs 10,000 plates when $k = 5$ (yellow line).
➤ The column needs 64,000 plates when $k = 0.5$ (blue line).

Only a capillary column can do that.

The column length

After calculating the required plate number (N_r) from Equation 5.5, it's easy to calculate the necessary column length from the plate height (H) of the column. If all else is constant,[3] the plate height doesn't change with column length. Rearranging Equation 2.3 gives the column length (L) as:

$$L = N_r \cdot H \tag{5.8}$$

Now let's see what it takes to get better resolution. We'll use the green line in Figure 5.1. This line is for a retention factor of $k = 2$, so the peaks are in the liquid phase for two-thirds of the time. Again, if $\alpha = 1.05$:

➢ For $R_s = 1.0$, the column needs 16,000 plates (from Figure 5.1).
It would be about 8 m long.

➢ For $R_s = 1.5$, the column needs 36,000 plates (calculated).
It would be about 18 m long.

➢ For $R_s = 2.0$, the column needs 64,000 plates (calculated).
It would be about 32 m long.

The column lengths noted above are just for illustration; they are minimal estimates from Equation 5.8 with an optimistic plate height $H = 0.5$ mm. The actual column lengths might be twice as long, depending on the measured plate number of the column.

We shall soon discover an optimum range for the retention factor: not too small and not too large. In the present example, we assumed $k = 2$, which is actually pretty good!

A deeper understanding

For a deeper understanding, we need to look more closely at the physics of distribution because it's the basic science behind gas chromatography. The following **SCI-FILE:** *On Distribution* defines the *distribution constant* and explores its relationship to retention factor and separation factor. The SCI-FILE is optional reading but highly recommended.

SCI-FILE: *On Distribution*

Retention volume

Chapter 2 describes a way to get column performance data directly from the chromatogram by measuring peak retention times and widths. It was convenient to measure those distances in millimeters, although they are really times. It works out fine because our main performance parameters – retention factor, separation factor, plate number, and resolution – are all ratios, so the units cancel out.

To understand the distribution constant and the variables derived from it, we go one stage further and convert the measured retention times to volumes. This is easy to do if you know the carrier

---

[3] In practice, of course, the carrier gas pressure cannot be the same for different column lengths working at the same carrier velocity which complicates the issue. See Chapter 9.

flow rate ($\dot{V}_o$) at the column exit. To convert the retention time (t_R) of a peak to its **retention volume** (V_R), multiply by the flow rate:

$$V_R = t_R \cdot \dot{V}_o \cdot \frac{T_c}{T_a} \quad (5.9)$$

If retention time is in seconds, the flow rate should also be in seconds (for example, 0.5 mL/s) and retention volume comes out in milliliters. Since it's necessary to measure carrier flow at ambient temperature (T_a), Equation 5.9 includes a factor to correct the measured flow rate to the column temperature (T_c). If using a soap-film flowmeter, a correction for the added water vapor saturation is also necessary.

Retention volume of a solute is the volume of carrier gas needed to elute an average molecule of that solute from the column. It's a more fundamental measure than retention time because it doesn't vary with carrier gas flow rate. Its independence from flow rate makes retention volume a better unit for reporting and cataloging experimental retention data.

Equation 5.9 will also convert other measured times to volume units, including holdup time, adjusted retention time, and peak width. For example, here's Equation 2.1 expressed in volume units:

$$V_R = V_M + V_R' \quad (5.10)$$

This shows that retention volume is the sum of **holdup volume** (V_M) and **adjusted retention volume** (V_R').

For accurate work, it's best to adjust the holdup volume by subtracting the extracolumn volume inside valves, connection tubes, and detectors. Generally, however, this level of detail is unwarranted for work on process gas chromatographs. The correction is significant only for low-volume capillary columns and to avoid band broadening a PGC designed to use those columns should already have negligible extracolumn volume.

It's necessary to use volume units in the discussion on the distribution constant that follows. In PGC practice, though, the calculations we need are mostly ratios that work equally well in time units, so retention volumes are not particularly useful to us on a daily basis.

Distribution constant

This is important. To become proficient at gas chromatography, it helps to understand how our chromatogram measurements relate to the underlying science. So let's look at what happens inside a column.

In Waters (2020, 36–37), the **SCI-FILE:** *On Solubility* introduced the dynamic equilibrium that exists between the concentration of a solute in the liquid phase $[A]_L$ and the concentration of that solute in the gas phase $[A]_G$. At constant temperature and pressure, the ratio of those concentrations is constant and *defines* the distribution constant (K_c):

$$K_c = \frac{[A]_L}{[A]_G} \quad (5.11)$$

The distribution constant is the fundamental science that all chromatography rests upon. Luckily, it is indeed *constant* for very dilute solutions and that's the reason why ideal chromatogram peaks are symmetrical. Yes, real peaks are a little asymmetric and that imperfection is the subject of Chapter 8.

In Equation 5.11, note that the distribution constant is a ratio of *concentrations*: the amount of solute per unit volume of each phase. There are two ways to measure the amount of solute; one can weigh it or count the number of molecules present.

Analytical chemists like to weigh stuff because it's the most accurate way to measure quantity in a laboratory. They often measure analyte quantities in nanograms (ng). A PGC can't weigh anything, so this book prefers to state quantities in moles. We express concentration as nanomoles-per-liter (nmol/L), or similar.

> *One mole is a count of 6.022×10^{23} molecules, so a nanomole is only 6.022×10^{14} of them!*
>
> *A mole is the number of molecules you get when their total weight in grams is equal to their molar mass. It's too many to count, so we usually weigh them first, and then convert their weight to a number of moles.*

As noted in Chapter 2, plate theory models a column as a series of plates. In a single plate at equilibrium, the solute distributes itself between the liquid and gas phases in the plate. Then, the solute concentration in the liquid phase is the number of moles in the liquid (n_L) divided by the volume of that liquid (v_L):

$$[A]_L = \frac{n_L}{v_L} \quad (5.12)$$

And the solute concentration in the gas phase is:

$$[A]_G = \frac{n_G}{v_G} \quad (5.13)$$

Inserting these concentrations into Equation 5.11 leads to a more useful statement of the distribution constant:

$$K_c = \frac{n_L}{v_L} \cdot \frac{v_G}{n_G} \quad (5.14)$$

Equation 5.14 contains the *phase ratio* (β), defined as the ratio of gas phase volume to liquid phase volume:

$$\beta = \frac{v_G}{v_L} \quad (5.15)$$

The volumes of liquid phase (v_L) and gas phase (v_G) are those contained in one plate. To get the volumes for the whole column, multiply by the plate number.

$$\beta = \frac{N \cdot v_G}{N \cdot v_L} = \frac{V_G}{V_L} \quad (5.16)$$

Then, V_G is the total gas volume in the column (equal to the holdup volume V_M) and V_L is the total liquid volume. As one might expect, the phase ratio is the same throughout the column. It's a function of the liquid loading in a packed column or internal diameter and liquid film thickness in a capillary column.

To calculate the phase ratio from Equation 5.16, it's necessary to adjust the value of V_G to allow for the average pressure in the column. Chapter 9 reviews the effects of pressure drop in a column and gives the appropriate equation.

Definition of retention factor

Equation 5.14 also includes the peak *retention factor*, although it's not immediately visible. The true definition of retention factor (k) is the ratio at equilibrium of the amount of solute in the liquid phase (n_L) to the amount of solute in the gas phase (n_G):

$$k = \frac{n_L}{n_G} \quad (5.17)$$

This definition is similar to the distribution constant but invokes the actual amounts of solute in one plate, rather than their concentrations.

We can use the theoretical plate model to reconcile this basic definition of retention factor with the calculation from chromatogram measurements done in Chapter 2. For one equilibrium, the k-value defines the number of solute molecules in each phase. When the carrier gas moves to the next plate, it leaves exactly n_L solute molecules behind in the liquid phase.

Clearly, a solute of higher k-value would leave more molecules behind. For example, a solute B with $k = 4$ will always leave twice as many molecules in the liquid phase than a solute A with $k = 2$. Then, the adjusted retention time of Peak B will be twice that of Peak A. This is why we can calculate the k-values from peak locations on the chromatogram.

We now have definitions for β and k. Inserting these in Equation 5.14 highlights a useful correlation between these three column design variables:

$$K_c = k \cdot \beta \quad (5.18)$$

Wherein:

- The k-value is the dependent variable that quantifies the retention of a peak and hence its separation from other peaks.

- The K_c-value is a measure of solute affinity for the stationary phase and is a function of the temperature of the column.

- The β-value is a function of the physical volumes of gas and liquid present and is constant for a given column.

These are the variables most used in column design. However, the absolute value of K_c would be difficult to determine so we use a relative value instead; this is the separation factor (α).

Definition of separation factor

The use of retention volumes also allows insight into the true nature of the separation factor. In Chapter 2, Equation 2.4 calculates separation factor (α) from the ratio of adjusted retention times of two chromatogram peaks, Peak B and Peak A. When converted to volume units, that equation is:

$$\alpha = \frac{V_{RB}'}{V_{RA}'} \tag{5.19}$$

Thus, separation factor is also the ratio of the adjusted retention volumes of the two peaks.

Similarly, Equation 2.3 is an equation for the adjusted retention time of a peak. In volume units, it becomes:

$$V_R' = k \cdot V_M \tag{5.20}$$

Combining Equations 5.19 and 5.20, we discover that separation factor is also the ratio of retention factors for Peak B and Peak A:

$$\alpha = \frac{k_B}{k_A} \tag{5.21}$$

Finally, the merging of Equations 5.18 and 5.21 shows that separation factor is also the ratio of distribution constants for Peak B and Peak A. This anchors the α-value to the fundamental science of chromatography and is the true definition of separation factor:

$$\alpha = \frac{K_{cB}}{K_{cA}} \tag{5.22}$$

In summary, there are several equalities for separation factor:

$$\alpha = \frac{K_{cB}}{K_{cA}} = \frac{k_B}{k_A} = \frac{V_{RB}'}{V_{RA}'} = \frac{t_{RB}'}{t_{RA}'} \tag{5.23}$$

And this is why we can calculate separation factor from simple chromatogram measurements.

Optimizing liquid phase performance

Adjusting the separation factor

The above SCI-FILE explains how the distribution constant (K_c) is the fundamental science behind chromatographic retention. Simply put, it's a measure of the solubility of an analyte in the liquid phase. For two peaks to separate, they must have different solubilities. The separation factor (α) is simply the ratio of those solubilities; that is, the ratio of their distribution constants.

Unfortunately, it's usually not possible to increase the separation factor on a given column. Most changes in operating conditions will affect both peaks equally and leave their α-value unchanged. To improve column selectivity, you'll need a different stationary phase and that means a new column. Refer to Chapter 4 for the many options available.

Adjusting the retention factor

The *resolution equation* tells us that peaks with low retention factors will be difficult to resolve. At the limit, when $k = 0$, the solutes don't enter the liquid phase at all and no resolution is possible, even if their separation factor is good. That might get you thinking that the k-value should be as high as possible; but then the analysis time would be too long. There must be an optimum value somewhere in between.

For liquid-phase columns, there's a universal retention factor that yields the best combination of resolution and analysis time. Therefore, since the retention factor increases with a peak's position along the chromatogram, there's a unique location on any chromatogram that delivers optimum resolution.

Of course, all the peaks cannot be at one location, but we can design and operate a column so the peaks that are most difficult to resolve are near that location. Here, we call it the **power region** of the chromatogram.

The optimization procedure focuses on the two peaks that are most difficult to resolve. By adjusting the column design and operating conditions, we move those two peaks into the power region of the chromatogram, thereby maximizing their resolution.

The *lower limit* of the power region is about twice the holdup time, where $k = 1$. A peak at this location has been in the liquid phase for half of its retention time, thereby allowing the liquid to have some effect. Since this procedure targets the pair of peaks that are most difficult to resolve, there may be other peaks closer to the air peak. In most applications, this is acceptable because their resolution must be better than the target pair.

The **SCI-FILE:** *On Retention Factor* explains that the fastest analysis occurs when the peaks are at three times the holdup time, where $k = 2$. Beyond that point, resolution increases faster than analysis time up to about four times the holdup time ($k = 3$), when the column delivers the maximum resolution per second – its optimum performance. Yet the curve is not steep and you won't gain much by having the peaks exactly at the optimum location; anywhere in the power region will suffice.

For process work, the *upper limit* of the power region is about six times the holdup time, where $k = 5$. Beyond this point, the resolution continues to improve, but the analysis time increases faster than the resolution. It's still common for some well-separated peaks to be out beyond the power region. There's nothing wrong with that unless they are too wide and low, and difficult to measure.

> *Maximum column performance occurs in the **power region** of the chromatogram, the time zone that lies between two times the holdup time and six times the holdup time.*

The following **SCI-FILE:** *On Retention Factor* explains the theoretical basis for the power region.

SCI-FILE: *On Retention Factor*

Optimum resolution

The optimization procedure discussed in the main text targets the most difficult pair of peaks to resolve and aims to move them to a position on the chromatogram where $k = 3 \pm 2$; the "power region."

The theory supporting the power region is complex and this SCI-FILE gives only a summary. Consult the cited sources for the full derivation of equations.

We start from the resolution Equation 5.6 previously discussed:

$$R_s = \frac{\alpha - 1}{\alpha} \cdot \frac{k_B}{1 + k_B} \cdot \frac{\sqrt{N}}{4}$$

Recall that the three terms in the equation are functions of separation factor (α), retention factor (k), and plate number (N). The complexity of the equation diminishes when you focus on the isothermal separation of two specified peaks by a given liquid phase. Then, we can consider the separation factor to be constant. For small changes in k, the plate number is also constant and the effect of retention factor on resolution is:

$$R_s = Q \cdot \frac{k_B}{1 + k_B} \quad (5.24)$$

where Q is a constant.

The blue line in Figure 5.2 plots this effect graphically. Notice that the resolution increases rapidly up to about $k = 5$, and not much thereafter. There is little gain in resolution when retention factors are high ($k > 5$).

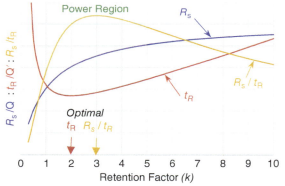

The red line shows that the retention time (and analysis time) necessary to achieve a required resolution reaches a minimum when $k = 2$. The blue line indicates that resolution improves rapidly with increased retention factor up to about $k = 5$ and not much beyond. The orange line traces the resolution per unit time, which peaks at $k = 3$ and then slowly declines. Thus, the "power region" of the chromatogram lies approximately between $k = 1$ and $k = 5$.

The chart assumes that Q and Q′ remain constant when k varies.

Figure 5.2 Optimum Retention Factor.
Source: Adapted from Skoog et al. (2007).

Optimum analysis time

Skoog et al. (2007, 767) derive a general equation for peak retention time (t_R):

$$t_R = \frac{16 \cdot R_s^2 H}{u} \cdot \left(\frac{\alpha}{\alpha - 1}\right)^2 \cdot \frac{(1 + k_B)^3}{k_B^2}$$

Again, for a specified pair of peaks on a given column, we consider the separation factor and plate height to be constant. If we also assume that the carrier velocity (u) doesn't change and we need a fixed resolution, the necessary retention time is:

$$t_R = Q' \cdot \frac{(1 + k_B)^3}{k_B^2} \quad (5.25)$$

where Q' is another constant.

The red line in Figure 5.2 plots the above relationship. The necessary retention time is very high for $k < 1$ and rapidly decreases to a minimum at $k = 2$.

By comparing the plot lines in Figure 5.2, you can see the disadvantage of operating a column with low or high k-values. It would be a mistake to set up a column with $k < 1$ because the resolution would be poor, and the analysis time would be long. Having $k > 5$ may also be an unattractive option because the analysis time increases faster than the resolution does.

In conclusion, the time zone on a chromatogram that lies between $k = 1$ and $k = 5$ generates the maximum resolution in the shortest analysis time.

Optimum retention factor

The practical optimization procedure described in the main text focuses on the pair of peaks that are the most difficult to resolve. The procedure adjusts the column operating conditions to move these two peaks into the "power region" between $k = 1$ and $k = 5$.

From Equation 5.18, only two variables are effective for adjusting the retention factor (k) of a peak:

$$k = \frac{K_c}{\beta} \qquad (5.26)$$

For example, to increase k:

- Increase the distribution constant (K_c) of the peak (that is, its "solubility") by reducing the temperature of the column.

- Reduce the phase ratio (β) of the column.

The main text describes this procedure in more detail and gives a worked example to illustrate its efficacy.

Exploring the power region

It's easy to locate the power region on a chromatogram. To do so, you will need to know the column *holdup time*: that is, the retention time of an "air peak." It's unlikely that your process sample will actually contain air, so to discover the holdup time you may have to inject a sample containing a gas that doesn't much dissolve in the liquid phase such as air, nitrogen, or hydrogen. Methane may be an adequate substitute if you're using a flame ionization detector (FID) that can't detect the diatomic gases. For a given column and carrier gas flow rate, you only need to do it once.

Consider the isothermal chromatograms in Figure 5.3. In each case, the holdup time is one minute, so the power region on all three chromatograms lies between two and six minutes, as the green shading indicates. The two peaks X and Y are the same components in each case but they appear in different locations on the chromatogram. The advantage of peaks in the power region (Figure 5.3b) is immediately clear:

➢ In Figure 5.3a, the peaks are too close to the air peak and their resolution is inadequate: $R_s = 1.2$.

➢ In Figure 5.3b, the same two peaks are in the power region and the resolution is excellent: $R_s = 3.1$.

➢ In Figure 5.3c, the same two peaks are beyond the power region and the resolution is excessive, leading to a longer analysis time and lower peak heights.

Given the above, you might like to practice your diagnostic skills. Why do the peaks appear at different locations on these three chromatograms? What could be causing them to arrive at the positions shown?

Before reading on, try to identify the variables that might be responsible for the different peak positions. For each variable, predict the direction of change that will move the peaks in Figure 5.3a and 5.3c into the power region. For the moment, ignore the effect of the column pressure; a later chapter will examine that more-complex subject.

132   PGC column design

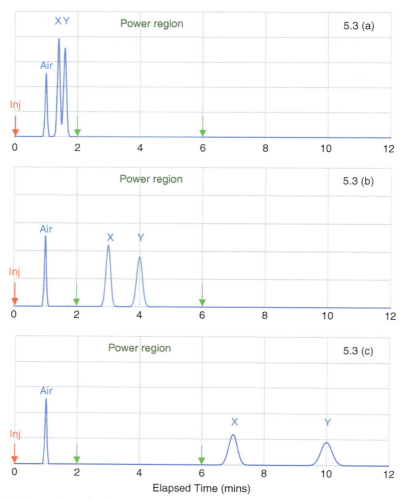

In these three isothermal chromatograms the power region is the same, between two and six minutes. Only the middle chromatogram (b) has peaks X and Y in the power region. In the top (a) and bottom (c) charts, what variables could we adjust to bring the peaks into the power region?

Figure 5.3 The Power Region.

Here's a place to record your suggestions:

Column Variable to Adjust	Direction of Change	
To move peaks into power region:	*For Figure 5.3a*	*For Figure 5.3c*
1.		
2.		
3.		
4.		

Stop for a moment to think. Are you a good troubleshooter? Write your answers above or on a separate notepad. Be sure to include all possible variables.

Optimizing retention factor

To move the target peaks into the power region, you'll need to make some changes to column design or operating variables that bring peak retention factors into the range: $k = 3 \pm 2$. There are not many variables that can do that.

Ineffective variables

A sure way to change the retention factor of a peak is to use a different liquid phase. Yet *"the chromatograms use different liquid phases"* is not an acceptable diagnosis of the above troubleshooting exercise. Why? Because the separation factor in Figure 5.3 is the same on each chromatogram ($\alpha = 1.5$), which strongly implies that they use the same liquid phase. Therefore, the different peak locations in Figure 5.3a and 5.3c are not due to different liquid phases. They must be due to different column designs or operating conditions.

To change the retention factor of a target peak, it's retention time must change relative to the air peak. A change in *column length* or *carrier gas flow rate* won't do it because those variables affect the retention times of all peaks, including the air peak. The retention factor is a ratio, so when a variable moves all peaks equally their retention factors are unaffected and they do not move into or out of the power region.

Ironically, people looking for more resolution often increase the column length, and it always seems to work because the additional column length increases the plate number. But the longer column will probably increase the analysis time too, so it may not be the best move. First, you need to shepherd the target peaks into the power region. Only then is it okay to increase the column length if necessary to do so. But you may be in for a pleasant surprise: the resolution may have improved so much that you can *reduce* the column length!

Effective variables

The **SCI-FILE:** *On Retention Factor* explains that only two variables affect the retention factor of a peak, *distribution constant* (K_c) and *phase ratio* (β), so to optimize liquid-phase performance these are the ones we need to manipulate. Here again is Equation 5.26 that expresses the relationship succinctly:

$$k = \frac{K_c}{\beta}$$

The distribution constant (K_c) of a solute is a precise way to express its solubility in the liquid phase and hence its position on a chromatogram. Temperature has a major effect on solubility, so it's a powerful tool for optimizing peak retention factors. Carrier gas pressure also has an effect but becomes important only when a column operates under elevated gas pressure, as may occur in a multiple-column system. There's more about that in Chapter 9.

The phase ratio (β) is a way to indicate the amount of liquid available in the column to dissolve the solutes. Clearly, if there were no liquid in the column there would be no retention, so retention factor depends on the quantity of liquid present. Perhaps less clearly, it also depends on the amount of gas in the column. The phase ratio incorporates both of these effects.

Optimizing the column temperature

Often, the easiest way to optimize the retention factor is to change the column temperature. Temperature has a large effect on retention time. A decrease in temperature of about 15 K (15 °C) is often sufficient to double the retention factor (Zeeuw 2021). This is why the precise control of oven temperature is so important! Yet it gives us a powerful way to optimize peak retention factors; a small change in temperature will often move the target peaks into the power region with little effect on their separation factor.

For isothermal PGC applications, we prefer to operate with low column temperature to reduce column bleed and extend column life. However, some liquid phases can solidify or gel at low temperatures resulting in reduced efficiency. Set the column temperature as low as the application will allow.

Temperature programming

When the column temperature starts low and gradually increases during analysis, each peak elutes at a different average temperature, increasing the retention factor of the more volatile solutes while reducing the retention factor for those less volatile. This reduced range of peak retention factors allows more peaks to elute within or close to the power region.

Capillary columns are typically much longer than packed columns yet they work at about the same carrier gas velocity. Therefore, the holdup time of a capillary column may be quite long, resulting in a wide power region that can accommodate many peaks.

When a capillary column employs temperature programming, the double advantage of a reduced range of retention factor and a wide power region allows many peaks to enjoy optimum resolution. Contrary to what we're often told, the fabulous resolution exhibited by capillary columns is not solely due to their high plate numbers. Recall the resolution equation. The resolution of peaks depends more on their retention factors than on the column plate number.

The rate of temperature rise also affects the retention factors. Peaks that elute quickly from the column do not experience the higher temperatures occurring towards the end of the program, so a slower ramp reduces their average temperature and increases their observed retention factors.

Optimizing the column phase ratio

The other way to move target peaks into the power region is to adjust the phase ratio. Equation 5.16 defines phase ratio (β) as the volume (V_G) of gas in a column relative to the volume (V_L) of liquid phase:

$$\beta = \frac{V_G}{V_L}$$

Thus, a column with a lower phase ratio has more liquid phase in it and will provide higher retention factors for all component peaks. To reduce phase ratio:

➢ Packed columns will require an increased percent loading of liquid phase on the solid support.

➢ Capillary columns will need an increased liquid film thickness or a reduced internal diameter, or both.

Phase ratio is not a useful concept for a solid-phase column.[4]

A worked example

It's fun to see how easy (and how effective!) it is to choose the liquid-phase loading and operating temperature of an isothermal column.

Evaluate the two chromatograms in Figure 5.4. You can assume they come from isothermal liquid-phase columns. The resolution of Peaks A and B is identical in each case, so either chromatogram is adequate for measuring those two peaks. Yet one of the chromatograms offers an

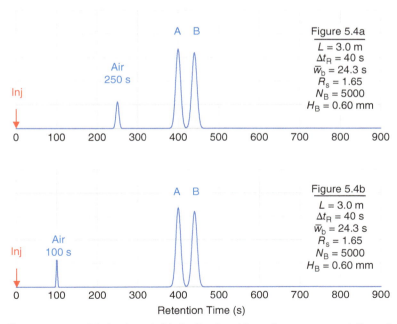

Chromatogram (a) (top) and (b) (bottom) achieve the same resolution of Peaks A and B. But which one offers the best opportunity for improvement?

Figure 5.4 Challenge Question.

---

[4] The retention factor of a component on a solid-phase column can be adjusted only by changing the column temperature.

opportunity for improvement that the other does not. Which chromatogram would you choose to improve?

Make your choice before reading on.

Evaluation of options

If your only interest is to measure the two peaks, both chromatograms offer the same resolution and the same analysis time, so they are equally good for the job. Pick any one you like.

If you want to optimize the peak resolution, though, there's a clear difference between the two chromatograms. The key, of course, is the air peak time. Figure 5.4b has the peaks in the power region and Figure 5.4a does not. Therefore, it's unlikely that you can improve Chromatogram 5.4b but Chromatogram 5.4a may have hidden potential not yet realized.

To evaluate the two options, calculate the peak separation factor (α) on each chromatogram using Equation 2.4:

$$\alpha = \frac{t'_{RB}}{t'_{RA}}$$

For Figure 5.4a:

$$\alpha_a = \frac{190 \text{ s}}{150 \text{ s}} = 1.27$$

For Figure 5.4b:

$$\alpha_b = \frac{340 \text{ s}}{300 \text{ s}} = 1.13$$

The separation factors are different, which implies that the chromatograms are from different liquid phases. The liquid phase used for Chromatogram 5.4a is capable of far superior separation than the liquid phase used for Chromatogram 5.4b (about double, actually).

So why are the two resolutions identical? It's because the column used for Chromatogram 5.4a does not allow the peaks to stay in the liquid phase long enough. The peaks are not in the power region so both their retention factors are less than optimum. This liquid phase is capable of much better performance. Let's see how we capture that potential.

Before getting into the calculations, here are the starting conditions we used to draw the chromatogram in Figure 5.4a:

CONDITIONS for FIGURE 5.4a:		Air	Peak A	Peak B
Retention Time	t_R	250 s	400 s	440 s
Peak Base Width	w_b		23.7 s	24.9
Average Base Width	$\overline{w}_b$		24.3 s	

How to optimize liquid phase performance

To improve the performance of Chromatogram 5.4a, we need to move the peaks into the power region. You should already know which variables to change. Reduce the *column temperature* or the *phase ratio*.

It's easier to lower the column temperature, but a liquid-phase column should not be too cold or its efficiency may suffer. If a lower column temperature is inadvisable, install a new column with a lower phase ratio.

These adjustments will have little effect on the column holdup time, so they don't shift the power region. That's why they can move peaks into the power region.

Note the effect of the variables:

➤ The effect of liquid phase loading (or liquid film thickness) is linear. If you have twice as much liquid in the column, the peaks will spend twice as long in the liquid phase.

➤ The effect of temperature on the adjusted retention time is highly non-linear. As noted above, a temperature reduction of about 15 K is usually enough to double the adjusted retention time.

For simplicity of argument, let's reduce column temperature or column phase ratio by just enough to double the peak retention factors. We are not changing the liquid phase, so separation factor remains the same.[5]

Figure 5.5 illustrates the result of this change when the column operates at the same carrier gas velocity (that is, the air peak is at the same position). Notice that the peaks are now in the power region.

The new chromatogram is impressive. The air peak time is the same, but the adjusted retention times have doubled; Peak A from 150 to 300 seconds, and Peak B from 190 to 380 seconds.

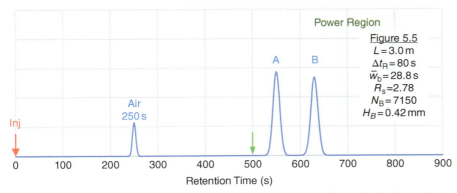

Starting from Figure 5.4a, this chromatogram illustrates the effect of reducing column temperature or phase ratio to move the peaks A and B into the power region. The resolution has improved by 68 % from 1.65 to 2.78, as evidenced by the flat baseline now between the peaks.

Figure 5.5 Peaks in Power Region.

---

[5] There are a few exceptions, but it's generally true that column temperature has only a small effect on separation factor.

Therefore, the new retention time for Peak A is:

$$t_{RA} = 250 + 300 = 550 \text{ s}$$

And the new retention time for Peak B is:

$$t_{RB} = 250 + 380 = 630 \text{ s}$$

Of course, the peaks are wider now, due to their additional time in the column, but they are not twice as wide. To estimate the new base widths, assume that the plate height has not changed[6] and apply Equation 2.23:

$$w_b = 4\sqrt{H_t \cdot t_R}$$

$$\frac{(w_b)_{\text{new}}}{(w_b)_{\text{old}}} = \sqrt{\frac{(t_R)_{\text{new}}}{(t_R)_{\text{old}}}}$$

The new base width of Peak A is:

$$w_{bA} = 23.7 \times \sqrt{\frac{550 \text{ s}}{400 \text{ s}}} = 27.8 \text{ s}$$

The new base width of Peak B is:

$$w_{bB} = 24.9 \times \sqrt{\frac{630 \text{ s}}{440 \text{ s}}} = 29.8 \text{ s}$$

In summary, here are the new conditions used to draw the chromatogram in Figure 5.5:

CONDITIONS for FIGURE 5.5:		Air	Peak A	Peak B
Retention Time	t_R	250 s	550 s	630 s
Peak Base Width	w_b		27.8 s	29.8 s
Average Base Width	$\overline{w}_b$		28.8 s	

Notice that the peak separation doubled from 40 to 80 seconds but the average peak width ($\overline{w}_b$) increased only 19 % from 24.3 to 28.8 seconds. From Equation 2.8, the new resolution is:

$$R_s = \frac{\Delta t_R}{\overline{w}_b}$$

$$R_s = \frac{80 \text{ s}}{28.8 \text{ s}} = 2.78$$

This is a massive increase in resolution. The column is now very efficient due to a significant reduction in plate height. Using Equation 2.26, the plate number for Peak B is now:

---

[6] The next chapter reveals that the plate height will not be exactly the same but it's close enough for the present discussion.

$$N_B = 16 \times \left(\frac{630 \text{ s}}{29.8 \text{ s}}\right)^2 = 7150$$

And from Equation 2.14, the plate height for Peak B is:

$$H_B = \frac{3000 \text{ mm}}{7150} = 0.42 \text{ mm}$$

Nice. The only negative outcome is the analysis time, which increased by 43 % from 480 to 680 seconds. The column is now producing more resolution than we really need. The column is too long!

The reduction in plate height means more plates per meter. It's the plate number that indicates the efficiency of a column, not its length. A shorter column can be more powerful if it has smaller plates.

Reducing the column length

The peaks are now in the power region with optimum retention factors. This makes the column very efficient, so it doesn't need to be so long. It's okay to reduce the column length as it won't move the peaks out of the power region nor will it affect the plate height.

Notice that we intend to reduce the column length, not increase it!

To start, let's reduce the column length enough to revert to the original analysis time, using the same carrier gas velocity. That means reducing the retention time of Peak B from 630 to 440 seconds. With the same carrier gas flow rate (and velocity), the retention time is directly proportional to the column length, so the new length (L) needs to be:

$$L = 3000 \text{ mm} \times \frac{440 \text{ s}}{630 \text{ s}} = 2100 \text{ mm}$$

Figure 5.6b shows the new chromatogram.

This is magic! Look closely at Figure 5.6b and compare it with the original chromatogram in Figure 5.6a. The new column is only 70 % of the original length and the analysis time is the same. Yet the separation between peaks has improved from 40 to 56 seconds and the resolution is up from 1.65 to 2.32, a significant improvement in performance. You don't need math to verify this: just look at the baseline now visible between the peaks! However, a few calculations will reveal *why* the resolution has improved. None are difficult to understand.

The retention time of Peak B on the shorter column is 440 seconds, as expected from the calculated change in length. The air peak time (t_M) also drops in proportion to the new column length:

$$t_M = 250 \text{ s} \times \frac{2100 \text{ mm}}{3000 \text{ mm}} = 175 \text{ s}$$

Therefore, the adjusted retention time of Peak B is now:

$$t'_{RB} = 440 - 175 = 265 \text{ s}$$

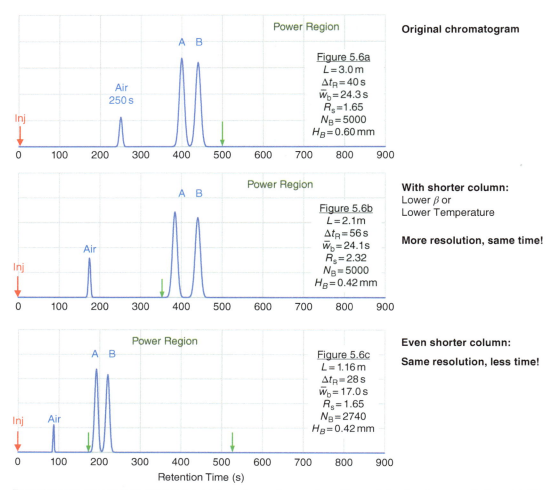

Figure 5.6 Optimum Performance.

For reference, Figure 5.6a (top) is the original chromatogram from Figure 5.4a showing a resolution of 1.65. Figure 5.6b (middle) uses a shorter column, yet the resolution improves to 2.32 because the peaks are now in the power region. Figure 5.6c (bottom) uses an even shorter column yet achieves the original resolution in far less time.

We can use this Peak B data to discover the exact position and width of Peak A. To find the adjusted retention time of Peak A, use the separation factor between the peaks that we calculated above for this liquid phase ($\alpha = 1.27$):

$$\alpha = 1.27 = \frac{t'_{RB}}{t'_{RA}}$$

$$t'_{RA} = \frac{265 \text{ s}}{1.27} = 209 \text{ s}$$

Therefore, the retention time for Peak A is:

$$t_{RA} = 209 + 175 = 384 \text{ s}$$

And the time between the peaks (Δt_R) is:

$$\Delta t_R = 440 - 384 = 56 \text{ s}$$

This is a significant improvement on the original 40 seconds. Since the operating conditions for the 2.1 m column are the same as the 3.0 m column, the plate height for Peak B has not changed, it's still 0.42 mm. From Equation 5.4, the new plate number (N_B) for Peak B is:

$$N = \frac{L}{H}$$

$$N_B = \frac{2100 \text{ mm}}{0.42 \text{ mm}} = 5000$$

Wow! This is the same plate number as the original, which means Peak B has the same width (24.9 seconds) as the original, yet the time between peaks has increased by 16 seconds. Sixteen seconds of additional separation is huge.

By again using Equation 2.23, you can estimate the base width of Peak A from the base width of Peak B and the retention times of the two peaks:

$$w_{bA} = 24.9 \text{ s} \times \sqrt{\frac{384 \text{ s}}{440 \text{ s}}} = 23.3 \text{ s}$$

From this, the average base width ($\overline{w}_b$) is now:

$$\overline{w}_b = \frac{23.3 \text{ s} + 24.9 \text{ s}}{2} = 24.1 \text{ s}$$

And the new resolution is:

$$R_s = \frac{56 \text{ s}}{24.1 \text{ s}} = 2.32$$

Let's pause to evaluate our success. In Figure 5.6b:

➢ The column is 30 % shorter than it was.

➢ The column temperature is about 15 °C lower, *or*
The phase ratio is half of what it was.

➢ The carrier gas flow rate is the same.

➢ The resolution has increased by 41 % from 1.65 to 2.32.

➢ The analysis time has not changed.

You might be wondering whether such impressive gains in performance are commonly achievable. Probably not. To illustrate the calculations, we started with a column operating a long way from optimum. Hopefully, your columns are not like that. If your peaks are already near the power region only minor improvements will be possible.

Getting a faster analysis

The resolution in Figure 5.4a was 1.65, which was enough to give a baseline separation between equal peaks. With the new column conditions, what length of column would be enough to regain that same resolution?

As before, the resolution is proportional to the square root of the column length. To change the present resolution ($R_s = 2.32$) back to the original resolution ($R_s = 1.65$), the new column length must be:

$$L = 2100 \text{ mm} \times \left(\frac{1.65}{2.32}\right)^2 = 1060 \text{ mm}$$

This is about one-third of the original column length, yet the resolution is the same! And the analysis time is now about half of the original. That's why we call it the *power* region.

Figure 5.5c shows the new chromatogram. Amazing, isn't it?

Feel free to practice the calculations on this chromatogram, following the above procedures. By doing so, you will start to become familiar with what to expect when working on column systems.

You're on your way to becoming an expert chromatographer!

Knowledge Gained

Theory

- Resolution is the combined effect of retention factor, separation factor, and plate number.

- The resolution equation quantifies the effects of these three factors on the achieved resolution.

- Peaks with small retention factors are not long in the liquid phase and achieve poor resolution.

- Peaks with small separation factors are close together and difficult to resolve.

- Analysis time drops rapidly as the retention factor increases, reaching a minimum at about $k = 2$.

- The maximum resolution per unit time occurs at four times the air peak time, when $k = 3$.

- Resolution continues to rise at higher k values, but at diminishing rates of improvement.

- Beyond $k = 5$, the analysis time increases more than resolution does.

- The power region of a chromatogram lies between two and six times the air peak time ($1 < k < 5$)

- To increase k, reduce the column temperature or phase ratio (increase the liquid-phase loading).

- To reduce k, increase the column temperature or phase ratio (reduce the liquid-phase loading).

- Retention volumes provide a deeper understanding of the relationship between variables.

- Volumetric retention data is useful in theoretical work, but not in practical process chromatography.

- Multiply a retention time by carrier gas flow rate (at column temperature) to get retention volume.
- Distribution constant (K_c) is the equilibrium ratio of solute concentrations in the liquid and gas phases.
- Phase ratio (β) is the ratio of liquid phase volume to gas phase volume within the column.
- Retention factor (k) is the equilibrium ratio of the amounts of solute in liquid phase and gas phase.
- Distribution constant is the product of retention factor and phase ratio.
- Separation factor (α) is the ratio of the distribution constants for Peaks B and A on a liquid phase.
- Separation factor (α) is also the ratio of the retention factors of Peaks B and A on a column.

Practice

- The main objective of column design is to achieve adequate resolution of two adjacent peaks.
- The secondary objective of column design is to achieve the desired resolution in minimum time.
- A column system must resolve the analyte peaks from all other components of the sample.
- The resolutions must be adequate for reliable operation and easy maintenance.
- The column designer needs to have an overall strategy that includes the goals for each column.
- Recognizing that the initial design might fail, the designer needs to keep many options open.
- Each column in a column system has a specific goal, a resolution it must achieve in minimum time.
- In any analysis, there's always one pair of peaks that are most difficult to resolve.
- Resolution is a function of separation factor (α), retention factor (k), and plate number (N).
- It's necessary to optimize all three factors to get the best resolution in minimum time.
- There is a minimum plate number to resolve peaks with known separation and retention factors.
- Resolution becomes difficult as the separation factor approaches 1.0.
- Resolution becomes difficult as the retention factor approaches zero.
- The "power region" of a chromatogram is between two and six times the air peak time.
- If peaks stay longer in the liquid phase, they get wider and lower, and more difficult to measure.
- To optimize the resolution, move the peaks into the power region of the chromatogram.
- To increase the retention factor, reduce the column temperature or its phase ratio.
- To reduce the retention factor, increase the column temperature or phase ratio.
- For a temperature program, a slow ramp increases retention factors, while a fast ramp reduces it.
- Then, adjust the column length and operating conditions to yield the necessary plate number.

Did you get it?

Self-assessment quiz: SAQ 05

This SAQ also uses your knowledge from previous chapters. Consider the chromatogram in Figure 5.3b:

Q1. *Estimate the retention factors for Peak X and Peak Y.*
Q2. *Estimate the separation factor for Peak X and Peak Y.*
Q3. *Given that the plate number for Peak Y is 3600, estimate in seconds the base width of Peak X and Peak Y.*
Q4. *What is the resolution of Peak X and Peak Y in Figure 5.3b?*
Q5. *Using a shorter column under the same operating conditions, what plate number would achieve a resolution of 2.0 between Peak X and Peak Y?*
Q6. *Given that the original column length was 1.6 m and the operating conditions are unchanged, what are the retention times of Peak X and Peak Y on the new column?*
Q7. *For the new resolution of 2.0, estimate in seconds the base widths of Peak X and Peak Y.*

Check your SAQ answers with those given in the back of the book.

References

Further reading

For more information about column tuning and performance refer to:

> *Guide to GC Column Selection and Optimizing Separations,* published by Restek (2021) and available online.

> *Optimizing GC parameters for faster separations with conventional instrumentation,* an article by Anila Khan (2013).

Cited

Annino, R. and Villalobos, R. (1992). *Process Gas Chromatography: Fundamentals and Applications.* Research Triangle Park, NC: Instrument Society of America.

Khan, A.I. (2013). Optimizing GC parameters for faster separations with conventional instrumentation. Thermo Literature: TN20743 E 02/13S. Runcorn, UK: Thermo Fisher Scientific.

Knox, J.H. (1961). The speed of analysis by gas chromatography. *Journal of the Chemical Society* 1961, 433–441.

Purnell, J.H. (1960). The correlation of separating power and efficiency of gas-chromatographic columns. *Journal of the Chemical Society* 1960, 1268–1274.

Purnell, J.H. (1962). *Gas Chromatography.* New York, NY: John Wiley & Sons, Inc.

Restek (2021). Guide to GC Column Selection and Optimizing Separations. *Lit. Cat.# GNAR1724A-UNV.* Bellefonte, PA: Restek Corporation.

Scott, R.P.W. (1998). *Introduction to Analytical Gas Chromatography.* Book 76 of Chromatographic Science Series. New York, NY: Marcel Dekker Inc.

Skoog, D.A., Holler, F.J., and Crouch, S.R. (2007). *Principles of Instrumental Analysis*, Sixth Edition. Thomson Publishing USA.

Waters, T. (2020). *Process Gas Chromatographs: Fundamentals, Design and Implementation.* Chichester, UK: John Wiley & Sons Ltd.

Zeeuw, J. de (2021). Impact of GC parameters on the separation, Part 5: Choice of column temperature. Accessed on 2024/10/30 at https://www.restek.com/resource-hub-search?s=Impact-of-GC-Parameters.

Figures

5.1 Plate Number for Resolution.
5.2 Optimum Retention Factor.
5.3 The Power Region.
5.4 Challenge Question.
5.5 Peaks in Power Region.
5.6 Optimum Performance.

Symbols

Note: the text may add subscripts A and B or subscripts X and Y to indicate the first and the second peak, respectively; "none" indicates a dimensionless quantity.

Symbol	Variable	Unit
$[A]_L$	Concentration of solute "A" in the liquid phase	mol/mL
$[A]_G$	Concentration of solute "A" in the gas phase	mol/mL
H	Plate height	mm
k	Retention factor of peak	none
K_c	Distribution constant	none
L	Column length	m
n_G	Amount of solute in gas phase per plate	mol
n_L	Amount of solute in liquid phase per plate	mol
N	Plate number (observed)	none
N_r	Plate number (required)	none
Q	A constant	none
Q'	Another constant	s
R_s	Resolution of adjacent peaks	none
t_M	Holdup time	s
t_R	Retention time of peak	s
t_R'	Adjusted retention time of peak	s
Δt_R	Time between adjacent peaks; that is, "separation"	s
T_a	Temperature of flow measurement	K

T_c	Temperature of column	K
v_G	Volume of gas phase in one plate	mL
v_L	Volume of liquid phase in one plate	mL
V_G	Column holdup volume	mL
V_L	Volume of liquid phase in column	mL
V_R	Retention volume of peak	mL
V_R'	Adjusted retention volume of peak	mL
$\dot{V}_o$	Flow rate of carrier gas at column outlet	mL/s
w_b	Base width of chromatogram peak	s
$\overline{w}$	Average base width of adjacent peaks	s
α	Separation factor between Peaks B and A	none
β	Phase ratio of column	none

Equations

Note: subscripts A and B refer to the first and second peak but the text sometimes substitutes X and Y, respectively.

5.1 $\quad R_s = \dfrac{t_{RB} - t_{RA}}{w_b}$ — Conservative resolution of Peaks A and B.

5.2 $\quad w_b = 4 \cdot \dfrac{t_R}{\sqrt{N}}$ — Base width of a peak.

5.3 $\quad \sqrt{N_r} = 4R_s \cdot \dfrac{t_M + t_R'}{t_{RB}' - t_{RA}'}$ — Required plate number for a selected peak.

5.4 $\quad \sqrt{N_r} = 4R_s \cdot \dfrac{t_M/t_{RB}' + 1}{1 - t_{RA}'/t_{RB}'}$ — Required plate number for Peak B using ratios of retention times.

5.5 $\quad N_r = 16R_s^2 \cdot \left(\dfrac{\alpha}{\alpha - 1}\right)^2 \cdot \left(\dfrac{1 + k_B}{k_B}\right)^2$ — The necessary plate number from Peak B data.

5.6 $\quad R_s = \dfrac{\alpha - 1}{\alpha} \cdot \dfrac{k_B}{1 + k_B} \cdot \dfrac{\sqrt{N}}{4}$ — The resolution equation based on Peak B.

5.7 $\quad N_r = 16R_s^2 \cdot \left(\dfrac{1}{\alpha - 1}\right)^2 \cdot \left(\dfrac{1 + k_A}{k_A}\right)^2$ — The necessary plate number from Peak A data.

5.8 $\quad L = N_r \cdot H$ — Column length from required plate number.

5.9 $\quad V_R = t_R \cdot \dot{V}_o \cdot \dfrac{T_c}{T_a}$ — Retention volume from retention time and corrected flow rate.

5.10	$V_R = V_M + V_R'$	Holdup volume and corrected retention volume.
5.11	$K_c = \dfrac{[A]_L}{[A]_G}$	Definition of distribution constant.
5.12	$[A]_L = \dfrac{n_L}{v_L}$	Molar concentration of solute in the liquid phase of one plate.
5.13	$[A]_G = \dfrac{n_G}{v_G}$	Molar concentration of solute in the gas phase of one plate.
5.14	$K_c = \dfrac{n_L}{v_L} \cdot \dfrac{v_G}{n_G}$	Distribution constant from molar concentrations.
5.15	$\beta = \dfrac{v_G}{v_L}$	Definition of phase ratio in one plate.
5.16	$\beta = \dfrac{V_G}{V_L}$	Phase ratio of whole column.
5.17	$k = \dfrac{n_L}{n_G}$	Definition of retention factor.
5.18	$K_c = k \cdot \beta$	Distribution constant from practical variables.
5.19	$\alpha = \dfrac{V_{RB}'}{V_{RA}'}$	Separation factor from adjusted retention volumes.
5.20	$V_R' = k \cdot V_M$	Adjusted retention volume per holdup volume.
5.21	$\alpha = \dfrac{k_B}{k_A}$	Separation factor as ratio of retention factors.
5.22	$\alpha = \dfrac{K_{cB}}{K_{cA}}$	Definition of separation factor.
5.23	$\alpha = \dfrac{K_{cB}}{K_{cA}} = \dfrac{k_B}{k_A} = \dfrac{V_{RB}'}{V_{RA}'} = \dfrac{t_{RB}'}{t_{RA}'}$	Equalities for separation factor.
5.24	$R_s = Q \cdot \dfrac{k_B}{1 + k_B}$	Effect of retention factor on resolution.
5.25	$t_R = Q' \cdot \dfrac{(1 + k_B)^3}{k_B^2}$	Effect of retention factor on retention time.
5.26	$k = \dfrac{K_c}{\beta}$	Retention factor from distribution constant and phase ratio.

New technical terms

When first introduced, these words and phrases were in bold type. You should now know the meaning of these technical terms

adjusted retention volume **power region**
column system **resolution equation**
critical pair **retention volume**
holdup volume

Refer to the Glossary at the end of the book for an explanation of these terms.

6

Optimizing performance

"There's always an optimum carrier gas flow rate that maximizes a column's resolving power. With the flow rate set at optimum and the peaks in the power region, the resolution you get is the very best that a column can do. Yet you might be surprised to find that process gas chromatographs rarely work under such ideal conditions ..."

Caution

Before we start, a word of caution. This chapter discusses methods for optimizing the performance of a single column. Our goal is to understand the effect of carrier flow rate and column length on the performance of a column. A real PGC column works in series with at least one other column so any change in flow rate or column length affects the performance of both columns, an additional complexity discussed in Chapter 9.

Your existing PGC column system should be a nuanced design where each column has a specific purpose. Don't attempt to change the column or its operating conditions unless you fully comprehend what each column is doing and why it's doing it:

Don't change a column or the settings of a PGC column system unless you know what you are doing!

If you decide to proceed, save all the settings before you change anything. Always have a way to get back to where you started.

All the examples given here relate to liquid-phase columns. The theory of solid-phase columns is different but the end results are similar.

Optimum flow rate

For every column, there is an optimum carrier flow rate that achieves the best resolution that the column can deliver. In this section, we describe a practical way to discover that optimum flow rate.

The optimizing procedure uses measurements from the chromatogram to calculate a column performance parameter at different carrier gas flow rates. Then, a graphical plot of that parameter immediately reveals the optimum flow rate. The graph also shows the performance to expect when the column runs at a non-optimum flow rate.

The procedure given here is for optimizing a single column, but most PGCs use multiple columns in series. Of course, the procedure also works for columns in series when they use the same stationary phase and the same liquid-phase loading.

For different columns connected in series the theory is less clear, but the procedure will still find their best overall performance. However, you'll need to be careful. Any change of flow rate will necessitate a change in carrier gas pressure and this change will have a different effect on each column. It may significantly alter the contribution of the first column to the overall separation, perhaps spoiling the resolution of peaks. For a detailed look at the properties of serial column arrangements, see Chapter 9.

Before proceeding, ensure that the most difficult peaks to resolve are in the power region of the chromatogram, following the procedure given in Chapter 5.

Experimental

Finding the optimum flow rate is an experimental procedure: inject the same sample several times, each time with the carrier gas set to a different flow rate. From each resulting chromatogram, record two values: an attribute of the flow rate and a measure of column efficiency. A graphical plot of this data displays the nonlinear relationship between these two variables and reveals the optimum flow rate.

In theory, column performance depends on the velocity of the carrier gas, rather than its flow rate. It would not be easy to measure the velocity of the carrier gas but we have a convenient surrogate – an unretained peak travels through a column at the same speed as the carrier gas, so its holdup time is a measure of carrier gas velocity.

Thus, to calculate the average carrier gas velocity ($\bar{u}$), measure the holdup time (t_M) of an unretained peak on each chromatogram. Then divide by the column length (L) using Equation 2.2:

$$\bar{u} = \frac{L}{t_M}$$

You'll need an unretained peak on each chromatogram, often called an "air peak." We don't really want to put oxygen into the column, so it's best to inject a sample containing a little nitrogen (for a thermal conductivity detector) or a little methane (for a flame-ionization detector).

This procedure assumes that the nitrogen or methane has essentially zero solubility in the liquid phase. It may not be applicable to a solid-phase column.

Working in velocity units (m/s) gives you the flexibility of applying your column efficiency curve to different column lengths but if you have no intention of changing the column length, it's okay to plot the curve in flow units. Measure the column exit flow rate ($\dot{V}$) for each chromatogram.

With open-tubular columns, you can compute the carrier velocity directly from the measured flow rate using the column internal diameter (d_c) and the liquid film thickness (d_f):

$$u = \frac{4\dot{V}}{\pi \cdot (d_c - 2d_f)^2} \tag{6.1}$$

This calculation uses the flow rate at vent pressure, so it will return the column exit velocity (u). To get the average velocity, apply the pressure correction given in Chapter 9.

The optimization procedure

The procedure for finding the optimum flow rate is simple and well-worth doing. This method is applicable to any liquid-phase column:

> Use a test sample that produces a fully separated analyte peak, preferably with a retention factor of $k = 4 \pm 1$.

> You can save time by including an unretained component (nitrogen or methane) in the test sample. Otherwise, it's okay to inject a second sample that contains that unretained component.

> If you prefer to measure the carrier flow rates, you won't need the unretained component but you'll have to measure the column exit flow rate for each chromatogram.

> Run several chromatograms at widely different carrier flow rates.

Then, for each chromatogram:

> If working in velocity units, measure the holdup time (t_M) of the unretained component.

> Calculate the average carrier gas velocity ($\bar{u}$) from the column length (L) and the air peak time (t_M) using Equation 2.2:

$$\bar{u} = \frac{L}{t_M}$$

> Measure the retention time (t_R) and width of the analyte peak. You can measure peak width at the baseline (w_b) or at half-height (w_h).

> Calculate the plate number (N) for the analyte peak using Equation 2.26 or Equation 2.27, as appropriate:

$$N = 16\left(\frac{t_R}{w_b}\right)^2$$

$$N = 5.54\left(\frac{t_R}{w_h}\right)^2$$

> Calculate the plate height (H) for the analyte peak using Equation 2.14:

$$H = \frac{L}{N}$$

Finally:

> Plot the **column efficiency curve** – plate height as a function of average carrier velocity (or exit flow rate). Your curve should look like the one in Figure 6.1. It has a similar characteristic shape on any column.

> If necessary, run a few more chromatograms at different carrier flow rates to clearly define the whole curve.

Notice that the dip in the curve gives the minimum plate height (H_{min}) and therefore the maximum plate number. This is the optimum velocity (or flow rate) for the best resolution of peaks. Of course, Figure 6.1 is just an example; your curve will have different values for the optimum velocity (or flow rate) and for the minimum plate height.

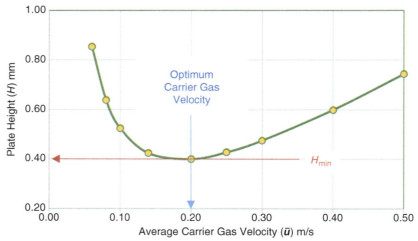

This is a useful plot, showing the performance of your column. In addition to finding the optimum flow rate for the column, you can use your graph to estimate the plate height at any flow rate.

Later, we will need this data in an exercise to optimize the length of a typical column and minimize the analysis time.

Figure 6.1 Typical Column Efficiency Curve.

Relation to theory

It's easy to mix up theory and reality here. Often, the graph in Figure 6.1 is called a *van Deemter curve* in honor of the leader of the team that originally developed a theory to explain it (van Deemter et al. 1956).

Be aware, though, that the well-known relationship between carrier gas velocity and plate height is not a theory; it's an empirical fact. That's easy to confirm, just do the experiment!

A knowledge of theory and practice are both important for success with gas chromatographs but be careful to distinguish between the two. The true **van Deemter curve** shown in Figure 6.2 is a theoretical construct, not an empirical one. As such, it's unlikely to exactly match your experimental findings.

Your experimental results are not a theory; they are real and specific to the column you are working with. So maybe you shouldn't label your graph a "van Deemter curve." To highlight the distinction between reality and theory, this text uses **column efficiency curve** to refer to the empirical relationship between plate height and carrier gas velocity (or flow rate) on a given column. Figure 6.1 illustrates a typical column efficiency curve.

That said, you can gain some useful insights from studying the theory of this famous curve. If you would find that interesting, read the following SCI-FILE *On Rate Theory* for a more detailed explanation.

SCI-FILE: *On Rate Theory*

Too much theory?

Before studying the lessons learned from rate theory, it will be good to clarify the role of theory in a book like this. We have no wish to undermine the great success of chromatographic theory; it has led to many amazing improvements in gas chromatography and has totally transformed liquid chromatography.

Yet it's still pertinent to ask; *what use is the theory to a modern process chromatographer?*

Let's be realistic; no worker outside of academia ever uses the theoretical equations to calculate column parameters from first principles. The usual way is to try a column and see how well it does the job; then modify that column as necessary to get an acceptable result.

After working with columns for many years, a column designer or troubleshooter may have significant experience of what works and what doesn't. Yet their working environment will always limit that experience to the norms of their workplace. A petrochemist and a biochemist have different chromatographic experiences. While both are experts in their own field, they don't know what they don't know.

That's where the theory comes in. We won't need all the math because you can find that in any standard textbook on chromatography. But it's vital to know the variables that affect column performance and how you can optimize them.

So this book focuses on how the theory can help with column design and troubleshooting. For those who enjoy the math the equations are given, but the emphasis is not on mathematical proof. Instead, we concentrate on what theory reveals about column performance and how to maximize it. With that proviso, let's take a look at the rate theory of chromatography.

Rate theory

Unlike plate theory, the rate theory of chromatography is not based on solute–solvent equilibrium. Contrarily, by alluding to the continuous movement of carrier gas, it discounts the possibility of an equilibrium occurring anywhere in a column. Instead, rate theory seeks to explain how dynamic processes inside the column cause peak broadening.

Rate theory adopts the same measure of broadening used by plate theory, the *plate height*. But it then goes beyond the mere existence of plates to discover mathematical relationships between column variables and the observed plate height. The identity of those variables and their effect on peak width is what makes rate theory so important to us.

The earlier plate theory started from the fundamental idea of theoretical plates and then figured plate height simply as the equivalent length of column occupied by one plate. But rate theory looks for causes. It starts from the realization that while a peak is in the column, it is inherently unstable; the higher concentration of molecules at its center is continuously spreading into the lower concentration gas on either side. Thus, the longer the peak is in the column, the wider it becomes.

The technical term for this linear spreading of molecules along the column axis is **dispersion**. Rate theory tries to identify and quantify the sources of dispersion.

To understand this approach, recall that variance is a statistical measure of dispersion and that the overall variance (σ^2) of a Gaussian peak might be the sum of several independent variances:

$$\sigma^2 = \sigma_1^2 + \sigma_2^2 \dots + \sigma_n^2 \qquad (6.2)$$

Recall from Equation 2.18 that the plate height (H) is the variance generated per unit length of column:

$$H = \frac{\sigma^2}{L}$$

The length (L) of a given column is constant, so the observed plate height is the sum of the plate heights of all the individual dispersive processes occurring in that column:

$$H = H_1 + H_2 \dots + H_n \qquad (6.3)$$

This is why all the equations developed by rate theory follow the form of Equation 6.3.

What we learn

We already know that to get the best performance, we should operate a column at its optimum flow rate, thereby achieving the minimum plate height, maximum plate number, and best resolution. We now learn from Equation 6.3 that plate height itself may have several independent contributors, so we can focus on reducing each one of those separately.

Rate theory is all about motion and several movements occur in gas chromatography. Let's start with the dominant pair: the incessant movement of the carrier gas and the gradual migration of the solute peaks along the column.

Carrier gas velocity

The movement of the carrier gas drives most dynamic effects occurring in a column, so many sources of peak broadening are functions of the carrier gas velocity.

We use the column length (L) and the holdup time (t_M) to estimate the average linear carrier gas velocity ($\bar{u}$) in the column. Per Equation 2.2:

$$\bar{u} = \frac{L}{t_M}$$

The carrier gas velocity is not the same at every place in the column because the carrier gas pressure drops continually from inlet to outlet. As the pressure drops, the gas expands and its velocity increases, so the inlet end of the column experiences a lower velocity than the outlet end does.

In Chapter 9, we shall find that a knowledge of pressure drop is important when designing a multiple column system for a process gas chromatograph. For a single column, though, we can simply evaluate its overall performance.

Solute band velocity

The other velocity in a column is the **migration rate** of a solute. Each band of solute molecules is moving along the column at a lower average velocity ($\bar{v}$) than the carrier gas:

$$\bar{v} = \frac{L}{t_R} \quad (6.4)$$

Clearly, the migration rate of a solute is also related to its retention factor (k):

$$\bar{v} = \frac{\bar{u}}{k} \quad (6.5)$$

The van Deemter equation

Van Deemter (1956) and his associates were the first of many scientists to offer progressively more complex theories to explain the observed dependence of plate height on carrier gas flow rate.

The original **van Deemter equation** has the advantage of being easy to understand and still gives the closest agreement with the experimental results for packed columns. For open-tubular columns, use the **Golay equation** reviewed below.

The van Deemter equation expresses the plate height (H) as a function of average carrier gas velocity ($\bar{u}$) in the sum of three terms:

$$H = A + \frac{B}{\bar{u}} + C\bar{u} \quad (6.6)$$

Figure 6.2 is a graphical representation of Equation 6.6. The green curve shows the nonlinear relationship between plate height and carrier gas velocity (or flow rate). The minimum point of the green curve marks the optimum carrier gas velocity for the column studied. The additional curves in Figure 6.2 show the individual contribution of the three terms in the van Deemter equation.

The original van Deemter equation applied to a packed column and still does. The values of the constants A, B, and C are dictated by column design and operating parameters. We are interested in discovering the effect of these parameters on column performance.

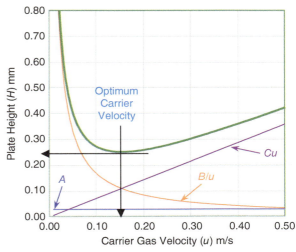

This graphical representation of the van Deemter equation (green curve), shows how column efficiency varies with carrier gas velocity. Plate height is the sum of three terms: the A-term (blue line) is constant, the B-term (orange line) declines exponentially with velocity, and the C-term (purple line) increases linealy with velocity. The optimum carrier velocity occurs at the lowest point of the combined curve.

Figure 6.2 van Deemter Curve.

The A-term

The first term in the van Deemter equation is a constant (A), which is unaffected by the flow rate. The blue plot line in Figure 6.2 represents its value.

The A-term applies solely to packed columns. It recognizes that there are multiple gas pathways through the packing bed, each of different length. Solute molecules randomly take longer or shorter pathways through the bed, so they travel different distances to reach the end of the column. Consequently, they arrive in the detector at different times, adding to the width of the peak.

We call this first dispersion process **eddy diffusion**. It causes the same degree of peak broadening at any carrier gas velocity.

The contributors to A are the particle diameter (d_p) and packing factor (λ):

$$A = 2\lambda d_p \quad (6.7)$$

What we learn

From Equation 6.7, we learn that smaller particles and more-perfect packing will improve column efficiency.

The use of smaller particles in liquid chromatography has provided enormous gains in column efficiency. In gas chromatography, though, the A-term is only a small contributor to plate height and the gains are less dramatic. While gains are sometimes possible, small particles will create more pressure drop, and this tends to limit column efficiency. In practice, packed columns for PGCs rarely use particles less than about 125 μm.

To minimize the packing factor (λ), we use particles of uniform size and endeavor to pack them tightly in the column. Unfortunately, it turns out that smaller particles are more difficult to pack uniformly.

The B-term

The second term in the van Deemter equation is a constant (B) divided by the average carrier gas velocity. In Figure 6.2, the orange curve represents the B-term.

The B-term recognizes that solute molecules in the gas phase are constantly diffusing away from the peak center, thereby causing a peak to continuously spread lengthwise in the column.

The rate of this **longitudinal diffusion** depends on the diffusion coefficient (D_M) of the solute in the chosen carrier gas:

$$B = 2\psi D_M \qquad (6.8)$$

In Equation 6.8, the **obstruction factor** (ψ) is applicable only to packed columns and allows for the permeability of the packed bed.

What we learn

The diffusion process is continuous, so the extent of the band spreading depends on how long the solute is in the gas phase. The orange plot in Figure 6.2 shows the effect of carrier velocity. When the carrier velocity is low, the solute spends a lot of time in the gas phase and the B-term is dominant, causing low column efficiency. When the gas velocity is high, the solute spends less time in the gas phase and the B-term becomes insignificant. Since PGCs tend to run fast, the B-term is rarely of concern in process gas chromatography.

A dense carrier gas, such as nitrogen or argon, would reduce the solute diffusion rate and thereby reduce the B-term. But we choose carrier gases mostly for other reasons (for example, to suit the detector), seldom to reduce the B-term.

The C-term

The third term in the van Deemter equation is a constant (C) multiplied by the average carrier gas velocity. This term is called the **resistance to mass transfer**. It allows for the peak broadening caused by the slow transfer of solute molecules into and out of the liquid phase.

Figure 6.3 illustrates this effect. In each diagram, the blue curve is a snapshot of the distribution of molecules in the gas phase, while the orange curve is the concurrent distribution in the liquid phase.

In Figure 6.3a, the two distributions are in equilibrium, but this can happen only if the carrier gas stops moving. Figure 6.3b illustrates what really happens when the gas phase is moving. The movement disturbs the equilibrium and makes the overall peak wider.

The next step is important. As indicated by arrows in Figure 6.3b, some of the leading molecules enter the liquid phase to reestablish the leading-edge equilibrium, and some of the trailing molecules enter the gas phase to reestablish the trailing-edge equilibrium. Given that the gas phase continues to move, a fast mass transfer will incur only a small increase in peak width, whereas a slow mass transfer will cause significant broadening.

The column variables associated with the rate of mass transfer in the liquid phase are the depth of the liquid film (d_f) and the rate of diffusion of the solute in the liquid phase (D_S).

The expression for C recognizes that a thick liquid film will slow down the mass transfer rate, while a high diffusion rate will speed it up:

$$C = f(k) \cdot \frac{d_f^2}{D_S} \qquad (6.9)$$

where $f(k)$ is a function of the retention factor (k).

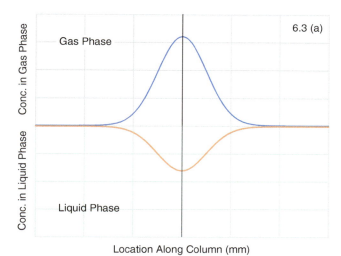

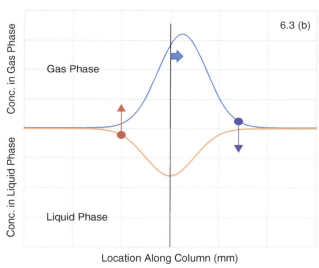

A static equilibrium (6.3a) would result in perfect alignment between the distributions of solute in the gas and liquid phases. But when the carrier gas moves (6.3b), the solute in the gas phase moves ahead of the solute in the liquid phase, making the peak wider. The effect is more pronounced when the rate of transfer between phases is slow.

Figure 6.3 Effect of Slow Mass Transfer.
Source: Adapted from Miller (2005, 78) for a solute peak having $K_c = 0.5$.

What we learn

Equation 6.9 contains the square of film thickness, so a thinner liquid film will give a significant improvement in column efficiency. Yet very thin films can be problematic. For some practical limits to the thinness of the liquid film, refer to that discussion in Chapter 4.

Reducing the viscosity of the liquid would allow more rapid diffusion of the solute. To that end, it seems that the column should operate at a high temperature, but Chapter 4 has several caveats to following that advice. Most PGC columns run best warm, not hot.

The full expression of Equation 6.9 also includes a function $f(k)$ of the peak retention factor. Here is the expansion of $f(k)$:

$$f(k) = \frac{8}{\pi^2} \cdot \frac{k}{(1+k)^2} \qquad (6.10)$$

Table 6.1 lists the numerical values of this function for typical values of k. The function reaches a maximum value at $k = 1$ and declines at higher values of k, which indicates that the C-term has less effect for later peaks on the chromatogram.

Note that a higher column temperature will reduce the peak retention factor and that will tend to negate any benefit gained from reduced liquid phase viscosity. One might propose to increase the film thickness to maintain constant retention factors, but a quick look at Equation 6.9 would quash that idea: film thickness is squared!

Looking again at Figure 6.3b, the time taken for mass transfer will cause more peak broadening at higher gas velocities than at low velocities. To capture this effect, the van Deemter equation multiplies the constant (C) by the carrier gas velocity.

Since PGC columns tend to run fast, the C-term has a large effect on plate height and is of most concern for PGC column design.

Table 6.1 Values of the k-Function in Equation 6.10.

k	1	2	5	10	20
$f(k)$	0.20	0.18	0.11	0.067	0.037

Extended van Deemter equation

The original van Deemter equation did not consider resistance to mass transfer in the gas phase, reasoning that this transfer would be very fast and thus negligible in a packed column. But the development of open-tubular columns quickly showed the necessity of including an extra term to account for the mass transfer rate in the gas phase.

So the extended van Deemter equation applicable to open-tubular columns (and to liquid chromatography) incorporates two mass transfer constants, one for the mobile phase (C_M) and one for the stationary phase (C_S):

$$H = A + \frac{B}{\overline{u}} + (C_M + C_S)\overline{u} \qquad (6.11)$$

As previously noted, the A-term allows for multiple pathways in the column packing and is not applicable to open-tubular columns. Thus, the Golay equation for capillary columns (reviewed below) does not include the A-term.

The Golay equation

Laminar flow in an open-tubular column sets up a wide range of carrier gas velocities, very slow near the walls due to friction and faster at the center. These different velocities would inevitably distribute the solute molecules randomly along the axis of the column, causing severe peak broadening. Fortunately, another effect limits the lengthwise distribution; the molecules also diffuse radially between the center of the tube and the tube wall. When this radial diffusion is fast, it mixes the high and low velocities and tends to negate the longitudinal broadening.

Golay (1958) developed an extended van Deemter equation for capillary columns that includes resistance to mass transfer in the gas phase (C_M) and is still valid today. Golay recognized the advantage of fast radial movement inside the column. His C_M term includes the internal diameter (d_c) of the column and the diffusion coefficient of the solute in the gas phase (D_M):

$$C_M = f(k) \cdot \frac{d_c^2}{D_M} \qquad (6.12)$$

where $f(k)$ is a new function of the retention factor.

What we learn

Equation 6.12 shows that a narrow column diameter and a fast rate of diffusion will improve the efficiency of a column.

The squared column diameter in the Golay equation predicted that thinner columns would return a large improvement in efficiency. The diameter of columns has been shrinking ever since. The remarkable efficiency of modern narrow-bore open-tubular columns confirms that prediction.

To improve the diffusion rate, use low-density carrier gases (hydrogen or helium). A higher column temperature might help, subject to the limitations discussed above. The full expression of Equation 6.12 includes a new function $f(k)$ of the peak retention factor. Here is the expansion of this function:

$$f(k) = \frac{1 + 6k + 11k^2}{96(1+k)^2} \qquad (6.13)$$

Numerically, this rather foreboding function of k is very simple and has little effect on plate height. Table 6.2 shows that its value varies by only a factor of two for normal values of k. Thus, later peaks experience a higher value of C_M and will be slightly less efficient.

Table 6.2 Values of the k-Function in Equation 6.13.

k	1	2	5	10	20
$f(k)$	0.047	0.066	0.089	0.100	0.107

Interestingly, this increase in C_M tends to compensate for the larger and opposite effect of k on the mass transfer rate in the liquid phase (C_S).

Table 6.3 summarizes the variables associated with rate theory and their effect on column performance, as discussed above.

Other effects

Over the years, researchers have proposed other modifications to the original van Deemter equation with the intent of more closely modeling the empirical efficiency curve. These modified equations are generally not applicable to process gas chromatography.

A source of variance not covered by column theory is the dispersion that occurs in the instrumentation, the so-called **extracolumn variance**. Dispersion occurs in the sample injection device, in unswept side volumes ("dead legs") of the flow path, in the tube connections between devices, and in the detector. We examine these issues in Chapter 8.

Also missing from the above discussion is the effect of pressure on column efficiency. In gas

Table 6.3 Quick Reference Guide to Rate Theory.

Column Variable	Symbol	Column Type	Action to Reduce Plate Height and Improve Column Efficiency	Equation Number
Average Carrier Velocity	$\bar{u}$	All	Run at optimum carrier gas velocity (or flow rate).	6.6
Particle Diameter	d_p	Packed	Use smaller particles but beware of creating too much pressure drop across column.	6.7
Packing Factor	λ	Packed	Use particles of uniform size. Ensure column is snugly packed with no voids.	6.7
Obstruction Factor	ψ	Packed	Use smaller particles to reduce the A term, but this will cause more pressure drop.	6.8
Diffusion Coefficient of Solute in Gas Phase	D_M	All	Run at higher carrier gas velocity to reduce the contribution of the B-term. Use a dense carrier gas (nitrogen or argon).	6.8
Liquid Film Thickness *(Squared!)*	d_f^2	All	Use a low liquid phase loading, but not too low! (see Chapter 4).	6.9
Diffusion Rate of Solute in Liquid Phase	D_s	All	Use a low-viscosity liquid phase. For some liquid phases, a higher temperature might help.	6.9
Retention Factor of Peak	k	All	Only a minor effect – and not easily adjusted. Focus on having the peaks in the power region.	6.9 & 6.12
Column Internal Diameter *(Squared!)*	d_c^2	Open Tubular	Reduce column diameter but beware of exceeding maximum sample volume (see Chapter 4). Check detector sensitivity to a reduced sample volume.	6.12
Diffusion Rate of Solute in Gas Phase	D_M	Open Tubular	Use a low-density carrier gas (hydrogen or helium).	6.12

chromatography, the gradual decompression of the carrier gas as it moves down the column length has a significant effect on peak broadening, particularly in packed columns. Refer to Chapter 9 for more information on the effects of pressure drop.

Getting a fast analysis

It may come as a surprise that PGC columns rarely run at their optimum flow rate. It's usually too slow. Often, the last step in PGC performance optimization is to optimize the analysis or cycle time.

From a practical point of view, the procedure for fast analysis is rather simple: install a long column and run it fast. To fully understand the procedure, though, it's best to follow the calculations given in the Worked Example below, so you don't waste your time with a column that's never going to work. The calculations are not difficult, and the results are astounding!

The Worked Example that follows starts with a fully optimized column and deliberately runs it under non-optimal conditions to gain some useful performance improvements, such as

- More resolution with the same analysis time.
- Faster analysis with the same resolution.

These are advantages that you don't want to miss.

A worked example

Introduction

This is a practical procedure for improving column performance in four steps.

The *first step* shows the calculations needed to fully understand the base case. It shows you how to use the efficiency curve of a column to calculate the resolution of adjacent peaks and the overall efficiency of the column.

The *second step* examines a longer column, but for ease of understanding, the other operating conditions remain unchanged. Naturally, the longer column gives a longer analysis time. Example calculations illustrate the inevitable changes to peak width and resolution.

The *third step* considers the possibility of increasing the carrier gas velocity in the longer column to regain the original analysis time. The calculations are similar, so we show them in summary form. The outcome of this change may surprise you.

The *final step* shows how it might be possible to achieve a faster analysis using the longer column at an even higher carrier gas flow rate. We will discuss this paradoxical result in detail but challenge you to do the calculations yourself. It's a good practice exercise for you.

Initial assumptions

Now for the actual procedure. We start from the following arbitrary assumptions:

- The column length is 10 m.
- The retention time of Peak A is 285 s.
- The retention time of Peak B is 300 s.
- The column efficiency curve displayed in Figure 6.1 applies to Peak B on this column.
- The column is running at its optimum flow rate.

Therefore, from Figure 6.1:

- The carrier gas velocity at optimum flow rate is 0.20 m/s.
- The plate height for Peak B at optimum flow rate is 0.40 mm.

Although we will consider four different column conditions, the above starting data, together with the efficiency curve in Figure 6.1, is all the information we need to calculate performance parameters for the entire exercise.

Step #1: Evaluate the existing column

Look at the starting chromatogram in Figure 6.4a.

Before changing anything, we first need to understand the base case. The column in this example is doing a good job. It's separating Peak A and Peak B in 300 seconds with a clear baseline resolution. It's evident that the column is operating under optimum conditions since both peaks are in the power region and the flow rate is set at the optimum velocity.

On the face of it, the performance of this column cannot be improved but we are going to improve it anyway!

To get a faster analysis, one could propose a shorter column or a higher flow rate and accept the loss of resolution incurred. In a real application, though, that's hardly likely to be an acceptable outcome. Therefore, let's place a constraint on this exercise; that any modified configuration shall produce at least as much resolution as before.

To ensure this, we need to know the starting resolution. You can calculate the resolution from the time separation of the peaks and their average width.

The time separation (Δt_R) is simply the difference between the retention times of Peak B and Peak A, and we have that data:

$$\Delta t_R = 300 \text{ s} - 285 \text{ s} = 15 \text{ s}$$

Now we need the peak widths. That's an easy calculation, but the units can be confusing. The plate height obtained from Figure 6.4a is in length units (0.4 mm), whereas the calculation of chromatogram peak widths must be in time units, normally seconds. So, before calculating the peak widths, convert the length-based plate height (H) to the time-based plate height (H_t) using Equation 2.21 to:

$$H_t = H \cdot \frac{t_R}{L}$$

From Figure 6.4a:

$$H_B = 0.40 \text{ mm}$$
$$t_R = 300 \text{ s}$$
$$L = 10 \text{ m}$$

Therefore,

$$H_t = 0.40 \text{ mm} \times \frac{300 \text{ s}}{10{,}000 \text{ mm}} = 0.0120 \text{ s}$$

162 Optimizing performance

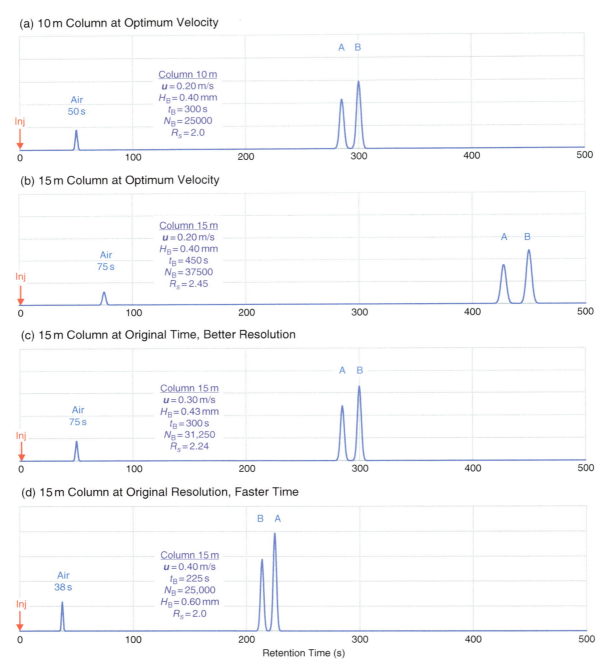

PGC columns tend to be longer and run faster than optimum, thereby gaining better time efficiency. In this sequential example, chromatogram:
(a) Uses a column running at optimum flow rate and achieving an adequate resolution;
(b) Uses a longer column running at optimum flow rate and achieves excess resolution in long time;
(c) Increasing the carrier flow yields the same analysis time but with better resolution than the original column – better time efficiency;
(d) Running even faster yields the same resolution as the original chromatogram in less analysis time.

Figure 6.4 Advantages of Using a Longer Column.

Our column is producing about 83 plates-per-second; one equivalent equilibrium occurs every 12 milliseconds. The phase distribution process is fast!

Now calculate the base widths (w_b) of Peaks A and B using Equation 2.23:

$$w_b = 4\sqrt{H_t \cdot t_R}$$

$$(w_b)_A = 4\sqrt{0.012 \times 285 \text{ s}} = 7.397 \text{ s}$$

$$(w_b)_B = 4\sqrt{0.012 \times 300 \text{ s}} = 7.589 \text{ s}$$

You may notice that the above calculations use the plate height derived from Peak B data to calculate the widths of both peaks. This is allowable because the plate height in time units is sensibly constant for all peaks on the chromatogram, as noted in the **SCI-FILE** *On Plate Theory* in Chapter 2.

To continue, calculate the average peak width ($\overline{w}$) of the two peaks:

$$\overline{w} = \frac{7.397 \text{ s} + 7.589 \text{ s}}{2} = 7.49 \text{ s}$$

Finally, get their resolution from Equation 2.8:

$$R_s = \frac{\Delta t_R}{w_a}$$

$$R_s = \frac{15 \text{ s}}{7.49 \text{ s}} = 2.00$$

As shown by the chromatogram in Figure 6.4a, this resolution produces a very good baseline separation.

To evaluate the changes that we are about to make, we will also need to know the separating power of the column, given by its plate number. The plate number (N) is not the same for both peaks, it comes out higher for Peak B than for Peak A. Using Equation 2.26:

$$N = 16\left(\frac{t_R}{w_b}\right)^2$$

$$N_A = 16 \times \left(\frac{285 \text{ s}}{7.40 \text{ s}}\right)^2 = 23{,}730$$

$$N_B = 16 \times \left(\frac{300 \text{ s}}{7.59 \text{ s}}\right)^2 = 25{,}000$$

If you think about it, this difference is not surprising. The width of a peak doesn't increase as fast as its retention time does; peak width is proportional to the square root of peak retention time. It's inevitable, therefore, that each peak on a chromatogram has a different plate number that gradually increases with peak retention time. That's why a longer column always improve resolution – the additional peak separation outweighs the increase in peak width. But this extra resolution is not free; the cost to you is lower peak heights and longer analysis time.

For a given peak, plate number always indicates the resolving power of the column. If the plate number increases, it's a certain indication that the resolution has also increased. For us, this is a good reason to keep an eye on the plate number for Peak B when making changes to the column.

Finally, recall that the power region lies between two times and six times the holdup time. For this chromatogram, that's from 100 to 300 seconds and both peaks are in the power region.

Summary of chromatogram 6.4a

In summary, these are the key starting conditions:

- The column is running at its optimum flow rate.
- Carrier gas velocity is 0.2 m/s.
- Plate height is 0.40 mm.
- Peak B plate number is 25,000.
- Peak B retention time is 300 s.
- Time between peaks is 15.0 s.
- Average peak width is 7.49 s.
- Resolution is 2.00.

This column is fully optimized and doing a fine job. It's the very best that the column can do. You would think that all columns would run like this – set at their optimum flow rate. But no, most PGC applications need the fastest analysis time and the optimum flow rate tends to be slow. It's counterintuitive, but a longer column may provide a faster analysis.

To examine this apparent paradox, let's take it one step at a time.

Step #2: Use a longer column

In the second chromatogram, we replace the 10 meter column with a 15 meter column of the same type and run it at the same average carrier velocity as before. Because of the additional carrier gas pressure, the column exit flow rate might not be the same; to obtain the same average velocity, adjust the flow rate until the air peak elutes at 75 seconds.

Figure 6.4b shows the new chromatogram.

Here's what we know:

- The column length is now 15 m.
- The carrier gas flow rate is about the same.

- The average carrier gas velocity has not changed – it's still optimized at 0.20 m/s.
- The plate height has not changed – it remains at 0.40 mm.

Here are the calculations to evaluate the new column condition.
The retention time of Peak A increases in proportion to column length:

$$(t_R)_A = 285 \text{ s} \times \frac{15 \text{ m}}{10 \text{ m}} = 427.5 \text{ s}$$

Similarly, the retention time of Peak B also increases:

$$(t_R)_B = 300 \text{ s} \times \frac{15 \text{ m}}{10 \text{ m}} = 450 \text{ s}$$

So, the time separation between peaks is more than it was before:

$$\Delta t_R = 450 \text{ s} - 427.5 \text{ s} = 22.5 \text{ s}$$

The column is still working at optimum velocity, so its plate height is the same as it was before, 0.012 s.

Again using Equation 2.23, the new base widths of Peaks A and B are:

$$(w_b)_A = 4\sqrt{0.012 \text{ s} \times 427.5 \text{ s}} = 9.060 \text{ s}$$

$$(w_b)_B = 4\sqrt{0.012 \text{ s} \times 450 \text{ s}} = 9.295 \text{ s}$$

And, the average peak width has increased to 9.18 s:

$$\overline{w} = \frac{9.060 \text{ s} + 9.295 \text{ s}}{2} = 9.18 \text{ s}$$

Yet the resolution is much better, as expected with the longer column. The resolution increased from 2.00 to 2.45:

$$R_s = \frac{22.5 \text{ s}}{9.18 \text{ s}} = 2.45$$

And from Equation 2.13, the plate number of the new column is:

$$N = \frac{L}{H} = \frac{15,000 \text{ mm}}{0.40 \text{ mm}} = 37,500$$

Summary of chromatogram 6.4b

Table 6.4 lists the values of column parameters for the longer column running under optimum conditions. Noted below are the most important results of this change:

- The column is 1.5 times longer.
- The carrier gas velocity is the same.

166 Optimizing performance

Table 6.4 Four-Step Optimization of Analysis Time.

Key Performance Data		Figure Number			
		6.4a	6.4b	6.4c	6.4d
Column Length	m	10	15	15	15
Carrier Gas Velocity	m/s	0.20	0.20	0.30	0.40
Plate Height	mm	0.40	0.40	0.48	0.60
Plate Number of Peak B		25,000	37,500	31,250	25,000
Resolution A–B		2.0	2.45	2.24	2.0
Retention Time of Peak B	s	300	450	300	225
Plates per Second	Hz	83	83	104	111
Peaks in Power Region		Yes	Yes	Yes	Yes

Plate number and resolution are the same in both original and final conditions (yellow highlight) but the time efficiency has improved (green highlight). However, it's fair to note this improvement in analysis time comes at the cost of twice the carrier gas consumption.

> - The plate height is the same.
> - The Peak B retention time increased from 300 to 450 s.
> - The time between peaks improved from 15.0 to 22.5 s.
> - The average peak width worsened from 7.49 to 9.18 s.
> - The Peak B plate number went up from 25,000 to 37,500.
> - The resolution improved from 2.00 to 2.45.
> - The analysis time is about 1.5 times longer than it was.

There should be no surprises here. As expected, the longer column provides a much better resolution at the expense of a longer analysis time.

Step #3: Regain the analysis time

The next step is to increase the carrier gas velocity. That will certainly speed up the analysis, but the column will no longer be operating at its optimum velocity. Figure 6.1 shows that plate height will increase when the carrier velocity increases and this will cause a loss of resolution. But that may not be a problem because we now have more resolution than we need.

In the third chromatogram, the carrier gas flow rate is just high enough to bring Peak B back to its original location, reducing its retention time from 450 to 300 s. To achieve this reduction in retention time, the carrier gas velocity ($\bar{u}$) must be:

$$\bar{u} = 20 \text{ m/s} \times \frac{450 \text{ s}}{300 \text{ s}} = 30 \text{ m/s}$$

Figure 6.4c shows the new chromatogram.

Here's what we know:

➢ The column length remains at 15 m.
➢ The column is not running at its optimum flow rate.
➢ The average carrier gas velocity is now 0.30 m/s.
➢ From Figure 6.1, the plate height has increased to 0.48 mm.

Calculate the new column parameters in much the same way as before, noting the effect of the changes:

The retention time of Peak A is now the same as the original:

$$(t_R)_A = 427.5 \text{ s} \times \frac{0.20 \text{ m/s}}{0.30 \text{ m/s}} = 285 \text{ s}$$

The retention time of Peak B is now the same as the original:

$$(t_R)_B = 450 \text{ s} \times \frac{0.20 \text{ m/s}}{0.30 \text{ m/s}} = 300 \text{ s}$$

The separation between peaks is now the same as the original:

$$\Delta t_R = 300 \text{ s} - 285 \text{ s} = 15 \text{ s}$$

The plate height is now 0.48 mm.
The plate height converted to time units is:

$$H_t = 0.48 \text{ mm} \times \frac{300 \text{ s}}{15{,}000 \text{ mm}} = 0.0096 \text{ s}$$

So, the new peak widths are:

$$(w_b)_A = 4\sqrt{0.0096 \text{ s} \times 285 \text{ s}} = 6.616 \text{ s}$$
$$(w_b)_B = 4\sqrt{0.0096 \text{ s} \times 300 \text{ s}} = 6.788 \text{ s}$$

And, the average peak width is:

$$\overline{w} = \frac{6.616 \text{ s} + 6.788 \text{ s}}{2} = 6.70 \text{ s}$$

The column is achieving a better resolution, with the same analysis time!

$$R_s = \frac{15 \text{ s}}{6.70 \text{ s}} = 2.24$$

And, the plate number of Peak B is higher than the original value of 25,000 – as one would expect with the increased resolution:

$$N_B = \frac{L}{H} = \frac{15{,}000 \text{ mm}}{0.48 \text{ mm}} = 31{,}250$$

Summary of chromatogram 6.4c

As noted in Table 6.4, this is a great improvement. Here are the most important performance parameters compared with their original values:

➤ The column is 1.5 times longer than the original.
➤ Carrier gas velocity is 1.5 times faster than optimum.
➤ Plate height is worse, increasing from 0.40 to 0.48 s.
➤ Both peaks are back to their original retention times.
➤ Both peaks are narrower.
➤ The resolution is better, improving from 2.0 to 2.24.
➤ Plate number went up from 25,000 to 31,250.
➤ The analysis time is the same as it was before.

Nice. Compare the new chromatogram in Figure 6.4c with the original in Figure 6.4a. Notice that the retention times of Peaks A and B are now the same as they were in the original chromatogram, but the peaks are narrower! That's why the resolution improved. This increased resolution explains why PGC columns are often longer and run faster than you might expect.

Actually, it's useful to know that your PGC column may have untapped separating power. If you need more resolution and you don't mind a longer analysis time, just reduce the flow rate a little. In most cases, the reduced carrier gas velocity will give you more resolution. It just takes time.

Step #4: Faster analysis with the same resolution

The final change aims to get the shortest analysis time. Our present column is still producing more resolution than the original, so it should be possible to increase the velocity until the resolution falls back to its original value. This would give the same resolution in less time, a worthy goal indeed.

Once again, more carrier gas pressure is necessary. With all three of the proposed changes needing more pressure, you may reach a limit to what is possible. It depends on the starting conditions and the carrier gas in use. The section on *Time Efficiency* below discusses the choice of carrier gas for fast analysis.

Figure 6.4d shows the final chromatogram. Here's what we know:

➤ The column length remains at 15 m.
➤ The column is running twice as fast as its optimum flow rate.
➤ The average carrier gas velocity is now 0.40 m/s.
➤ From Figure 6.1, the plate height is now 0.60 mm.

We calculated the new column parameters in much the same way as before, and briefly note the changes below. You can write in the calculation of each item yourself; it's a good training exercise:

The retention time of Peak A is now 71 seconds less than the original:

$$(t_R)_A = \underline{\hspace{2cm}} = 213.8 \text{ s}$$

The retention time of Peak B is now 75 seconds less than the original:

$$(t_R)_B = \underline{\hspace{2cm}} = 225 \text{ s}$$

The separation between peaks is now 3.8 seconds less than the original:

$$\Delta t_R = \underline{\hspace{2cm}} = 11.2 \text{ s}$$

The plate height converted to time units is:

$$H_t = \underline{\hspace{2cm}} = 0.0090 \text{ s}$$

So the new peak widths are:

$$(w_b)_A = \underline{\hspace{2cm}} = 5.549 \text{ s}$$

$$(w_b)_B = \underline{\hspace{2cm}} = 5.692 \text{ s}$$

And the average peak width is less than on the original chromatogram:

$$\overline{w} = \underline{\hspace{2cm}} = 5.62 \text{ s}$$

The resolution is the same as the original, with 75 seconds faster analysis time:

$$R_s = \underline{\hspace{2cm}} = 2.0$$

And the plate number of Peak B is back to its original value of 25,000:

$$N_B = \underline{\hspace{2cm}} = 25{,}000$$

Summary of chromatogram 6.4d

Table 6.4 lists most of this new data. Here's a summary that compares the final column performance with the performance of the original column:

- The column is 1.5 times longer than it was.
- Carrier gas is now flowing at twice the optimum velocity.
- Plate height is much worse, having increased from 0.40 to 0.60 mm.
- Peak B is now 75 s faster than in the original chromatogram.
- The peaks are closer together.

- Both peaks are narrower.
- Resolution is the same as the original column.
- Plate number is the same as it was in the original column.
- The analysis time is about 75 s faster than it was before.

Look again at Figure 6.4d. It's an amazing improvement in the speed of analysis.

> *A good thing to remember:*
>
> *Optimum column performance is usually slow. A better resolution or a faster analysis is often possible by running a longer column faster than its optimum flow rate.*

Time efficiency

This Worked Exercise illustrates the possibility of using a longer column running fast to get an improved resolution (per Figure 6.4c) or a reduced analysis time (per Figure 6.4d). To get these advantages, PGC columns tend to run fast. It's not unusual for the carrier gas to be set at about twice the optimum flow rate.

In the end, the best performance comes from generating the most theoretical plates in the shortest time. So, when analysis time is an issue, as it often is with a PGC, the best measure of performance is **time efficiency**, measured in plates-per-second. Table 6.4 includes the values of this new parameter, confirming that Figure 6.4d provides the best time efficiency of the four examples examined.

An interesting property of time efficiency is its constancy. Unlike plate number, which increases with retention time, time efficiency is about the same for every peak on the chromatogram. Because of its independence from the peak used to measure it, time efficiency is an ideal parameter for evaluating the performance of the column itself.

Designing for fast analysis

The efficiency curve

The ability to drive a column faster and yet get more plates depends upon the slope of the efficiency curve. The above worked example adopted the efficiency curve in Figure 6.1. If you look closely at that curve, you will see that any increase in carrier gas velocity beyond its optimum value results in a smaller increase in plate height. For instance, at twice the optimum velocity, the plate height increases by only 50 %.

This is the key to fast analysis; it relies on the slope of the right incline of the efficiency curve. Long columns with shallow curves lose little efficiency when running fast. Their time efficiency in plates-per-second is higher than a shorter column running at optimum speed. This advantage is lost for steeper efficiency curves, and for very steep curves, the time efficiency of the column may decline as the velocity increases. That kind of column must operate at or near its optimum flow rate.

The **SCI-FILE** *On Rate Theory* discusses the factors that determine the slope of the efficiency curve at high flow rates. In brief, a shallow slope is desirable and more likely to occur with reductions in column diameter, carrier gas density, liquid phase loading, or liquid phase viscosity.

Narrow column diameter

It's well known that narrower column diameters yield flatter efficiency curves and thus allow faster analysis. For PGC columns, though, the need for stability and reliability for long periods in an industrial environment will often outweigh the advantage of using a very narrow column with a very thin liquid film. Modern PGC columns aim to last more than one year in continuous operation.

Choice of carrier gas

The rate theory predicts that lower-density carrier gases will exhibit shallower efficiency curves than higher-density gases, thus allowing higher velocities with minimal loss of column efficiency. This is true in practice. Figure 6.5 illustrates typical efficiency curves for hydrogen, helium, and nitrogen carrier gas.

Hydrogen is best for fast analysis on two counts:

➢ The efficiency curve for hydrogen is much flatter than for nitrogen, and this favors high-speed analysis with little loss of efficiency. Helium is similar in this regard.

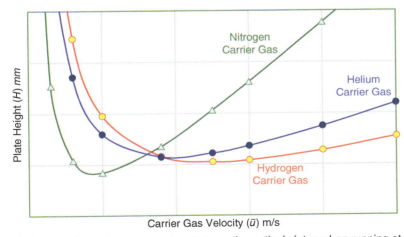

Higher density carrier gases generate more theoretical plates when running at optimum velocity but yield slow analyses. Because of the severe slope of their van Deemter curve, running at a faster flow rate is unlikely to improve resolution or analysis time. Hydrogen is the best carrier gas for fast analysis. Its curve is flatter and it requires only half the pressure drop of helium.

Figure 6.5 Effect of Carrier Gas Density.
Source: Adapted from Miller (2005, 86).

> Hydrogen is a low-density and low-viscosity gas and flows through a column at about twice the flow rate of helium, when at the same inlet pressure. The ability to flow fast is a strong advantage of hydrogen carrier when designing for fast analysis, particularly with packed columns.

Historically, most PGCs in the USA used helium as carrier gas because it was safe and inexpensive. But helium is now sometimes in short supply and is costlier than it was. So, many US plants are converting their PGCs to use hydrogen carrier gas, sometimes achieving faster analysis. Helium has always been expensive outside the USA, so most of the PGCs at other jobsites use hydrogen carrier gas.

Note, however, that hydrogen may react with unsaturates in your sample, creating extra peaks and reducing the analyte measurement. It's well known that hydrogen reacts with butadiene at temperatures above 100 °C to form several additional C4 peaks, probably paraffins and olefins. If you need to measure butadiene at high injection temperature, it's best to use helium carrier gas.

Liquid phase loading

The depth of the liquid phase layer has a major effect on the slope of the efficiency curve at high carrier gas velocities. Whenever possible, avoid deep liquid layers. It takes time for solute molecules to diffuse through a deep layer and this results in a steep efficiency curve. Figure 6.6 shows that a low liquid load (or thin liquid film) is much better for fast analysis because it allows rapid mass transfer and leads to a flatter efficiency curve.

Low liquid loadings are not always possible. If the solutes are not very soluble in the chosen liquid phase, a thicker liquid film may be necessary to bring

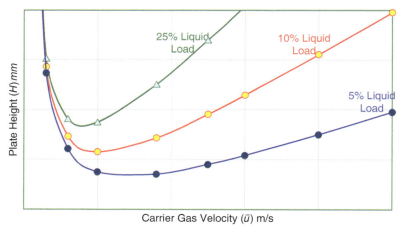

Figure 6.6 Effect of Film Thickness or Liquid Loading. *Source: Annino et al., 1992/with permission of International Society of Automation.*

It takes time for the solute to penetrate the liquid layer so a thicker liquid film loses efficiency more rapidly at higher carrier gas flow rates. The effect is worse for high-viscosity liquid phases. Thin liquid films (low loading) provide the best efficiency, but first confirm your closest analyte peaks are in the power region.

the peaks into the power region. Chapter 5 explains why the best peak retention time for fast analysis is three times the gas holdup time (i.e., $k = 2$).

Also, a large sample volume may overload a lightly loaded column, yet a large sample may be necessary to get the needed detector sensitivity.

Liquid phase viscosity

A liquid phase with low viscosity allows faster diffusion of the solute molecules when they are in the liquid phase, which then allows a rapid transfer of solute molecules between phases. The fast mass transfer minimizes band spreading in the column and produces a flatter efficiency curve.

It's useful to know that low viscosity is a desirable property of a liquid phase. In practice, though, we choose liquid phases for their ability to separate the peaks, not for their low viscosity.

As noted in the **SCI-FILE** *On Rate Theory*, a higher column temperature will reduce the viscosity of many liquid phases and may help to flatten the efficiency curve. Even so, the advice sometimes given to run a column hot is shortsighted. It's much more important to set the column temperature so the peaks are in the power region. Refer to Chapter 4 for a more detailed discussion of column temperature limitations.

Knowledge Gained

Practice

- On every column, there is an optimum carrier gas flow rate that delivers the best resolution.

- A simple experimental procedure discovers the optimum carrier flow rate.

- The procedure is applicable to a single column or to similar columns connected in series.

- Before starting, ensure that the peak pair that are hardest to resolve is in the power region.

- Measure the plate number and plate height of a peak at several different carrier flow rates.

- Record the flow rates, or optionally calculate the carrier gas velocity from the holdup time.

- Plot a column efficiency curve of plate height against carrier gas velocity (or flow rate).

- Note the optimum velocity (or flow rate) that gives the minimum plate height.

- PGC columns tend to be longer and run faster than their optimum flow rates.

- It's possible to achieve better resolution or faster analysis on a longer column running fast.

- Hydrogen is usually the best carrier gas for fast analysis.

- A thin film of low-viscosity liquid phase allows fast analysis with little loss of resolution.

Theory

- Rate theory assumes that equilibrium cannot form because the mobile phase is always moving.

- Rate theory seeks to understand the causes of molecular dispersion within the column.

- *Variance is a statistical measure of the dispersion of solute molecules in the column.*
- *The variance of a peak is the sum of the variances of all independent dispersions.*
- *Plate height is a measure of the variance generated per unit length of column.*
- *Rate theory expresses plate height as the sum of several independent sources of dispersion.*
- *The van Deemter equation equates plate height to the sum of three terms: A, B, and C.*
- *The A-term is the dispersive effect of multiple gas pathways through a packed column bed.*
- *The A-term is lower for small particles of uniform size, tightly packed in the column.*
- *The B-term is due to the longitudinal diffusion of solute molecules away from the peak center.*
- *The B-term is lower with a high-density carrier gas or increased carrier gas velocity.*
- *The C-term is due to slow transfer of solute molecules into and out of the liquid phase.*
- *The C-term is lower with a thin liquid film, a low-viscosity liquid phase, or a slow carrier gas.*
- *The extended van Deemter has an extra C-term for slow transfer of solute in the gas phase.*
- *The extra C-term applies mainly to capillary columns.*
- *The extra C-term is lower for smaller diameters or lighter carrier gases.*

Did you get it?

Self-assessment quiz: SAQ 06

On practice

For questions 6Q1–6Q5, assume you are working with Peaks A and B on an 8 m column, for which Figure 6.1 is the Peak B column efficiency curve. The retention time of Peak A is 185 s. The retention time of Peak B is 200 s.

Q1. What is the maximum plate number for Peak B on this column?
Q2. What is the holdup time of this column when it is generating the maximum plate number for Peak B?
Q3. Are the Peaks A and B in the power region?
Q4. What is the base width of Peak B?
Q5. What is the plate number for Peak B when the column is operated at twice its optimum flow rate?

On theory

Q6. Which of the following statements is/are true of the van Deemter curve shown in Figure 6.2?

    A. The $B/\bar{u}$ term contributes less to the minimum plate height than the $C\bar{u}$ term does.

- **B.** *The $B/\bar{u}$ term contributes about the same to the minimum plate height as the $C\bar{u}$ term does.*
- **C.** *The $B/\bar{u}$ term contributes more to the minimum plate height than the $C\bar{u}$ term does.*

Q7. *For the van Deemter curve shown in Figure 6.2, which of the following changes might be effective to reduce the slope of the curve at high carrier gas velocities?*

- **A.** *Reduce the carrier gas flow rate.*
- **B.** *Increase the liquid loading.*
- **C.** *Use a lower-viscosity liquid phase.*
- **D.** *Increase column temperature.*
- **E.** *Change to hydrogen carrier gas.*
- **F.** *Use a longer column.*
- **G.** *Use a wider-bore column.*

References

Further reading

The helpful little reference book by Leslie Ettre and John Hinshaw (1993) is a *must have* guidebook on the equations of theoretical gas chromatography.

Cited

Annino, R. and Villalobos, R. (1992). *Process Gas Chromatography: Fundamentals and Applications*. Research Triangle Park, NC: Instrument Society of America.

Ettre, L.S. and Hinshaw, J.V. (1993). *Basic Relationships of Gas Chromatography*. Cleveland, OH: ADVANSTAR Communications, Inc.

Golay, M.J.E. (1958). Theory of chromatography in open and coated tubular columns with round and rectangular cross-sections. In: *Gas Chromatography 1958 (The Amsterdam Symposium)* (ed. D.H. Desty), 36–55. London, UK: Butterworths.

Miller, J.M. (2005). *Chromatography: Concepts and Contrasts*. Hoboken, NJ: John Wiley & Sons.

Van Deemter, J.J., Zuiderweg, F.J., and Klinkenberg, A. (1956). Longitudinal diffusion and resistance to mass transfer as causes of nonideality in chromatography. *Chemical Engineering Science* 5, 271–289.

Tables

6.1   Values of the *k*-Function in Equation 6.10.
6.2   Values of the *k*-Function in Equation 6.13.
6.3   Quick Reference Guide to Rate Theory.
6.4   Four-Step Optimization of Analysis Time.

Figures

6.1 Typical Column Efficiency Curve.
6.2 van Deemter Curve.
6.3 Effect of Slow Mass Transfer:
   a. Distribution of solute at equilibrium.
   b. Band broadening due to gas movement.
6.4 Advantages of Using a Longer Column:
   (a) 10 m Column at Optimum Velocity.
   (b) 15 m Column at Optimum Velocity.
   (c) 15 m Column at Original Time, Better Resolution.
   (d) 15 m Column at Original Resolution, Faster Time.
6.5 Effect of Carrier Gas Density.
6.6 Effect of Film Thickness or Liquid Loading.

Symbols

Symbol	Variable	Unit
a	Effective cross-sectional area of gas pathway in column	m
A	van Deemter constant – eddy diffusion term	m
B	van Deemter constant – longitudinal diffusion term	m^2/s
C	van Deemter constant – resistance to mass transfer term	s
C_M	Expanded van Deemter – mass transfer into mobile phase	s
C_S	Expanded van Deemter – mass transfer into stationary phase	s
d_c	Internal diameter of open tubular column	m
d_f	Liquid phase film thickness	m
d_p	Diameter of particles in a packed column	m
D_M	Diffusion coefficient of solute in gas phase	m^2/s
D_S	Diffusion coefficient of solute in liquid phase	m^2/s
H	On-column plate height in length units	m
H_{min}	Minimum plate height in length units	m
H_t	Chromatogram plate height in time units	s
k	Retention factor	none
K_c	Distribution coefficient	none
L	Length of column	m
N	Plate number	none
R_s	Resolution of adjacent peaks	none
t_R	Peak retention time	s
u	Carrier gas velocity at column exit	m/s
$\bar{u}$	Average carrier gas velocity	m/s
$\bar{v}$	Average migration rate of a solute	m/s
$\dot{V}$	Carrier gas flow rate	mL

$\bar{w}$	Average base width of two adjacent peaks on chromatogram	s
w_b	Base width of a solute peak on chromatogram	s
w_h	Width at half height of a solute peak on chromatogram	s
Δt_R	The time separation between adjacent chromatogram peaks	s
λ	Packing factor for the uniformity of column packing	none
σ	Standard deviation of an on-column peak in length units	m
ψ	Obstruction factor for diffusion in a packed column	none

Equations

6.1 $\quad u = \dfrac{4\dot{V}}{\pi \cdot (d_c - 2d_f)^2}$ 	Open tubular column exit velocity.

6.2 $\quad \sigma^2 = \sigma_1^2 + \sigma_2^2 \ldots + \sigma_n^2$ 	Addition of variances for a Gaussian peak.

6.3 $\quad H = H_1 + H_2 \ldots + H_n$ 	Plate height as sum of independent variances.

6.4 $\quad \bar{v} = \dfrac{L}{t_R}$ 	Migration rate of a solute along column.

6.5 $\quad \bar{v} = \dfrac{\bar{u}}{k}$ 	Relation between velocities and retention factor.

6.6 $\quad H = A + \dfrac{B}{\bar{u}} + C\bar{u}$ 	The original van Deemter Equation.

6.7 $\quad A = 2\lambda\, d_p$ 	Van Deemter A-term: eddy diffusion.

6.8 $\quad B = 2\psi\, D_M$ 	Van Deemter B-term: longitudinal diffusion.

6.9 $\quad C = f(k) \cdot \dfrac{d_f^2}{D_s}$ 	Van Deemter C-term: resistance to mass transfer in liquid phase.

6.10 $\quad f(k) = \dfrac{8}{\pi^2} \cdot \dfrac{k}{(1+k)^2}$ 	$f(k)$ for the van Deemter C-term in Equation 6.9.

6.11 $\quad H = A + \dfrac{B}{\bar{u}} + (C_M + C_S)\bar{u}$ 	The extended van Deemter Equation.

6.12 $\quad C_M = f(k) \cdot \dfrac{d_c^2}{D_M}$ 	Golay expression for resistance to mass transfer in the gas phase.

6.13 $\quad f(k) = \dfrac{1 + 6k + 11k^2}{96\,(1+k)^2}$ 	$f(k)$ for the Golay C_M term in Equation 6.12.

New technical terms

If you read the whole chapter, you should now know the meaning of these technical terms:

column efficiency curve **migration rate**
dispersion **obstruction factor**
eddy diffusion **resistance to mass transfer**
extra-column variance **time efficiency**
Golay equation **van Deemter curve**
longitudinal diffusion **van Deemter equation**

For more information, refer to the Glossary at the end of the book.

7

Extracolumn broadening

"Peaks also get wider as they travel through flow passages outside the column. Nearly every process gas chromatograph incorporates at least one sample injection valve, one column switching valve, one detector, and various lengths of connecting tubing that run between these devices and the columns. These extracolumn flow paths cause some additional peak broadening."

Introduction

When our process chromatographs used quarter-inch outside diameter columns, we didn't have to worry about peaks getting wider as they traveled through the one-sixteenth-inch connecting tubes or became exposed to the internal volumes of sample valves and detectors. That was a long time ago.

By 1962, many PGCs were using three-sixteenth-inch columns, and soon thereafter one-eighth-inch columns became the *de facto* standard for all brands of PGC. This quiet change reduced the internal volume of columns by a factor of 6.8, yet we still had no concern about external volumes. Everything worked well because the volume of external devices, notably detectors, had also changed.

Turns out, we were lucky back then. Later, when we tried to use capillary columns in PGCs, we discovered that the old way of connecting columns to clunky valves and detectors was inadequate. The whole instrument had to change.

Capillary columns can produce astounding separations, but only with very small samples. As peak volumes became smaller, the internal volume of the flow-path external to the column became critical. There's no point in generating 100,000 plate peaks if they mix with the carrier gas on their way to the detector, or if the detector is so big that adjacent peaks are both in the detector at the same time!

A modern PGC, even one using micropacked columns, will inject a very small sample, so peak broadening in the flow passages external to the

---

Process Gas Chromatography: Advanced Design and Troubleshooting, First Edition. Tony Waters.
© 2025 John Wiley & Sons Ltd. Published 2025 by John Wiley & Sons Ltd.

column is always a concern. To ensure the most efficient system design and to properly maintain that system on-site, it's important for everyone to understand the deleterious effect of extracolumn volumes on the efficiency of separation.

Peaks get wider because their molecules drift apart, or *disperse*, with time. Two sources of **peak dispersion** account for most of the broadening experienced by a peak while it's outside the columns:

> the sample volume and its injection profile.

> the geometry of the flow passages that the sample molecules encounter between and within injectors, valves, and detectors.

In older equipment, the response time of the detector and its signal-processing circuits also distorted the peaks. Modern detectors respond much faster than their ancestors did, but detection speed may still limit the measurement of very fast peaks from capillary columns.

Sample injection

Sample injection is a crucial function of a PGC – the reliability, repeatability, and accuracy of the entire analyzer depend on it. The effect of sample injection on column performance rests on two significant variables: the **sample volume** and the **injection profile**.

The injection profile is an imagined plot of sample concentration versus time, as seen at the column entry point. Figure 7.1 is an idealized example. After the sample injection event, there's a short transport delay for the sample to travel to the head of the column, arbitrarily shown as 400 ms in the figure. Ideally, the sample concentration in the carrier gas then rises instantly from 0 to 100 % and soon thereafter falls rapidly back to 0 %. We call this ideal condition a rectangular injection profile, or more colloquially a **plug injection**.

A plug injection has the shortest duration possible for a given volume of sample and has two distinct advantages:

> The narrow injection profile adds little to the inevitable dispersion of molecules in the column, thereby ensuring narrow chromatogram peaks.

> For a given amount of solute, narrow peaks generate more detector signal than wide peaks do, thereby improving signal-to-noise ratio and analyte detectability.

Clearly, the duration of a plug injection depends on the volume of sample injected. A large volume of sample gas takes more time to enter the column than a small volume does. This extended entry time distributes the analytes over a longer stretch of column, making the chromatogram peaks wider than they would otherwise be.

It's wise to keep the sample volume small and the injection fast.

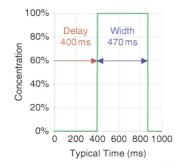

Figure 7.1 An Ideal Injection Profile.

Sample volume

The injected sample volume must be sufficiently small to avoid overloading the column, yet large enough for the detector to provide the desired sensitivity. A large sample volume may overload the stationary phase, distorting peaks on the chromatogram. This distortion can take many forms, including fronting, tailing, or even flat-topped peaks as seen in Chapter 8.

All columns have a maximum injection volume to avoid peak distortion. It depends partly on the limits to solute solubility (or adsorption) and partly on the width of the injection profile.

Solubility limit

Symmetrical peak shapes in gas chromatography occur only when the ratio of analyte concentrations in the liquid and gas phases is constant. This ratio is the distribution constant (K_c). Yet there's always an upper limit to the injected amount of an analyte. This upper limit is the *sample capacity* of the column, a concept first introduced in Chapter 3. Injecting more analyte than the sample capacity of a column exceeds the linear range of the distribution constant and causes peak distortion.

Sample capacity is a column property that is specific to a given analyte and may vary widely according to the solubility of the analyte in the liquid phase, the amount of liquid present in the column, and the linearity of the analyte partition isotherm. Refer to Chapter 8 for more discussion on the latter.

Recall that the *retention factor* (k) of a peak is a function of the solubility of that analyte in the liquid phase and the amount of liquid phase present in the column. Therefore, you can improve the sample capacity of your column for a specific peak by increasing its retention factor. Your options are:

➢ Choose a liquid phase that strongly retains the analyte.
➢ Use a higher liquid loading or thicker film thickness.
➢ Operate the column at a lower temperature.

Peak width

The second factor affecting sample capacity is the width of the analyte peak. Think of a peak like a constant-area isosceles triangle: a wider base means a lower apex. Therefore, a wider peak can accept a larger quantity of analyte without exceeding the concentration limit.

> *Since sample capacity is proportional to peak width, very narrow peaks always have low sample capacity.*

In Chapter 2, we found that the base width of a peak (w_b) depends on the plate height (H) and the column length (L) per Equation 2.17:

$$w_b = 4\sqrt{H \cdot L}$$

And from Equation 2.14:

$$H = \frac{L}{N}$$

Therefore,

$$w_b = \frac{4L}{\sqrt{N}} \tag{7.1}$$

Thus, the width of a peak varies inversely by the square root of its plate number (N). A capillary column with a high plate number will provide excellent resolution but the trade-off for those narrow peaks is a low sample capacity; highly efficient columns always need miniscule samples. This is why PGCs using capillary columns tend to employ the less efficient "megabore" columns having thicker liquid films.

Overall effect

We have seen that the permissible injected volume (V_{max}) of an analyte increases with peak retention volume and reduces with the square root of its plate number. Hence, we can write:

$$V_{max} = \frac{a \cdot V_R^\circ}{\sqrt{N}} \tag{7.2}$$

Wherein:

> The **corrected retention volume** (V_R°) of a peak is its retention volume (V_R) corrected for the average pressure in the column. This procedure uses the *gas compressibility factor* (j) as fully explained in Chapter 9.

> The empirical constant (a) relates to the pairing of an analyte with a liquid phase and is a function of analyte solubility and the linear extent of its partition isotherm. Reported values of (a) range between 0.02 and 0.6 (Annino and Villalobos 1992, 60).

To discover the constant (a) for a specific application, reduce the sample size until there is no further improvement in resolution. After finding the optimum sample volume, though, it's unlikely that anyone would want to go back to the equation!

Before moving on, realize that sample capacity calculated by Equation 7.2 is due to *solubility limits* and is applicable only to a specified analyte on a specified stationary phase at a specified temperature. Moreover, sample capacity limits only the quantity of that analyte in the injected sample; if a sample contains a lower percentage of the analyte, the injected sample volume can be larger.

Worked example

Sample capacity

As an example, consider a peak with a retention volume of 200 mL eluting from a column having 10,000 plates.

Using the most conservative value cited for the constant (a), Equation 7.2 indicates a maximum injection volume of 0.04 mL:

$$V_{max} = \frac{0.02 \times 200 \text{ mL}}{\sqrt{10{,}000}}$$

$$= 0.04 \text{ mL}$$

Note that this is the injected volume of a particular component. If the concentration of that component in the sample gas is 10 % by volume, it would be okay to specify a sample volume of 0.4 mL.

Feed volume

The technical literature often quotes Equation 7.2, but not always for the right reason. A similar equation can calculate the maximum sample volume from an entirely different criterion: the number of plates (and therefore the length of column) that the injected sample occupies. This is the *feed volume* limit for a column also introduced in Chapter 3.

Although the feed volume equation looks very much like our Equation 7.2, it isn't based on solubility and has nothing in common with sample capacity. Instead, it uses a principle from statistical theory:

> To have an insignificant effect on a peak, the average duration of the injection profile should not exceed one-half of the standard deviation of the chromatogram peak: that is, one-eighth of its base width (Purnell 1962, 110).

Applying this constraint to the theoretical plate equation yields an expression for the maximum duration (t_{max}) of the injection profile.

Recall Equation 2.26, which computes plate number (N) from retention time (t_R) and base width (w_b) in time units:

$$N = 16 \left(\frac{t_R}{w_b} \right)^2$$

Therefore,

$$w_b = \frac{4 t_R}{\sqrt{N}}$$

Apply the constraint ($t_{max} = w_b/8$):

$$t_{max} = \frac{t_R}{2\sqrt{N}} \tag{7.3}$$

Therefore,

$$\frac{t_{max}}{t_R} = \frac{0.5}{\sqrt{N}} \tag{7.4}$$

The ratio (t_{max}/t_R) is the maximum duration of the injection profile as a fraction of the retention time. Therefore, the maximum number of plates (N_{max}) that the injected sample volume should occupy is:

$$N_{max} = \frac{t_{max}}{t_R} \cdot N$$

Substituting from Equation 7.4:

$$N_{max} = 0.5 \sqrt{N} \tag{7.5}$$

This is a quick way to estimate the maximum feed volume of a column.

When expressed as a volume, the duration of a plug injection is the feed volume of the column. Although calculated for a particular peak, it's about the same for any peak on the chromatogram. To calculate the maximum feed volume (V_{max}) of a column, apply the fraction (t_{max}/t_R) to the corrected retention volume (V_R°) of the peak:

$$V_{max} = \frac{t_{max}}{t_R} \cdot V_R^\circ$$

Substituting again from Equation 7.4:

$$V_{max} = \frac{V_R^\circ}{2\sqrt{N}} \quad (7.6)$$

Clearly, the above equation is equivalent to Equation 7.2 with the constant set at ($a = 0.5$), yet it is based on quite different constraints. Both equations are valid.

Note that the feed volume equations calculate the *maximum sample volume* from the duration of a plug injection, regardless of the analyte concentrations. They neither consider the limits to solubility nor calculate the sample capacity for individual solutes.

Practical outcomes

Purnell examined the permissible sample volume (V_{max}) calculated from Equation 7.6 and concluded that it would overload any column (*ibid*, 167). Therefore, analyte solubility in the liquid phase is more likely to be the controlling factor, not the feed volume.

In practice, the two constraints are inseparable. Experimental estimates of the constant (a) in Equation 7.2 automatically include the influences of solubility *and* feed volume, so it's the preferred equation to use.

You can use the calculated feed volume as a starting point for estimating the permissible sample volume, but it might be excessive. Usually, a smaller sample volume will be necessary in practice.

All the equations confirm that high-efficiency columns demand smaller samples. They predict that the maximum injection volume decreases in proportion to the square root of increasing plate number. This concurs with experience: a highly efficient capillary column can accept far less sample than a typical packed column can.

In addition, the solubility equations show that the maximum sample volume of a substance depends on the retention volume of that substance. Therefore, when you need a high detector sensitivity it might help to increase the retention volume of an analyte, so the column can accept a larger sample.

One way to increase the retention volume is to increase the internal diameter of the column. A wider column tube will allow a larger sample and increase the overall sensitivity of the PGC. Unfortunately, however, the wider column may be less efficient, making the desired resolution more difficult to achieve.

Luckily, there are some other ways to increase sample capacity. The retention volume of a peak also depends on the chosen liquid phase, the volume of that liquid in the column, and the column operating temperature. To increase the sample capacity of the column, use a higher liquid loading, a lower oven temperature, or even a different liquid phase.

For a given column, the only way to be sure about your sample volume is to run several chromatograms with different sample volumes and observe the detector sensitivity and peak resolution accomplished. For the latter, it would be sufficient to look at the depth of the valley (or the extent of baseline separation) between the closest analyte peaks on the chromatogram and note the sample volume that significantly degrades their resolution. Then, choose a sample volume that provides good detector sensitivity and adequate resolution between the measured peaks.

> *Since each peak may have its own sample capacity, ensure that each analyte peak can reach its full-scale concentration without significant distortion.*

When measuring very low analyte concentrations, it's inevitable that the large sample size will distort the *unmeasured* major component peak. This distortion is normal and is acceptable provided that the injected volume of each analyte doesn't exceed its own maximum.

Chapter 11 introduces column-switching techniques to circumvent the major peak distortion.

In summary, a _smaller sample volume_ is appropriate when:

- The analyte has a higher concentration.
- The analyte peaks appear earlier on the chromatogram.
- The column has a higher efficiency.
- The column has a narrower bore (for example, a capillary column).
- The column is providing inadequate resolution.

Whereas a _larger sample volume_ is appropriate when:

- The detector has insufficient sensitivity to measure the analytes.
- The detector signal-to-noise ratio is too low.
- The analyte peaks appear later along the chromatogram.
- The column has a wider bore (for example, a packed column).

Sample size calculations

Duration of injection

The duration of a plug injection (t_i) is easy to compute. It's a simple function of the vapor sample volume (V_i) entering the column and the volumetric flow rate ($\dot{V}_c$) of the carrier gas inside the column. If the measurements

Worked example

Duration of a plug injection

Consider a gas sample volume of 0.30 mL filled at atmospheric pressure, then injected into a packed column at 100 °C. The carrier gas flow rate is 30 mL/min measured at atmospheric pressure and 20 °C.

By Equation 7.8, the plug injection width (t_i) is:

$$t_i = \frac{V_s}{\dot{V}_c} \cdot \frac{P_i}{P_f} \cdot \frac{T_f}{T_i}$$

The pressure is constant, so:

$$t_i = \frac{0.30 \text{ mL}}{30 \text{ mL/min}} \times \frac{293 \text{ K}}{373 \text{ K}}$$
$$\times \frac{60 \text{ s}}{1 \text{ min}}$$
$$t_i = 0.47 \text{ s}$$

Figure 7.1 illustrates this injection profile.

---

of sample gas volume and carrier gas flow rate are at the same temperature and pressure, no correction is necessary:

$$t_i = \frac{V_i}{\dot{V}_c} \qquad (7.7)$$

Otherwise, correct the sample volume for differences in pressure or temperature inside the column entry (P_i, T_i) and inside the flowmeter (P_f, T_f).

$$t_i = \frac{V_i}{\dot{V}_c} \cdot \frac{P_i}{P_f} \cdot \frac{T_f}{T_i} \qquad (7.8)$$

Gas sample injections are often close to ideal, provided the analyte molecules don't adsorb onto the tube walls. Upon injection, the carrier gas compresses the sample gas and then pushes it onto the column.

Liquid sample injections tend to be slower because the liquid must vaporize and this takes time. The injection profile for a liquid sample is more likely to resemble a tailing chromatogram peak than a square wave. The slower injection inevitably causes wider peaks. The following SCI-FILE gives an equation to calculate the dispersion caused by the injection profile, whether it's rectangular, exponential, or Gaussian.

In summary, the width of the injected band of molecules usually decreases with decreasing sample size, thereby reducing chromatogram peak widths. Although column efficiency may improve with smaller sample size, the dispersion occurring in the connecting tubes and detector will remain the same – and will be responsible for a larger percentage of the chromatogram peak width.

An example

The worked example on the duration of a plug injection (see margin) may help you to visualize the sample injection profile for a 300 µL plug injection on a typical column and illustrates the calculations involved. The previously shown Figure 7.1 illustrates the outcome: after a short transport delay, the sample enters the column as a compact 470 ms pulse.

The academic theories of chromatography assume a plug injection no wider than a single plate (Perry 1981, 27). Clearly, this is an ideal notion not achievable in practice. Yet it's evident that the plug duration should be as short as possible to get the narrowest peak widths on the final chromatogram.

Let's see how close our worked example gets to meeting the theoretical ideal. How many plates does the 470 ms sample occupy?

For that, we need the plate height. An average plate height for a packed column is about 0.6 mm. To convert the plate height from length units (H) to time units (H_t), use Equation 2.21:

$$H_t = H \cdot \frac{t_R}{L}$$

For example, say the column length (L) is 6 m and our peak of interest has a retention time (t_R) of 300 s. Its plate height in time units is:

$$H_t = 0.6 \text{ mm} \times \frac{300 \text{ s}}{6{,}000 \text{ mm}} = 30 \text{ ms}$$

Now we see the effect of the 470 ms plug injection. The plate height is 30 ms, so the injected sample occupies about 16 plates. While this is certainly not a single-plate injection, it's very narrow.

From Equation 2.24, the plate number (N) of this peak is 10,000:

$$N = \frac{t_R}{H_t} = \frac{300 \text{ s}}{30 \text{ ms}}$$
$$N = 10{,}000$$

So from Equation 7.5, the permitted occupation is 50 plates:

$$N_{max} = 0.5\sqrt{N} = 0.5 \times \sqrt{10{,}000}$$

Therefore,

$$N_{max} = 50$$

In conclusion, the plug injection spread of 16 plates easily complies with the theoretical limit for maximum feed volume, but the injected sample volume may still exceed the sample capacity of the column. That can be determined only by experiment.

Flow path geometry

Each device and each connecting tube in the carrier flow path allows more dispersion of the sample molecules, increasing the width of the chromatogram peaks. Because the extracolumn volumes tend to be constant in each PGC analyzer, they become increasingly significant when you reduce the column bore and sample size.

> *Narrow-bore columns require smaller sample volumes. But the peaks are then smaller and need even smaller volumes in connecting tubes and detectors!*

As the sample molecules pass through the system from injector to detector, they may encounter three distinct flow-path geometries that worsen their dispersion:

➤ Small-diameter open tubing and borings that carry the flow between devices.

➤ Wider flow passages and larger volumes where mixing can occur.

➤ Unswept side volumes where static gas can collect – often called *dead volumes*.

Open tubing

The dispersion in open tubing is not severe, provided that the tubing does not have a larger internal diameter than the column tube. You can easily calculate the time delay (t_{ot}) in the tubing by assuming **plug flow** prevails and adapting Equation 7.7:

$$t_{ot} = \frac{V_{ot}}{\dot{V}_c} \tag{7.9}$$

A worked example at the end of this chapter illustrates the use of this calculation.

In practice, the gas in the tubing is in laminar flow, not plug flow. The laminar flow profile causes additional dispersion of the peaks due to the wide range of carrier gas velocities occurring across the radius of the tube. In most cases, this additional dispersion is insignificant.

The following SCI-FILE shows that peak dispersion in empty tubes is proportional to the length of the tube, but it's also proportional to the *fourth power* of the diameter. Don't use large tubing!

The important points to note are:

➤ The internal diameter of flow passages should be small and never greater than the internal diameter of the column tube.

➤ However, the connecting tubes should not be small enough to cause blockage or a noticeable pressure drop.

➤ The connecting tubes should be as short as possible.

Mixing chambers

When a peak flows into a larger chamber, it tends to mix with the carrier gas causing an exponential delay. This mixing causes 12 times as much dispersion than caused by plug flow in tubes of constant diameter.

In addition, the gas entering the **mixing chamber** is moving at higher velocity than the gas in the chamber. This may cause a complex turbulent flow pattern that drastically multiplies the peak width, causing gross peak distortion and loss of resolution.

To avoid these adversities, it's wise to avoid any change in diameter along the flow path. Try to maintain a constant diameter from the point of injection to the point of detection.

Dead volumes

Another very poor design feature is a side volume that is not swept clean by the flow of carrier gas. Usually, a **dead volume** is a tee junction in the flow passage where one leg of the tee is not flowing. Such visible dead volumes were present in older detector designs, when they were not a big problem because of the enormous sample volumes used. As samples became smaller, the old detectors proved inadequate and had to change. The effect of a dead volume always worsens as the peak volume becomes smaller.

Visible dead volumes no longer occur in modern PGC equipment, but small pockets may persist at tube connections, inside valves and detectors, or even in the walls of the tubing. Unswept pockets are particularly important because their effect on a peak depends on the *fourth power* of the depth of the pocket. For the smallest peaks, the connecting tubing should be internally smooth, silicon coated, and butt jointed to valves and devices using special fittings.

Detectors

Three very different effects contribute to peak broadening in detectors.

The *first contribution* is the dispersion that occurs in the detector flow passages. The internal flow paths in a detector are much the same as the connecting tubes, mixing chambers, and dead volumes already discussed. These detrimental effects all depend on the geometry of internal volumes. Consequently, the development of detectors has focused on simplifying their flow paths and reducing their volumes.

The *second contribution* to peak broadening is quite different and is an inherent feature of **concentration-sensing detectors** like the thermal conductivity detector (TCD). While detecting a peak, the concentration of analyte entering the detector is rapidly changing, but the detector can respond only to the *average concentration* in its sensing volume. Therefore, the internal volume of the detector causes a lag in response that extends the true peak width. To faithfully track a peak, the sensing volume must be small in comparison to the peak volume. This source of peak broadening doesn't occur in **mass-sensing detectors** like the flame ionization detector (FID). These detectors respond to the instantaneous mass flow of analyte exiting the column and no averaging occurs. It's even possible to add a diluent gas to the column effluent to accelerate its passage through the detector with no effect on the true peak width.

The *third contribution* to peak broadening originates from the limited speed of response that to some extent afflicts all detectors. In addition to the sensor speed of response, the time constant of a detector includes the ability of the electronic circuits to follow the rate of change of the signal. When the electronic response lags the true peak profile, it increases the width and asymmetry of the peak and reduces its resolution from adjacent peaks. In older equipment, the signal-processing circuits may respond too slowly to faithfully follow fast peaks, even those from packed columns. Electronic circuits are now fast enough to follow the response from any detector; so fast, in fact, that the response may be dampened to reduce detector noise. Then, the dampened signal may not be as fast as expected (Waters 2020, 283).

The theory

A little more math is necessary to figure out the effect of extracolumn dispersion on the chromatogram peak widths. The theoretical approach is to treat all sources of dispersion outside the column in the same way as we treat sources inside the column. To accomplish that, the theory assigns the statistical parameter *variance* to the dispersion that peaks experience when they pass through each type of flow passage, including the column itself. Since multiple and independent variances are additive, the total variance foretells the width of the final chromatogram peak.

Any dispersion occurring external to the column increases the width of the peaks and spoils resolution. The SCI-FILE *On Extracolumn Variance* provides some procedures for estimating the effect of sample injection profiles and flow path geometries on the final resolution of peaks. Feel free to follow along; the math is not difficult.

SCI-FILE: *On Extracolumn Variance*

Introduction

Inside the column, a peak is an unstable concentration of molecules surrounded by pure carrier gas, so it's continuously spreading and getting wider as it travels along the flow path from injector to detector. Various diffusion and velocity effects accentuate that dispersal.

In Chapter 2, we noted that the width of a peak is a function of standard deviation (σ), a measure of the distribution of its molecules. By definition, the *variance* of the distribution is equal to the square of the standard deviation (σ^2). Thus, variance is a measure of dispersion.

In Chapter 6, we discovered that rate theories of chromatography identify several causes of peak dispersion within the column and then rely on the statistical principle that independent variances are additive. From this, Equation 6.2 expresses the total peak variance occurring in the column as the sum of the contributing variances:

$$\sigma^2 = \sigma_1^2 + \sigma_2^2 \cdots + \sigma_n^2$$

In a similar way, we express dispersion occurring *external to the column* as variances and then add them to those caused by the column itself. Some external sources include the sample injection valve, the column switching valves, the detector, and the tubing that connects these devices.

Collectively, these deleterious effects are known as extracolumn sources of peak broadening. They derive from the physical design of the PGC instrument itself.

In the end, it doesn't matter whether the dispersal occurs inside the column or outside the column. The effect is exactly the same, so extracolumn variances and intracolumn variances are equally important. Both increase the width of the chromatogram peaks and reduce resolution.

Acceptable dispersion

To evaluate the effect of additional variance, adapt Equation 2.18 to show that peak variance (σ^2) is the product of column length (L) and plate height (H):

$$\sigma^2 = L \cdot H \qquad (7.10)$$

From this, a 10 % increase in the variance of a peak will also increase plate height by 10 %.

Equation 2.19 gives the base width (w_b) of a peak as:

$$w_b = 4\sqrt{H \cdot L}$$

Therefore, a 10 % increase in plate height widens the peak by a factor of $\sqrt{1.1}$, which in turn reduces the resolution by the same factor; that is, about 5 %.

Klinkenberg (1960, 182-183) was the first to suggest that 10 % extra plate height is a sensible limit for extracolumn variance, a limit generally accepted by laboratory workers. Additional variance is never welcome in a PGC because it results in longer analysis times, particularly in applications using efficient capillary columns. Although some extracolumn variance is inevitable, we must strive to minimize it.

Sources of variance

There are several places in a PGC system where additional peak dispersion can occur. Here are the five main sources:

- The sample injection.
- Flow through empty tubes and tube fittings.
- Mixing volumes in the flow path.
- Dead volumes along the flow path.
- Response speed of detectors and signal-processing circuits.

To quantify the effect of these dispersions on the chromatogram peak shape, it's more convenient to work in time units. By analogy with Equation 6.2, the total peak variance in time units (τ^2) is the sum of the independent time variances experienced by the peak, whether they occur inside or outside of the column:

$$\tau^2 = \tau_1^2 + \tau_2^2 \cdots + \tau_n^2 \qquad (7.11)$$

Warning

The concept of variance is challenging, and not all published works are correct. Part of the problem lies in the units of measure. Variance can be expressed in units of length (mm^2), volume (mL^2), or time (s^2). When reading the literature be careful to note the units used in equations, as different units change the equations given.

Sample injection profile

Sternberg (1966, 231) derived equations for calculating the variance added by sample injection. His equations use the shape of the injection profile and it's time constant. He identified six different shapes for the injection profile, of which only three are likely to occur in a PGC.

Plug injection

A valve injection of a gas or a rapidly vaporizing liquid usually results in a plug injection and the time constant is simply the duration (t_i) of the inlet plug per Equation 7.7:

$$t_i = \frac{V_i}{\dot{V}_c}$$

Then, the additional variance (τ_i^2) generated by the injection profile is:

$$\tau_i^2 = \frac{1}{12} \cdot \left(\frac{V_i}{\dot{V}_c}\right)^2 \qquad (7.12)$$

Exponential injection

During the injection of a slowly evaporating liquid, the injection profile will decline exponentially. The time constant (t_{63}) is the time for the sample concentration to drop by 63 %. This time is more difficult to estimate, but it's clearly longer than a plug injection.

The variance (τ_i^2) generated by the injection profile is then:

$$\tau_i^2 = t_{63}^2 \qquad (7.13)$$

Gaussian injection

An injection profile that is Gaussian in shape adds the least variance. Such an injection occurs when a peak enters the column from another column.

The additional variance (τ_i^2) generated by the injection profile is:

$$\tau_i^2 = \frac{1}{36} \cdot \left(\frac{V_i}{\dot{V}_c}\right)^2 \qquad (7.14)$$

If the columns are close coupled, the additional variance is often negligible.

What we learn

As expected, Equation 7.14 predicts that smaller sample volumes will provide narrower peaks. But it's the ratio between sample size and the carrier gas flow rate that ultimately determines the incremental peak width due to sample injection.

This shows why capillary columns need very small samples. If running at the same velocity, a column that has half the bore size has only one-fourth of the flow rate. Therefore, to maintain the same injection variance, the smaller column can accept only one-fourth of the sample volume.

Even then, the smaller sample size is only enough to maintain the *same* injection variance. In practice, the smaller column will produce narrower peaks, so the same amount of variance may cause an excessive widening of the peaks. The worked examples at the end of this chapter explore this critical relationship between additional variance and chromatogram peak width.

Dispersion in open tubes

A peak also suffers dispersion when it passes through an empty tube or a tubular volume within the injector, column valve, or detector.

A simple approach is to assume plug flow occurs in the tubes, and then adapt Equation 7.12 to estimate the additional variance caused by the plug flow (Annino & Villalobos 1992, 67).

Then, the variance (τ_{ot}^2) caused by the open tube depends on its internal volume (V_{ot}) and the volumetric flow rate ($\dot{V}_c$):

$$\tau_{ot}^2 = \frac{1}{12} \cdot \left(\frac{V_{ot}}{\dot{V}_c}\right)^2 \quad (7.15)$$

Laminar flow

The assumption of plug flow is reasonable for most purposes, but the actual flow profile is laminar at the low carrier flow rates used in chromatography.

Laminar flow is quite different from plug flow; in laminar flow, the gas velocity at the center of the tube is twice the average. It then drops to near zero at the walls. This difference in velocity tends to spread the molecules over time and distance, thereby increasing peak width.

Dispersion also occurs by diffusion of molecules from regions of high concentration to regions of low concentration. In an empty tube, dispersion occurs in two directions, by **axial diffusion** and **radial diffusion**. To a small extent, the mixing induced by radial diffusion tends to diminish the effect of velocity differences.

These velocity and diffusion effects are the same as those that occur in open-tubular columns, so the gas diffusion terms of the Golay equation give a better estimate of the open-tube variance.

To derive a Golay expression for the plate height (H) of the open tube, start with Equation 6.11 and omit the A term and the C_S term as these don't apply to an open tube:

$$H = \frac{B}{u} + uC_M \quad (7.16)$$

Then, substitute for B from Equation 6.8 omitting the inapplicable packing factor (ψ) and substitute for C_M from Equations 6.12 and 6.13 noting that the retention factor (k) is zero in an open tube:

$$H = \frac{2 D_M}{u} + \frac{ud^2}{96 D_M} \quad (7.17)$$

where:

> d is the internal diameter of the open tube.

> D_M is the coefficient of diffusion of the solute in the gas phase.

> u is the carrier gas velocity at column exit.

In Equation 7.17, the first term accounts for dispersion due to longitudinal (axial) diffusion, the second term for dispersion caused by a combination of radial diffusion and the wide range of gas velocities in laminar flow.

In small empty tubes, we find that $u \gg D_M$, so the first term is negligible, leaving:

$$H = \frac{ud^2}{96 D_M} \quad (7.18)$$

This gives the plate height of the tubing in length units. To convert the plate height to variance (σ_{ot}^2) per Equation 2.18, simply multiply by the tube length (L_{ot}):

$$\sigma_{ot}^2 = L_{ot} \cdot \frac{ud^2}{96 D_M} \quad (7.19)$$

Then, to get the variance in time units (τ_{ot}^2) per Equation 2.10, divide by velocity squared:

$$\tau_{ot}^2 = \frac{L_{ot}}{u} \cdot \frac{d^2}{96 D_M} \quad (7.20)$$

To rewrite this equation as a function of the flow rate, substitute $u = 4\dot{V}_c/\pi d^2$:

$$\tau_{ot}^2 = \frac{\pi d^2 L_{ot}}{4 \dot{V}_c} \cdot \frac{d^2}{96 D_M}$$

Hence:

$$\tau_{ot}^2 = \frac{\pi d^4 L_{ot}}{384 \dot{V}_c D_M} \quad (7.21)$$

Giddings (1963) takes another approach to the variance added by connecting tubes, deriving the plate height in open tubing (H_{ot}) as:

$$H_{ot} = \frac{L_c}{N_{ot}} \cdot \frac{V_{ot}}{V_R^\circ} \quad (7.22)$$

By introducing the column length (L_c) and the corrected retention volume (V_R°), the equation allows for the peak shape passing through the open tube. This tactic is of theoretical interest, but it doesn't help with practical PGC applications.

What we learn

The variance caused by tubing increases by the *fourth power* of the tube diameter, but only by the first power of its length.

Beware of wide tubing! The bore of the connecting tubes should be narrow, never wider than the column tubing, but not so narrow that it causes pressure drop or blockage. Keep the tubes short.

Dispersion by mixing

A mixing chamber is an undesirable larger-diameter tube or vessel in the flow path, where mixing occurs. Solute molecules disperse when they mix with carrier gas, increasing the peak width. The exit profile from a mixing chamber is exponential.

Sternberg derives an equation for the variance due to an inline mixing chamber. The time constant is the volume of the chamber (V_{mc}) divided by the flow rate passing through it, and the variance (τ_{mc}^2) is:

$$\tau_{mc}^2 = \left(\frac{V_{mc}}{\dot{V}}\right)^2 \quad (7.23)$$

But a mixing chamber implies a change in diameter of the flow path. When a gas flows from a narrow passage into a wider one, turbulence occurs in the larger tube. The turbulence persists for a distance of about eight tube diameters, after which laminar flow resumes (Maynard and Grushka 1972).

Based on this information, one might consider the first eight diameters of the larger tube to be a mixing chamber, followed by normal laminar flow.

Clearly, the behavior at a diameter change is rather complex. Sternberg identifies four possible behaviors, depending on tube layout and flow rate. We have already discussed three of these behaviors: laminar flow, mixing effect, and peak diffusion. The fourth behavior is dispersion, which is dominated by lateral diffusion between the inlet jet and the more static gas around it. This kind of dispersal would atypically *multiply* the entering peak variance (τ_{in}^2) by the fourth power of the diameter change:

$$\tau_{out}^2 = \tau_{in}^2 \cdot \left(\frac{d_2}{d_1}\right)^4 \quad (7.24)$$

Accordingly, a peak entering a tube that is twice as wide would suffer a 16-times increase in variance and a fourfold increase in width. This is enormous and illustrates the potential disaster of using wide connecting tubes.

Luckily, the actual behavior is likely to be a complex mix of the four limiting cases. But seeing that fourth power *multiplier* should convince you to avoid an increase in diameter at any point in the flow path.

Mixing chambers have the same effect wherever they are in the flow path.

What we learn

The additional dispersion due to a simple mixing chamber creates 12-times the variance found in open tubes.

The concomitant diameter change has a complex effect on variance and might cause a disastrous increase in peak width.

Mixing chambers are highly undesirable in the flow path and should be eliminated from PGC equipment.

Dispersion in dead volumes

A dead volume is an unswept cavity in the flow path that traps analyte molecules and holds them for a while. The molecules exit the dead volume exponentially and thereby increase peak tailing and asymmetry.

Sternberg gives an equation for the variance added by a dead space, showing that it depends upon the pocket depth (d_{dv}), and the diffusion coefficient (D_M) of the solute in the carrier gas:

$$\tau_{dv}^2 = \frac{d_{dv}^4}{4\,D_M^2} \qquad (7.25)$$

Interestingly, the variance caused by a dead volume is independent of the carrier gas flow rate, whereas the variance caused by a mixing volume lessens with increased flow. These different behaviors might offer a practical diagnostic test.

Carrier gas density is higher at the beginning of the column, reducing the diffusion rate. A dead volume at the column inlet therefore has a more severe effect than a similar dead volume at the column outlet.

What we learn

A dead volume in the flow path is one of the most damaging deficiencies found in PGC design, particularly when located at the column inlet.

The devastating effect of a dead volume comes from its fourth-power dependency on the depth of the cavity. A deep pocket or side volume along the flow path will severely distort the peaks and lead to loss of resolution.

For example, the very small dead volumes in standard tube fittings had a minimal effect on the efficiency of packed columns, but when capillary columns came along, special tube connectors became necessary.

Dispersion in detector

Mass flow detectors

A mass-sensing detector like the FID has virtually zero sensor volume. If the internal tubing to the jet is not wider than the column and connection tubing, it will have almost no effect on the peak shape. If hydrogen is added to the column effluent, the much-increased flow rate will effectively eliminate any open-tube or mixing chamber dispersion.

The integrated area of a peak from a mass-flow detector is unaffected by peak shape because the total mass entering the sensor is the same, no matter how it's distributed over time.

Concentration detectors

A concentration-sensing detector like a TCD often has wide flow passages and the sudden increase in width can cause catastrophic peak dispersion, as previously discussed. A high-performance TCD has no dead volumes and a small diameter flow path.

A TCD responds to the moving-average concentration of analyte molecules in contact with its sensor. The variance so caused depends on the effective volume of gas (V_{de}) that continually acts on the sensing element. The variance (τ_{de}^2) added by this surrounding volume of gas follows the form of a plug injection:

$$\tau_{de}^2 = \frac{1}{12} \cdot \left(\frac{V_{de}}{\dot{V}_c}\right)^2 \qquad (7.26)$$

What we learn

Ideally, the detector flow passages should be no larger than the columns and connecting tubes.

The FID comes close to achieving this goal, but it's more difficult for a TCD to comply.

Table 7.1 lists four examples of detector volume necessary to get a detector variance equal to 5 % of the peak variance. Few PGC detectors can boast internal volumes as small as these.

Response speed

The net response time of detection is a combination of sensor reaction time and signal processing time.

Sensor reaction time depends on the rapidity of the detecting principle, which might include thermal or capacitive time delays. For instance, older TCDs were too slow to accurately follow a fast hydrogen peak, but extensive development has improved the TCD response time to rival that of an FID.

Signal processing time depends on the dynamic response of electronic or pneumatic signal transmission and recording mechanisms. Digital processing is very fast, but PGC peak integration and noise cancellation algorithms may dampen the final response.

The combined sensing response characteristic is typically exponential, so it increases the asymmetry and width of a peak, bringing down its height.

This loss of height is of concern in older equipment that measures analyte peak height. For a peak with a base width of one second, a detector time constant of 50 ms is enough to drop the peak height by about 2 % (Sternberg 1966, 258).

The increased peak width doesn't affect peak area measurement, but it does reduce peak resolution.

The sensing time lag also increases the observed retention time. It may be useful to know that the top of a delayed peak always falls on the original peak curve (albeit on the trailing side), regardless of the degree of signal distortion.

What we learn

For very fast peaks, there may be a problem with the electronic speed of response. Be careful when setting noise suppression parameters to avoid excessively dampening the response.

Practical limits

From the theory, several authors estimate the limiting conditions for each source of variance that will ensure acceptable contributions to the total variance (for example, Scott 1998, 96).

McGuffin (2004, 34) accepts 12 % of the total variance as a reasonable budget for extracolumn effects. She then assigns 5 % each to the injector and detector, and 1 % each to the connecting tubes and electronics. Table 7.1 lists a small selection of her calculated data on four typical columns working at their optimum flow rates.

In the table, the optimum plate height (H_{min}) was assumed to be equal to the particle diameter for a packed column, or equal to the internal diameter of a capillary column. This simple way to predict the ultimate efficiency of a column may overestimate the performance of a real column.

These data confirm that the internal diameter of connecting tubing is far more important than its length. In the table, the connecting tubes used for capillary columns are the same bore as the column, to avoid any turbulence due to diameter change.

Table 7.1 Extracolumn Limit Parameters.

Variable and Unit		Packed	Open Tubular		
L	m	2.0	10	10	10
d	mm	4.0	0.50	0.32	0.20
$\dot{V}$	mL/min	64	1.2	2.1	2.0
H_{min}	mm	0.25	0.50	0.32	0.20
N		8,000	20,000	31,000	50,000
Acceptable parameter limits					
V_i	µL	110	6.2	2.0	0.6
d_{ot}	mm	1.0	0.5	0.32	0.2
L_{ot}	m	0.65	1.9	2.0	1.9
V_{de}	µL	110	6.2	2.0	0.6
τ_{el}	ms	22	71	38	18

The acceptable limits would cause 5 % additional variance each in injector and detector, and 1 % additional variance each in connecting tubes and electronics.
Source: Adapted from McGuffin (2004, 34).

The allowable volumes for the injector and detector are interesting. These maxima would add 5 % each to the variance of a peak, together increasing its width by about 5 % as noted above.

Also note the detector volume and response time requirements. For many PGCs, these may be the most difficult parameters to satisfy.

More detail

For those who yearn for a rigorous mathematical discourse on extracolumn variance, the article by Sternberg 1966 (205–270) is recommended. This excellent paper is *required reading* for anyone involved in PGC equipment design.

Worked examples

A few calculations will illustrate the application of the theory to practical PGC design. These examples use equations from the above SCI-FILE.

Sample injection

Let's start with a plug injection and compare two sample volumes, 75 and 300 µL, injected into a packed column, with a corrected flow rate of 15 mL/min.

From Equation 7.12, the plug injection variance is:

$$\tau_i^2 = \frac{1}{12} \cdot \left(\frac{V_i}{\dot{V}_c}\right)^2$$

For a 75 µL sample:

$$\tau_i^2 = \frac{1}{12} \cdot \left(\frac{0.075 \text{ mL}}{15 \text{ ml/min}} \cdot \frac{60 \text{ s}}{\text{min}}\right)^2 = 0.0075 \text{ s}^2$$

For a 300 µL sample:

$$\tau_i^2 = \frac{1}{12} \cdot \left(\frac{0.3 \text{ mL}}{15 \text{ ml/min}} \cdot \frac{60 \text{ s}}{\text{min}}\right)^2 = 0.12 \text{ s}^2$$

The larger sample volume creates 16 times the variance as the smaller one does, but its ultimate effect on the chromatogram will depend on the magnitude of the column variance. Let's see how the change in sample volume would affect a typical eluted peak with a base width of eight seconds.

> You should immediately see[1] that our 8 s wide peak has a standard deviation of 2 s and a variance of 4 s².

Consider the 75 µL injection and assume that no other extracolumn variances apply. From Equation 7.11, the observed chromatogram peak variance (τ^2) is equal to the sum of the column variance (τ_c^2) and the injection variance (τ_i^2):

$$\tau^2 = \tau_c^2 + \tau_i^2$$

---
[1] If not, see Figure 2.1

Therefore, the column variance (τ_c^2) is:

$$\tau_c^2 = 4 - 0.0075 = 3.9925 \text{ s}^2 \qquad \text{For 75 µL}$$

Then, from Equation 2.16, the base width of a column peak is four times its standard deviation:

$$w_b = 4\tau$$

$$w_b = 4 \times \sqrt{3.9925} = 7.99 \text{ s} \qquad \text{For 75 µL}$$

The 75 µL injection volume has added 0.01 seconds to the peak width, which is clearly negligible. Now, check what happens if you inject 300 µL:

$$\tau^2 = 3.9925 + 0.12 = 4.1125 \text{ s}^2 \qquad \text{For 300 µL}$$

$$w_b = 4 \times \sqrt{4.1125} = 8.11 \text{ s} \qquad \text{For 300 µL}$$

The contribution of the injection to peak width is now 0.12 seconds (about 1.5 %), thereby decreasing resolution, also by about 1.5 %. Such a loss in resolution is unlikely to be a problem in most applications, but it's starting to notice.

The effect of these sample volumes would be much greater for a narrower peak. Had we started with a four-second peak, the change from 75 to 300 µL would have increased the peak width and reduced the resolution by nearly 6 %, becoming quite significant.

Open tubing

The plug flow equation for empty tubes is functionally the same as the equation for plug injection.

For example, consider a plan to add 500 mm of one-sixteenth-inch tubing to connect the column to the detector. The volume of this tube is 0.30 mL.

Given a flow rate ($\dot{V}_c$) of 20 mL/min, Equation 7.14 returns the additional variance (τ_{ot}^2) due to plug flow in the open tube:

$$\tau_{ot}^2 = \frac{1}{12} \cdot \left(\frac{V_{ot}}{\dot{V}_c}\right)^2$$

$$\tau_{ot}^2 = \frac{1}{12} \cdot \left(\frac{0.30 \text{ mL}}{20 \text{ ml/min}} \cdot \frac{60 \text{ s}}{\text{min}}\right)^2 = 0.0675 \text{ s}^2 \qquad \text{For 20 mL/min}$$

Consider a peak that had a base width of two seconds before installing the additional tube.

➤ You should immediately see that this peak had a standard deviation of 0.5 s and a variance of 0.25 s².

Upon installing the connecting tube, the total variance rises to 0.3175 s² and the peak width increases to:

$$w_b = 4 \times \sqrt{0.3175} = 2.254 \text{ s} \qquad \text{(Equation 2.16)}$$

The increase is about 13 %, which is a significant loss of resolution. But the loss of performance would be much worse with a capillary column. The flow rate in a capillary column might be only one-tenth of the flow rate used in the example above.

Dropping the flow rate to 2 mL/min increases the variance in the connecting tube by a factor of 100:

$$\tau_{ot}^2 = \frac{1}{12} \cdot \left(\frac{0.30 \text{ mL}}{2 \text{ ml/min}} \cdot \frac{60 \text{ s}}{\text{min}}\right)^2 = 6.75 \text{ s}^2 \qquad \text{For 2 mL/min}$$

Adding the 6.75 s² of extracolumn variance totally overwhelms the original peak, whose variance increases to 7.0 s².

From Equation 2.16, the peak width (w_b) is now:

$$w_b = 4 \times \sqrt{7.0} = 10.58 \text{ s}$$

The peak is now over 10 s wide, and the resolution is ruined! Adding 500 mm of one-sixteenth-tube has increased the peak width by 500 %.

Although these examples use the approximate plug-flow equation, they clearly show that capillary columns will not work in a PGC designed for packed columns. Capillary columns need different equipment.

Knowledge Gained

Practice

- ❖ Peaks get wider as they pass through flow passages external to the column.

- ❖ Additional dispersion occurs during sample injection, and in connecting tubes and detector.

- ❖ The finite response time of electronic circuits may also distort the peak width and shape.

- ❖ Column overload may cause fronting, tailing, or even flat-topped peaks.

- ❖ Sample capacity is the maximum volume of an injected substance to avoid peak distortion.

- ❖ Sample capacity decreases in proportion to the square root of the plate number.

- ❖ Sample capacity is larger for later peaks on the chromatogram.

- ❖ Increase sample capacity by using another liquid phase, higher liquid load, or lower temperature.

- ❖ Use the smallest sample volume that provides adequate detector sensitivity.

- ❖ The duration of the sample injection profile affects the width and shape of chromatogram peaks.

- ❖ The ideal injection profile is rectangular but has a finite width that increases the peak width.

- ❖ The sample feed volume should not occupy more than 50 plates.

- ❖ The feed volume limit is statistical, not based on solubility or linearity, but the equation is similar.

- ❖ A smaller sample has a shorter injection profile, leading to narrower peaks.

- ❖ Gas sample injections often come close to the ideal plug injection profile.

- ❖ Liquid sample injection profiles may be rectangular or exponential in shape.

- *Extra dispersion from flow passages comes from open tubes, mixing chambers, or dead volumes.*
- *The connecting tubes should be as short as possible, but the internal diameter is far more important.*
- *The bore of connecting tubes should be as small as possible and never wider than the column tube.*
- *Mixing occurs when the carrier gas flows from a smaller tubular conduit into a larger one.*
- *The entry to a wider-bore chamber may cause a disastrous dispersion of narrow peaks.*
- *Unswept cavities cause extreme dispersion and must be assiduously avoided in analyzer design.*

Theory

- *Peaks suffer dispersion of their molecules as they travel in flow paths external to the columns.*
- *Variance is a measure of dispersion in flow paths both inside and outside the column.*
- *Independent variances are additive and increase the width of chromatogram peaks.*
- *Academics propose that 10–12 % of extracolumn variance is acceptable.*
- *But PGC design must always seek to minimize extracolumn variance.*
- *Dispersal occurs at injection and in empty tubes, dead volumes, mixing chambers, and detectors.*
- *The diameter of connecting tubing is far more important than its length.*
- *Dispersion in empty tubes increases with the fourth power of tube diameter.*
- *A sudden increase in flow-path diameter causes severe peak distortion and may ruin resolution.*
- *Highly efficient capillary columns require small injections and low-volume flow paths.*
- *Sample injectors and detectors may contribute the most to extracolumn broadening of peaks.*
- *Simple calculations can estimate the severity of extracolumn dispersion.*

Did you get it?

Self-assessment quiz: SAQ 07

On practice

Q1. In a well-constructed PGC, what are the two main sources of peak dispersion outside the column?

Q2. Give two independent reasons why a large sample volume can increase peak width.

Q3. Which symptom in the list below is the best reason for using a smaller sample volume?
   A. The analyte peak is fronting.
   B. The analyte has a very low concentration in the sample.
   C. The peaks are appearing late in the chromatogram.

On theory

Q4. Explain the difference between plate height and variance.

Q5. Consider a peak having a base width of four seconds:
   A. Calculate the extra variance caused by installing an additional connecting tube that has a volume of 0.2 mL. Assume plug flow at a rate of 2 mL/min in the connecting tube.
   B. Calculate the new peak base width, after the additional connecting tube is installed.

Q6. Which has the largest effect on peak width?
   A. Long connecting tubes between devices.
   B. Wide connection tubes between devices.

Q7. Consider two peaks exiting a capillary column both having a base width of two seconds and a separation of three seconds. If they enter a connecting tube that has twice the internal diameter, what happens to their resolution?

References

Further reading

The recent chapter by Vicky McGuffin includes an academic, yet practical review of extracolumn effects in chromatography (McGuffin 2004, 33–38).

A short book by Scott is available online and provides a good summary of extracolumn effects in gas and liquid chromatography (Scott 2003).

Those involved in the design or application of process gas chromatographs should study the original article by Sternberg (1966).

Cited

Annino, R. and Villalobos, R. (1992). *Process Gas Chromatography: Fundamentals and Applications*. Research Triangle Park, NC: Instrument Society of America.

Giddings, J.C. (1963). Principles of column performance in large scale gas chromatography. *Journal of Gas Chromatography* **1**, 12.

Klinkenberg, A. (1960). Prepared contribution. In: *Gas Chromatography* (ed. R.P.W. Scott). London, UK: Butterworths Scientific Publications, Ltd.

Maynard, V. and Grushka, E. (1972). Effect of dead volume on efficiency of a gas chromatographic system. *Analytical Chemistry* **44**, No. 8, 1427–1434.

McGuffin, V.L. (2004). In: *Chromatography, 6th edition. Journal of Chromatography Library*, Vol. **69A** (ed. E. Heflmann). Amsterdam, Netherlands: Elsevier B.V.

Perry, J.A. (1981). *Introduction to Analytical Gas Chromatography*. New York, NY: Marcel Dekker Inc.

Purnell, H. (1962). *Gas Chromatography*. New York, NY: John Wiley & Sons, Inc.

Scott, R.P.W. (2003). *Extra column dispersion*. Book 10: Chrom-Ed Book Series © LIBRARY4SCIENCE, LLC. Accessed 2022-05-22 at: https://www.academia.edu.

Scott R.P.W. (1998). *Introduction to Analytical Gas Chromatography*, Second Edition. New York, NY: Marcel Dekker, Inc.

Sternberg, J.C. (1966). Extracolumn contributions to chromatographic peak broadening. In: *Advances in Chromatography*, Vol. **2**. (eds. J.C. Giddings and R.A. Keller), 205–270. New York, NY: Marcel Dekker Inc.

Waters, T. (2020). *Process Gas Chromatographs: Fundamentals, Design and Implementation*. Chichester, UK: John Wiley & Sons, Ltd.

Table

7.1 Extracolumn Limit Parameters.

Figure

7.1 An Ideal Injection Profile.

Symbols

Symbol	Variable	Unit
a	An empirical constant	None
B	Golay constant for axial diffusion	None
C_{sub}	Golay constant for resistance to mass transfer	None
d	Internal diameter of tube or pocket depth	mm
D_{sub}	Diffusion coefficient of solute (see subs below)	None
H	On-column plate height in length units	mm
H_t	Chromatogram plate height in time units	s
j	Gas compressibility factor	None
k	Retention factor	None
K_c	Distribution constant	None
L	Length of column	m
L_{sub}	Length of open tube, etc. (see subs below)	m
N	Plate number	None
P_i	Pressure of injected gas volume	kPa
P_f	Pressure of carrier gas at flowmeter	kPa
N_{max}	Maximum plates occupied by injected sample	None
R_s	Resolution of adjacent peaks	None
t_{sub}	Time delay (see subs below)	s
t_i	Duration of a plug injection	s
t_R	Peak retention time	s
T_{sub}	Temperature (see subs below)	K
$\boldsymbol{u}$	Carrier gas velocity at column exit	m/s
$\dot{V}_c$	Carrier gas flow rate at column inlet	mL/min
V	Volume (see subs below)	mL
V_{max}	Maximum feed volume	mL
V_R	Peak retention volume	mL
V_R°	Retention volume, corrected to average column pressure	mL
w_b	Base width of a solute peak	s
σ	Standard deviation of peak in length units	mm
τ	Standard deviation of peak in time units	s
ψ	Obstruction factor for diffusion in a packed column	None

Subscripts

Sub$_{63}$		Exponential time constant (time for 63% change)
Sub$_c$		Pertaining to the column
Sub$_{de}$		Pertaining to the detector

Sub$_{dv}$		Pertaining to a dead volume
Sub$_{el}$		Pertaining to the signal-processing circuits
Sub$_f$		Pertaining to the flowmeter
Sub$_i$		Pertaining to injection
Sub$_{in}$		Going in
Sub$_{mc}$		Pertaining to a mixing chamber
Sub$_M$		Pertaining to the mobile phase
Sub$_{ot}$		Pertaining to open tubing
Sub$_{out}$		Coming out
Sub$_S$		Pertaining to the stationary phase

Equations

7.1	$w_b = \dfrac{4L}{\sqrt{N}}$	Peak width as a function of plate number.
7.2	$V_{max} = \dfrac{a \cdot V_R^\circ}{\sqrt{N}}$	Sample capacity of a column for a specified substance.
7.3	$t_{max} = \dfrac{t_R}{2\sqrt{N}}$	Maximum plug injection duration to limit peak broadening.
7.4	$\dfrac{t_{max}}{t_R} = \dfrac{0.5}{\sqrt{N}}$	Maximum plug injection time as a fraction of retention time.
7.5	$N_{max} = 0.5\sqrt{N}$	Maximum number of plates occupied by the feed volume to limit peak spreading.
7.6	$V_{max} = \dfrac{V_R^\circ}{2\sqrt{N}}$	Maximum sample feed volume to limit peak spreading.
7.7	$t_i = \dfrac{V_i}{\dot{V}_c}$	Width in time units (duration) of a plug injection.
7.8	$t_i = \dfrac{V_i}{\dot{V}_c} \cdot \dfrac{P_i}{P_f} \cdot \dfrac{T_f}{T_i}$	The injection duration corrected for differences in pressure and temperature.
7.9	$t_{ot} = \dfrac{V_{ot}}{\dot{V}_c}$	Time delay in open tubing assuming plug flow.
7.10	$\sigma^2 = L \cdot H$	Relation between variance and plate height.
7.11	$\tau^2 = \tau_1^2 + \tau_2^2 \ldots + \tau_n^2$	Variance in time units as the sum of individual variances, both intracolumn and extracolumn.
7.12	$\tau_i^2 = \dfrac{1}{12} \cdot \left(\dfrac{V_i}{\dot{V}_c}\right)^2$	Additive time variance of a plug injection due to sample volume and carrier gas flow rate.
7.13	$\tau_i^2 = t_{63}^2$	Additive time variance of an exponential sample injection profile.
7.14	$\tau_i^2 = \dfrac{1}{36} \cdot \left(\dfrac{V_i}{\dot{V}_c}\right)^2$	Additive time variance of a Guassian sample injection (a peak from another column).

7.15 $\quad \tau_{ot}^2 = \dfrac{1}{12} \cdot \left(\dfrac{V_{ot}}{\dot{V}_c}\right)^2$ — Additive time variance due to open tubing, assuming plug flow obtains.

7.16 $\quad H = \dfrac{B}{u} + uC_M$ — Golay equation for plate height of an open tube.

7.17 $\quad H = \dfrac{2D_M}{u} + \dfrac{ud^2}{96 D_M}$ — Plate height of an open tube derived from the Golay equation for wall-coated capillary columns.

7.18 $\quad H = \dfrac{ud^2}{96 D_M}$ — Simplified Golay equation for plate height in length units of an open tube.

7.19 $\quad \sigma_{ot}^2 = L_{ot} \cdot \dfrac{ud^2}{96 D_M}$ — Additive length variance due to an open tube, from the simplified Golay equation.

7.20 $\quad \tau_{ot}^2 = \dfrac{L_{ot}}{u} \cdot \dfrac{d^2}{96 D_M}$ — Additive time variance due to an open tube, from the simplified Golay equation.

7.21 $\quad \tau_{ot}^2 = \dfrac{\pi d^4 L_{ot}}{384 \dot{V}_c D_M}$ — Additive time variance due to an open tube, as a function of tube dimensions and flow rate.

7.22 $\quad H_{ot} = \dfrac{L_c}{N_{ot}} \cdot \dfrac{V_{ot}}{V_R^\circ}$ — Giddings equation for the plate height of an open tube.

7.23 $\quad \tau_{mc}^2 = \left(\dfrac{V_{mc}}{\dot{V}}\right)^2$ — Additive time variance due to an inline mixing chamber.

7.24 $\quad \tau_{out}^2 = \tau_{in}^2 \cdot \left(\dfrac{d_2}{d_1}\right)^4$ — Multiplicative time variance caused by a sudden diameter change in the flow path.

7.25 $\quad \tau_{dv}^2 = \dfrac{d_{dv}^4}{4 D_M^2}$ — Additive time variance due to the fourth power of the depth of an unswept pocket (dead volume).

7.26 $\quad \tau_{de}^2 = \dfrac{1}{12} \cdot \left(\dfrac{V_{de}}{\dot{V}_c}\right)^2$ — Additive time variance due to the sensing volume of a detector.

New technical terms

If you read the whole chapter, you should now know the meaning of these technical terms:

axial diffusion
concentration-sensing detector
corrected retention volume
dead volume
laminar flow
injection profile
mass-sensing detectors

mixing chamber
peak dispersion
plug flow
plug injection
radial diffusion
sample volume

For more information, refer to the Glossary at the end of the book.

8

Evaluating peak shape

"Chromatogram peaks are rarely symmetrical as some deviation from the exact Gaussian shape often occurs. The art of troubleshooting is to recognize the different forms of deviation and to know which forms are normal and thus acceptable and which forms are abnormal and need attention."

Real chromatogram peaks

We have seen that the theories of chromatography assume chromatogram peaks are Gaussian in shape or close enough to Gaussian to make no difference. This assumption is convenient for the theorist but not for the practitioner, since real chromatogram peaks (particularly those obtained from packed columns) often show a noticeable deviation from the pure Gaussian shape.

As a PGC column designer or troubleshooter, you will encounter misshapen peaks and will need to know the many possible causes. You'll need to recognize the peak shapes that cause imprecise or inaccurate measurement. A peak shape that can't be precisely integrated, or has a nonlinear response to analyte concentration, or interferes with the measurement of another peak is unacceptable.

Three peak shapes

Perfect Gaussian peaks are rare on process chromatograms, although the very narrow peaks from capillary columns may come close. When examined closely, most real peaks are somewhat distorted. They come in three distinct shapes as illustrated in Figure 8.1.

The majority of chromatogram peaks look slightly skewed toward their direction of travel as illustrated by the blue peak in Figure 8.1a. Such peaks fall more slowly than they rise and many authors call them "tailing peaks."

Process Gas Chromatography: Advanced Design and Troubleshooting, First Edition. Tony Waters.
© 2025 John Wiley & Sons Ltd. Published 2025 by John Wiley & Sons Ltd.

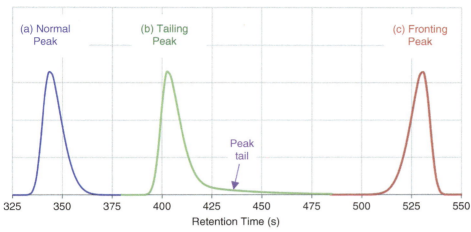

Real chromatogram peaks typically deviate from the ideal symmetrical shape. Three common peak shapes are (a) the normal peak, (b) the tailing peak, and (c) the fronting peak (sometimes called a leading peak).

Figure 8.1 Three Asymmetric Chromatogram Peaks. (a) Normal Peak. (b) Tailing Peak. (c) Fronting Peak.

But it's also possible for a nearly symmetrical peak to form a distinct tail as it returns to the baseline. The green peak in Figure 8.1b shows this undesirable effect, also called "tailing."

These two symptoms, skewing and tailing, are due to quite different underlying causes, so clear thinking requires us to distinguish between them. Calling both of them "tailing" can lead to confusion.

The blue peak in Figure 8.1a is certainly a skewed peak but is not tailing; herein, we call it the **normal peak** shape. Such peaks return promptly to the baseline and rarely cause any trouble with following peaks. Unless the skew becomes extreme, this is normal behavior of a chromatogram peak and does not require corrective action.

We reserve the term "tailing" for a peak followed by a low-level tail that approaches the baseline exponentially and often has a duration longer than the peak itself. The green peak in Figure 8.1b fits our definition of a *tailing peak*. If your chromatogram includes a peak like this, it may be a fault condition requiring some attention. Tailing is normally due to the strong adsorption and slow release of a few solute molecules on contact surfaces along the carrier flow path. Adsorption can occur inside a sample injector, column, or detector, or even in the tubing connections between them. An earlier cause of tailing, now mostly eliminated, was the presence of unswept side volumes (called **dead legs**) in the flow path.

A less common kind of peak is the *fronting peak*, sometime called a **leading peak**, one leaning backward away from the direction of travel as illustrated by the red peak in Figure 8.1c. Fronting peaks rise more slowly than they fall. A fronting peak might be normal behavior for some stationary phases or can be due to injecting a sample volume too large for the column in use.

It's also possible for a single peak to split into two partially separated peaks – sometimes called a **bimodal peak**. This is more likely to happen with capillary columns. The two potential causes of a split are:

> Partial condensation of sample components in the injector or column entrance followed by their evaporation. A higher temperature might be necessary.

> Partial chemical reaction in the injector or column creates a second component, particularly with hydrogen carrier. A lower temperature or helium carrier might be necessary.

If you are certain that the bimodal peak is a single analyte, it would be possible to calibrate its integrated area, but it's best to find the cause and fix the problem.

Normal asymmetric peaks

Measuring peak asymmetry

A measure of peak asymmetry is useful in tracking the shape of a peak suspected of abnormal behavior. It may also allow the more accurate calculation of column efficiency, as noted in the **SCI-FILE:** *On Asymmetric Peaks* that follows.

To measure the **asymmetry ratio** of a chromatogram peak body, excluding any tailing effects close to the baseline, first bisect the peak with a vertical line through its apex and then measure the leading and trailing widths at 10 % of the peak height. Figure 8.2 illustrates this procedure.

Then, Equation 8.1 defines peak asymmetry (A_s) as the ratio of trailing width (b) to leading width (a):

$$A_s = \frac{b}{a} \quad (8.1)$$

The asymmetry of *normal peaks* is slightly higher than one, whereas the less-common *fronting peaks* have an asymmetry of less than one. For the peak in Figure 8.2:

$$A_s = \frac{13.7 \text{ s}}{9.3 \text{ s}} = 1.47$$

A peak asymmetry of 47 % may be somewhat excessive.

Most chromatographers accept a skew of ±20 % as normal, requiring no remedial action. To quantify an analyte, modern PGCs measure the area under the peak rather than the peak height, and the area measurement is generally unaffected by a small degree of skew.

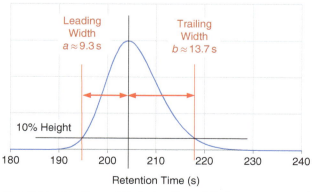

The ratio of trailing width to leading width is a measure of peak asymmetry. It's conventional to measure the leading and trailing widths at 10 % of the peak height.

Figure 8.2 Measuring Peak Skew.

A skew of greater than 20 % may be cause for concern. It will require some judgment to determine whether the observed asymmetry is a symptom of a fault or just a normal benign condition. The following subsection discusses some causes of skew, which may be helpful.

What causes asymmetry?

There are many reasons why a peak might not be perfectly symmetrical, so a real chromatogram peak shape is due to a combination of causes, making diagnosis difficult.

Some of the reasons for peak distortion are fundamental to the chromatographic process and are unavoidable. The **SCI-FILE:** *On Asymmetric Peaks* discusses these rather esoteric causes. Mostly, they have only a minor effect and tend to skew the peak in a trailing direction, resulting in the normal peak shape seen on many chromatograms. Other causes of peak distortion may be correctible and of more general interest so we examine them here.

Limited isotherm linearity

It's useful to remember that the chromatographic peak shape is a concentration profile. As the band of molecules passes a designated location in the column, the concentration of the solute rises from zero to a maximum and then declines back to zero again. To maintain a symmetrical peak shape, the fraction of molecules held by the stationary phase must be constant for all of the solute concentrations occurring within the peak.

In other words, the *distribution constant* should indeed be constant.

If the "distribution constant" is *not* constant, a smaller percentage of the molecules might dissolve at higher concentrations. Then, the top of the peak would move faster than its base, creating a normally skewed peak. Conversely, should a higher percentage of molecules dissolve at higher concentrations, the top of the peak would move slower than its base, creating a fronting peak. For a symmetrical Gaussian peak to form, the distribution constant must not change with concentration.

For a liquid-phase column, the distribution of an analyte between the gas phase and the liquid phase is known as partition. The **partition isotherm** is a graphical plot at constant temperature of the equilibrium concentration of solute in the liquid phase versus its concentration in the gas phase. To get a symmetrical peak, it must be a perfectly straight line.

The left side of Figure 8.3 illustrates the partition isotherm typical of most liquid phases. At the low concentrations usually occurring in gas chromatography, the isotherm is perfectly linear and small peaks like the one shown on the right of the figure are symmetrical, or nearly so.

But larger peaks may exceed the linear region. As the solute concentration increases, a limit to linearity is inevitable. Beyond this limit, the liquid phase may be unable to dissolve the same fraction of molecules. The isotherm then droops away from the desired straight line – like the blue dashed line in Figure 8.3.

208 Evaluating peak shape

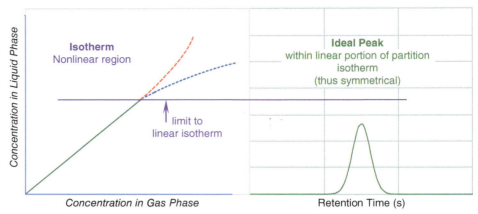

Figure 8.3 Typical Partition Isotherm.

This isotherm is linear at low concentrations indicating a constant distribution constant. A peak that stays under the linearity limit is symmetrical.

If so, consider what happens in the column as a large solute peak passes by. The concentration of solute molecules rises from zero and initially stays within the linear region of the isotherm. But if the solute concentration rises above the linearity limit, the solubility starts to fall and the solute molecules travel a little faster, thereby distorting the peak in the normal direction. Figure 8.4a illustrates this effect, common with liquid-phase columns.

It's possible but uncommon for the isotherm to rise above the expected straight line at high solute concentrations, as shown by the red dashed line in Figure 8.3. This occurs when the presence of a few solute molecules in or on the stationary phase encourages more molecules to attach. If that happens, the top of the peak will travel more slowly than its base and distort the peak in the fronting direction, as shown in Figure 8.4c.

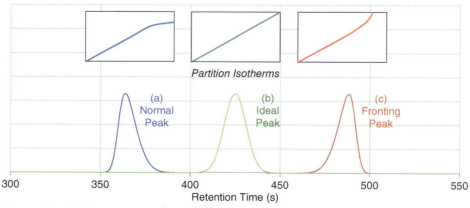

Figure 8.4 Isotherms and Peak Shape. (a) Normal Peak. (b) Ideal Peak. (c) Fronting Peak.

The partition isotherm determines peak shape. A symmetrical peak (green) is possible only when the isotherm is linear, often true at low analyte concentrations. Above the linearity limit, the isotherm is likely to droop downward (blue) causing normal skew but might occasionally soar upward (red) to yield a fronting peak.

Limited linearity will distort a peak, but it cannot cause tailing. The peak will quickly return to the baseline as its concentration declines. Also, it cannot cause a nonlinear calibration when the PGC measures peak area because all the analyte molecules reach the detector. For more information about the effect of excessive sample volume, colloquially known as **column overload**, refer to the sections on *Sample Size*, *Feed Volume*, and *Sample Capacity* in Chapter 3 (pp 67–71).

Curved isotherms

When a column uses a solid stationary phase, a linear isotherm is unlikely. Solid columns separate the solutes by their different affinity for a solid surface, which is the process of adsorption. Most adsorption isotherms are curves, indicating nonlinearity over the whole range of solute concentrations.

It's also possible to get a curved isotherm for certain combinations of solute and liquid phase, perhaps because of an electrochemical attraction between them.

With a curved isotherm, it's impossible to get a symmetrical peak. The reason should be obvious:

> *When the distribution constant is different at each solute concentration, every elevation of the peak will be traveling at a different speed.*

If the isotherm curves downward at higher concentrations, the peak molecules will migrate faster and faster as their concentration rises and the resulting peak will be severely asymmetric in the normal direction.

The chromatogram from a severely downward-curving isotherm exhibits peaks that look triangular, with a rapid rise and a slow decline, but without significant tailing as they return to baseline. If the isotherm is highly curved, the peak front will be almost vertical, rising sharply from the baseline like the front of a square wave. Such peaks are known as **self-sharpening peaks**. The propylene peak on the real chromatogram in Figure 8.5 shows a clear example of this effect.

An isotherm that curves upwards will have the opposite effect, generating a severely fronting peak. Again, the peak may be triangular, rising as a slow ramp, then descending rapidly to the baseline.

A second retention mechanism

We have seen how large peaks become distorted when they enter a nonlinear portion of the partition isotherm. Yet experimenters have found that even the smallest peaks are often asymmetric in the normal direction and they can't explain this asymmetry by nonlinearity.

To explain this unexpected result, theorists invoke a **second retention mechanism** operating in parallel to partition; the entrapment of solute molecules on scarce adsorption sites within the column. The basic idea is that desorption is a much slower process than partition, so the active sites

210 Evaluating peak shape

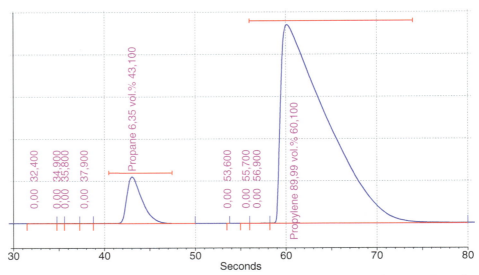

Figure 8.5 Typical Adsorption Peaks.
Source: Courtesy of Slovnaft Refinery.

This real chromatogram is typical of light gas separations on porous polymers, where the separation is mainly due to adsorption rather than partition. Clearly, this porous polymer has a significantly curved adsorption isotherm, but peak asymmetry doesn't affect peak area.

retain molecules adsorbed from the leading half of a peak long enough to release them into the trailing half of the peak, thereby creating peak asymmetry. Figure 8.6 illustrates this mechanism.

Note that in this type of adsorption, the analyte molecules are not strongly held and are released into the back of the peak as soon as its apex passes and the number of analyte molecules declines. Later, we shall see that some analytes suffer a stronger form of adsorption that doesn't release their molecules until after the peak has passed, causing a peak tail to form.

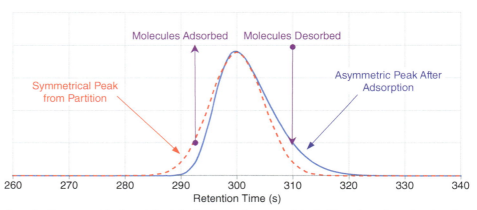

Figure 8.6 A Second Retention Mechanism.
Source: Giddings, J.C. (1965)/ with permission of Taylor & Francis Group.

A perfectly symmetrical peak (red dashes) may become distorted by adsorption of molecules at active sites on contact surfaces. The active sites subtract molecules from the peak front, hold them for random times, then release them into the peak back thereby forming a normal asymmetric peak shape (blue).

This peak-shaping process does *not* imply a nonlinear adsorption isotherm; it requires only that the desorption rate is slower than the adsorption rate. A nonlinear adsorption isotherm would further distort the peak, as discussed above.

It turns out that even the smallest amount of adsorption will distort a symmetrical peak. Giddings (1965, 77) found that the active sites have only to adsorb the average solute molecule once during its passage through the column to cause significant peak asymmetry.

In Chapter 4, we emphasized the importance of eliminating adsorption sites from liquid-phase columns and discussed the techniques available for doing so. The manufacturers of PGC columns routinely follow these techniques but it's not practical to eliminate all adsorption sites from a PGC column, especially a packed column. Thus, it's likely this second retention mechanism affects the shape of many solute peaks and is probably the main architect of the normal peak shape; slightly asymmetric in the trailing direction as exemplified by the blue peak in Figure 8.6.

The ultimate effect of the adsorption mechanism depends on the time delay that occurs before the active sites release the molecules back into the gas stream. If the delay is less than the peak width, the molecules adsorbed from the front of the peak tend to add to the back of the peak, causing the asymmetry discussed above. But if the delay is longer than the peak width, they return to the gas phase after the peak has moved on, forming a true tail like the one on the green peak in Figure 8.1c. Since the activity of adsorption sites varies, they will release the molecules after a wide range of delay times thereby forming a long exponential tail that interferes with the measurement of subsequent peaks. This desorption delay is the primary cause of peak tailing.

Again, note this tailing mechanism is due only to time delay and can occur even when the adsorption isotherm is perfectly linear.

Column overload

The most common cause of a fronting peak is column overload due to an excessive sample injection volume.

Generally, a fronting peak occurs when the distribution constant increases with solute concentration. While this can be due to an upward-curving isotherm, it's more likely to be a symptom of column overload. When the injected sample volume is too high, the large number of solute molecules immobilized by the stationary phase can attract and hold additional molecules of the same kind, reducing the rate of migration for high analyte concentrations. For more on fronting peaks consult the section on sample size in Chapter 3, pages 67–71.

SCI-FILE: *On Asymmetric Peaks*

Natural deviations from perfection

There are so many potential causes of chromatogram peak asymmetry that you might expect nice Gaussian peaks to be a rarity. Yet most of our peaks turn out to be only slightly distorted, which is a credit to the skill of the instrument manufacturers.[1]

The main text discusses some causes of peak asymmetry that might be amenable to improvement, including the effect of nonlinear isotherms or slow desorption processes.

This SCI-FILE focuses on the reasons for peak asymmetry inherent in the chromatographic process itself. These mechanisms skew the peaks in the normal direction ($A_s > 1.0$) and generally do not cause tailing. Mostly, they are not amenable to mitigation, although the magnitude of their effect may depend on column operating conditions.

Poisson distribution

Processes distributing solute molecules along the column are often assumed to be random, leading to the formation of Gaussian peaks. Yet the plate theory of chromatography predicts a peak shape that follows the **Poisson distribution** rather than the classic Gaussian shape (Giddings 1965, 35; Scott 2003, 31).

For a column having a plate number greater than a few hundred, there is no practical difference between the two distributions; both are symmetric. But for a column with very few plates, the Poisson distribution generates asymmetric peaks. So chromatogram peaks from low-efficiency packed columns may tend toward the Poisson shape illustrated in Figure 8.7, slightly skewed in the normal direction.

Since this contribution to asymmetry only affects low-efficiency columns, it rarely occurs. Most of our PGC columns have enough theoretical plates for their peaks to become Gaussian in shape.

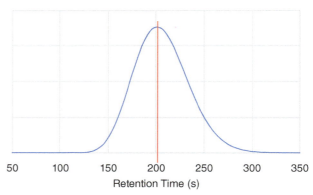

The plate theory of chromatography predicts a Poisson distribution rather than a Gaussian one. As shown here, the Poisson distribution is slightly asymmetric for low plate numbers, but becomes essentially Gaussian for plate numbers above a few hundred.

Figure 8.7 The Poisson Distribution.

Transformation to the time domain

As noted before (see Figure 2.6), the distribution of solute molecules in the column is spatial, which means some molecules in a peak have traveled further along the column than the average molecule, and some not so far.

For the moment, assume the on-column spatial distribution is perfectly Gaussian with a variance of σ^2 and a peak base width of 4σ mm. Consider what happens when the leading edge of the peak reaches the detector. At that instant, the peak starts to appear on the chromatogram. Yet the center of the peak must still travel a distance of about 2σ mm (half the peak base width) in the column before it reaches the detector, and the trailing edge of the peak must travel yet another 2σ mm, an additional distance of 4σ mm.

Since the trailing edge of a peak stays longer in the column than the leading edge does, it follows that the molecules toward the back of the peak will

---

[1] And to the theoretical principles elucidated by researchers over several decades of patient work. Their vital contributions are acknowledged by reference in this book.

suffer progressively more dispersion as the peak gradually elutes from the column. This additional dispersion makes the trailing half of a chromatogram peak wider than the leading half, so all chromatogram peaks are to some extent asymmetric.

While there is no way to change the peak skewing inherent in transforming a spatial separation into a temporal one, we can be sure narrow peaks will be less affected than wide peaks.

To get an estimate of the extent of this effect, consider each half of the Gaussian peak as if it were a separate peak. Recall that the base width of a Gaussian chromatogram peak is equal to 4τ, so the leading half-peak elutes from the column at about $(t_R - \tau)$ and the trailing half-peak elutes at about $(t_R + \tau)$.

For a peak whose base width is one-tenth of its retention time, $\tau = 0.025\, t_R$; so the leading half-peak elutes at $0.975\, t_R$ and the trailing half-peak at $1.025\, t_R$.

From Equation 2.23, we know the base width of a peak is proportional to the square root of its retention time.

Therefore, the ratio of their base widths is:

$$\frac{w_{\text{trailing}}}{w_{\text{leading}}} = \sqrt{\frac{1.025\, t_R}{0.975\, t_R}} = 1.025$$

This asymmetry (2.5 %) would be hardly visible.

Narrower peaks would exhibit even less asymmetry so, once again, this type of peak distortion occurs only with wide peaks from low-efficiency packed columns.

Equation 2.26 gives a plate number of 1,600 for the peak used in this example:

$$N = 16\,(10)^2 = 1{,}600$$

A wider peak might have a base width equal to one-fourth of its retention time. Then:

$$N = 16\,(4)^2 = 256$$

This chromatogram peak has a standard deviation $\tau = t_R/16$, so the width ratio of its trailing and leading half-peaks is:

$$\frac{w_{\text{trailing}}}{w_{\text{leading}}} = \sqrt{\frac{1.0625\, t_R}{0.9375\, t_R}} = 1.065$$

An asymmetry of 6.5 % is visually noticeable and confirms low-efficiency peaks suffer more skew.

This contribution to peak asymmetry is entirely due to the transformation of the on-column peak to the time-based domain of the chromatogram. As demonstrated here, even a perfectly symmetrical peak will suffer the distortion. To minimize this effect, increase column efficiency to reduce the peak widths. The peaks from highly efficient capillary columns exhibit very little skew.

Thermal peak distortion

When the solute dissolves into (or adsorbs onto) the stationary phase, it releases heat energy, increasing the local temperature. Then, when the solute returns to the gas phase, it adsorbs an equivalent amount of heat energy, reducing the local temperature.

Since the leading half of a peak is continuously dissolving as it travels through the column, it migrates at a slightly higher column temperature than the oven set point. Conversely, the trailing half of the peak migrates at a slightly lower temperature.

The temperature change depends on the size of the peak. For a very small peak it's negligible, but for a large peak it might reach ± 0.5 K (Scott 2014, 64).

We know from Chapter 4 that retention times are highly dependent on column temperature and may change as much as 4 % per kelvin. Clearly, if the trailing half-peak travels an average of 4 % slower than the leading half-peak, the chromatogram peak will be asymmetric in the trailing direction. And because the temperature difference is proportional to the size of the peak, large peaks will suffer more distortion than small peaks.

The contribution of the heat of solution (or heat of adsorption) to overall peak asymmetry is another natural consequence of the chromatographic process. There is no way to avoid the heat of solution, but a smaller sample volume will minimize its effect on peak shape.

A second retention mechanism

As noted in the main text, even very small peaks from liquid phase columns can exhibit a normal asymmetric shape. Theorists now believe this distortion is due to the combined effect of two separate retention mechanisms: rapid gas–liquid partition and slow gas–solid desorption at active sites on all contact surfaces.

In support of this hypothesis, skewed peaks often follow the **exponentially modified Gaussian** (EMG) function, which combines the Gaussian equation with an exponential decay function (see Foley & Dorsey, 1983, 1984; Foley 1987). If you are interested in this approach, you may wish to review the many references provided by these authors.

Figure 8.8 compares the pure Gaussian peak shape to the EMG peak shape. From this, you can imagine how adsorption removes molecules from the front of the peak and releases them into the back of the peak after a short delay.

The EMG function also provides a more accurate estimation of chromatographic variables from asymmetric peaks. In particular, our standard equations for plate number (Equations 2.26 and 2.27) are inaccurate when applied to significantly asymmetric peaks. Using the measurements shown in Figure 8.2, Foley and Dorsey (op. cit.) proposed this more-accurate equation for plate number:

$$N = \frac{41.7(t_R/w_{0.1})^2}{b/a + 1.25} \quad (8.2)$$

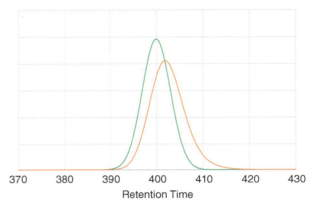

The chart compares a perfectly symmetrical Gaussian peak (green) with a "normal" chromatogram peak (orange) as modeled by the EMG function. The orange peak has an asymmetry ratio (b/a) of 1.35 and is typical of chromatogram peaks from packed columns.

Figure 8.8 The EMG Peak Model.

Problems with asymmetric peaks

Most chromatogram peaks are slightly asymmetric and this small deviation from perfection has no effect on the measurement of peak area and the computation of analyte concentration.

The height of an asymmetric peak may not be linear with analyte concentration, but PGCs are usually set up to measure peak area and will reliably measure even the most severely distorted peak. Yet grossly asymmetric peaks can cause problems of peak detection during automatic operation. Of main concern are:

> The retention time of an asymmetric peak may change when its concentration changes.

➤ The resolution of adjacent peaks may change when the area of the asymmetric peak changes.

➤ The wide integration window makes peak area measurement more susceptible to baseline error.

Retention time shift

Figure 8.9 illustrates (in a rather exaggerated way) that the observed retention time of a severely asymmetric peak can change with its size. This illustration is typical of a substance with a curved-down isotherm on the stationary phase in use. As the peak grows larger, the additional molecules experience a lower distribution constant and move a little faster through the column. The additional molecules then make the peak wider by adding to its front, while the peak tail stays about the same shape. Of course, the mirror image of this effect occurs when the substance has a curved-up isotherm on the stationary phase.

If the peak measurement window is manually set for the smaller peak, a larger concentration of the analyte may push the front of the peak out of the window, causing a measurement error. And in systems that automatically identify the peak by its retention time, the time shift of the peak apex may cause an incorrect identification of the peak.

Figure 8.9 also illustrates the self-sharpening effect occurring at the leading edge of the peak, which is a sure symptom of a curved isotherm.

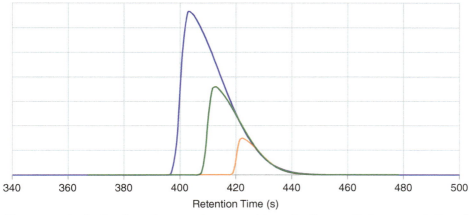

The apparent retention time of a severely asymmetric peak will vary with the concentration of the analyte. The peak shape shown indicates a downward-curved isotherm. As its concentration increases, the stationary phase holds a smaller percentage of the analyte allowing the additional molecules to move faster through the column.

Figure 8.9 Retention Shift with Concentration.

Evaluating peak shape

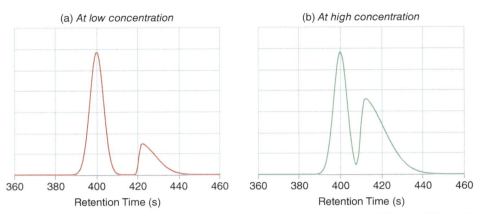

Figure 8.10 Effect of Peak Shape on Resolution. (a) At Low Concentration. (b) At High Concentration.

An asymmetric peak grows wider as its concentration increases and this additional width might spoil the resolution of an adjacent peak.

Loss of resolution

The varying width of a severely asymmetric peak can cause loss of resolution. At higher concentrations, the additional width of the peak will reduce the resolution between the two peaks and may even cause them to merge, as shown in Figure 8.10. It follows that application testing must use samples containing the maximum expected concentration of the asymmetric peak, whether measured or not.

Tailing peaks

The peaks on your chromatogram should be more-or-less symmetric and quickly return to the baseline like those in Figure 8.11a. Ideally, they should not form tails although some tailing is unavoidable in some applications.

A tailing peak will take up more space on the chromatogram baseline, spoiling the resolution, and making it difficult to get an accurate measurement of the following peak. It will also make peak-end detection more difficult and more variable, thereby reducing the precision of peak area measurement.

Different troubleshooting symptoms include:

➢ All onscale peaks are tailing.

➢ Some onscale peaks are tailing.

➢ Offscale peaks are tailing.

Let's consider these.

All peaks are tailing

When a chromatogram includes polar and nonpolar substances and all peaks are tailing, the analyzer may need attention. If both polar and nonpolar peaks tail, the columns are unlikely to be the cause. The observed tails may indicate a slow sample injection, dead legs in the analytical flow path, or a detector malfunction.

Injecting a wide-boiling liquid sample can cause a slow sample injection due to the time taken for the liquid to vaporize. Increasing the injector temperature or reducing the sample volume might be an effective remedy.

Dead legs are small upswept volumes that can harbor any kind of peak molecules for a while, then release them to follow the main peak. Examine the flow paths in your injector, detector, and tube connections in between. Even a simple tube union can have minute unswept volumes around the tube ends and every side-volume adds to the peak tail. Special tube connectors are available to eliminate such dead legs.

In older PGCs, thermal detectors and their electronics were slow to respond, but modern detector systems are less likely to cause tailing.

Some peaks are tailing

Polar solutes are much more likely to form tailing peaks than nonpolar solutes. In Figure 8.11b, the first two peaks are nonpolar and are not tailing, whereas the last two peaks are polar solutes with tails. This selective tailing is usually due to the adsorption of analyte molecules on active metal surfaces. To reduce this kind of tailing, replace all tubing and fittings in the flow path with surface-deactivated components.

Tailing of polar peaks may also occur when using a nonpolar liquid phase. A very nonpolar phase might not completely neutralize the active sites on a capillary-column wall or packed-column support. Sometimes a small increase in the polarity of the liquid phase is enough to minimize the tailing, with little affect on the separation of the analytes.

Occasionally, tailing might be due to deposits of nonvolatile solids or oils within the injector, connecting tube, or column inlet. When such deposits come from substances dissolved in a liquid sample, it will be necessary to clean or replace the injector and its connecting tube, or even replace the first column. It's also possible to cut a short length from the front of a long capillary column.

When the deposits are polymer oils due to the polymerization of a gas sample, a lower injector temperature will reduce the accumulation rate. However, this is often impractical as PGC gas injectors are normally at column-oven temperature.

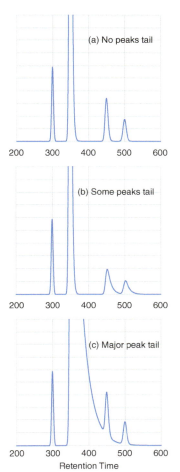

The normal peaks in chromatogram (a), are slightly asymmetric but exhibit no tailing. In chromatogram (b) the last two peaks suffer moderate tailing – a more polar column might help. In chromatogram (c) the second peak is so large it has formed a tail which overlaps the third peak.

Figure 8.11 Normal and Tailing Peaks. (a) No Peaks Tail. (b) Some Peaks Tail. (c) Major Peak Tail.

Offscale peaks are tailing

When a large peak goes offscale, it may seem wider than the onscale peaks. You can see this effect on the second peak in Figure 8.11a. Yet the triangulated base width of the peak hasn't changed. If the second peak was onscale, it wouldn't look much different than the first peak.

To understand this effect, look back at the perfect Gaussian peak in Figure 2.5 and imagine how it would appear at a higher sensitivity. Most of the peak is now offscale. The increased sensitivity amplifies the small tails on either side of the peak (those outside the triangulated base width) eventually driving them offscale. The peak then appears wider because you're seeing only its magnified base. The peak itself hasn't changed.

An offscale peak looks wider than onscale peaks, but if it quickly returns to the baseline it's not a problem. Yet when a large peak tails exponentially it can be a real challenge. Figure 8.11c reveals how a major peak tail can interfere with the measurement of a following peak. However, the figure shows only the first stage of tailing when the analyte peaks are a few hundred parts-per-million (ppm) and are still visible on the chromatogram. To display lower analyte concentrations the sensitivity would be perhaps 100 times higher, which would bury the analyte peaks in the magnified major-peak tail. Analysis is impossible under such conditions.

All columns suffer this shortcoming, ultimately caused by the large volume of sample needed to detect and measure a few ppm. The only solution is to remove most of the major peaks by column switching. The heartcut technique discussed in Chapter 11 is a powerful way to measure these very low concentrations.

Diagnosis of deviant peak shapes

Reading the chromatogram

The chromatogram is the fundamental readout of a PGC and contains all the information necessary for calculating the analyte concentrations. It also exhibits all symptoms necessary to diagnose potential or actual problems with the analysis.

> *The essential skill for troubleshooting a PGC is your ability to read information from a chromatogram. Everything else is just electronics.*

One of your chromatogram-reading skills should be to notice and to correctly diagnose the cause of misshapen peaks. It requires a keen eye and a logical brain. In this chapter, we have seen how peaks may not be perfectly symmetrical. This is true, but even asymmetric peaks must display a smooth curve without sudden change in direction or slope. Look very carefully at your chromatogram peaks. Even the slightest deviation from the expected curvature is a vital clue; it's telling you something isn't right.

Observed deviations from the expected peak curvature may indicate:

➢ An unknown peak is overlapping the analyte peak and causing a high measurement value.

➢ A slice of the peak is missing, cut by inaccurate timing of a column valve and causing a low measurement value.

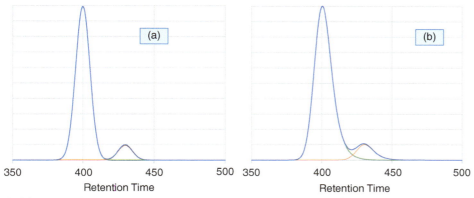

In (a), the resolution of 1.5 is adequate for perfectly symmetric peaks even when the second peak is only one tenth as tall as the first peak. Yet for the "normal" asymmetric peaks in (b), a resolution of 1.5 is not enough for accurate measurement.

Figure 8.12 Resolution of Normal Asymmetric Peaks.

Overlapping peaks

An asymmetric peak may be a composite of two or more peaks merged together. When a small partly resolved peak closely follows a larger peak, the combination of the two peaks may look like a single asymmetric peak on the chromatogram. But it's rare for the second peak to be exactly the right size and position to completely disappear into the trailing edge of the first peak. If you look carefully at your chromatogram, you may see the telltale signs of overlapping peaks.

The classic signs of a merged peak are the strange shapes generated when two peaks overlap. Published examples typically show the chromatogram generated by the overlap of two equal ideal peaks at various resolutions. For example, see Waters (2020, 70).

Those curves are revealing and give a good first lesson in diagnosis, but we now know that real peaks are slightly asymmetric and this changes the game. Figure 8.12 compares unequal symmetric and asymmetric peaks at a resolution of 1.5. The baseline separation of ideal symmetric peaks is adequate but look what happens when the more realistic "normal" peaks overlap!

It gets worse as resolution declines. Figure 8.13 shows the merged peak obtained at a resolution of 1.0 – the second peak is still noticeable if you have your eyes open. At even lower resolutions, it's just a minor inflection in the side of the peak and easily missed.

On the chromatogram in Figure 8.14 are two symmetrical peaks partly merged. The first peak is three times the size of the second peak. At a resolution of 0.75, chromatogram (a) clearly shows the presence of the two peaks but reducing the resolution to 0.50 results in a single asymmetric peak.

It's important to recognize the symptoms of peak asymmetry caused by merged peaks because any measurement of the composite peak would include two different substances. This is not unusual in a laboratory analysis separating dozens of components but it's unexpected in a PGC; the extra

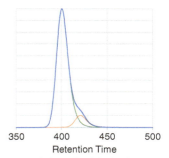

A resolution of 1.0 is inadequate for the measurement of normal asymmetric peaks. Carefully note the resulting shape of the blue curve. The deviation from a smooth curve indicates another peak is hiding there (orange curve).

Figure 8.13 Normal Peaks at Low Resolution.

220 Evaluating peak shape

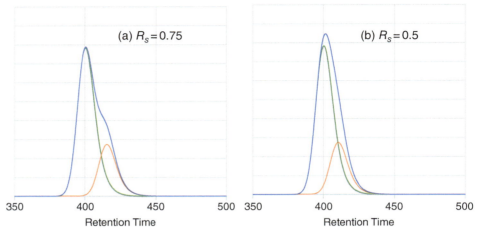

Figure 8.14 Asymmetry due to Merged Peaks. (a) $R_s = 0.75$. (b) $R_s = 0.5$.

When two or more normal peaks merge together, the resulting chromatogram (blue) may show a partial separation (a) or just a single asymmetric peak (b) depending on the resolution. To diagnose the presence of two peaks in (b) realize the composite peak is wider than a normal peak. Hint: look at the individual peaks (green and orange).

peak is probably a fault condition causing a large error in the analyte measurement. A PGC troubleshooter must closely inspect the gradual change of slope of each analyte peak, searching for any inconsistency. To get a better look at its shape, try stretching a narrow peak by using a fast chart recorder or by expanding the display time base.

Figure 8.15 illustrates this technique. In chromatogram (a), the single peak gives little indication that it's really two merged peaks. In chromatogram (b), the stretched baseline reveals a second peak. Can you see it?

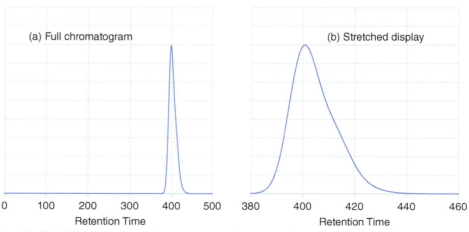

Figure 8.15 Looking for Merged Peaks. (a) Full Chromatogram. (b) Stretched display.

In (a), it's difficult to detect the presence of a merged peak in the full chromatogram. Stretching the timebase in (b) visually reveals an inflection on the trailing side of the peak due to a 20 % second peak hiding there. To see it more clearly hold the page at an angle with your eye aligned with the trailing edge of the peak.

In this book, we show simulated chromatograms with wide peaks to clearly show their characteristic shapes. The peaks from high-resolution capillary columns may be too narrow to evaluate – to see more, stretch their time base!

Some merged peaks show no irregularity whatsoever. For instance, the blue peak in Figure 8.14b apparently has a normal shape. The ultimate test is to compare the peak width with the width of nearby peaks. This is easy to do in Figure 8.14 because the graphic includes the green and orange traces of the individual peaks. A real chromatogram would show only one curve like the blue line in the figure. Yet an expert troubleshooter would notice that the peak is too wide and would deduce the combination of real peaks necessary to generate the observed shape.

An easy clue is when an onscale peak is wider than the peaks on either side of it. If all the peaks are from the same column, it's a sure sign that the wide peak is a composite of two or more peaks.

Offscale peaks

There are two kinds of chromatogram display:

> A time-based plot of the raw detector signal at constant sensitivity, showing the relative size of each peak. Often, the tops of one or more peaks will be offscale.

> A time-based plot of the modified detector signal, showing each measured peak at a sensitivity selected to keep it onscale. The top of any unmeasured peak may be offscale.

Figure 8.16 illustrates the effect of different size peaks on peak separation. The blue line shows the chromatogram at constant sensitivity, so the large peak goes offscale. The green line traces the shape of the large peak when attenuated to bring it back onscale.

In chromatogram 8.16a, the two peaks have a concentration ratio of about 2:1. Their resolution is excellent.

In chromatogram 8.16b, the concentration ratio has increased to 100:1 and the first peak is now offscale and wider than before. The green peak shows the shape of this peak when attenuated by a factor of 100. Notice that its retention time and width have not changed much. When looking at an offscale peak remember that its retention time is not at the center of its width; it's much closer to the front of the peak. Realize that the extra width of the blue peak is mainly due to amplifying its base and is mostly on its trailing side because the peak is asymmetric. However, the large peak has not formed a tail and the resolution is still good; both peaks are measurable.

In chromatogram (c), the first peak is much larger and is way offscale. Again, its retention time has not changed but the peak is wider and has formed a tail that runs into the second peak, interfering with its measurement. The tailing is due to an increase in the *second retention mechanism*,

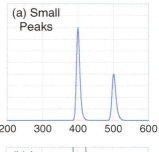
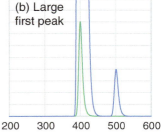
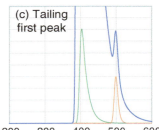

Simulated chromatograms in blue, same peaks at reduced sensitivity in green. In (a), two peaks of similar size exhibit excellent resolution. In (b), the first peak is offscale but shows no significant tailing. In (c), the first peak is so large that it has overloaded the column forming a tail that overlaps the second peak.

Figure 8.16 Effect of Peak Size. (a) Small Peaks. (b) Large First Peak. (c) Tailing First Peak.

more molecules have adsorbed on contact surfaces because of the larger number of molecules present. At an even higher concentration, the peak tail will totally obliterate the second peak. This illustrates the classic problem of trying to measure ppm of an analyte that follows a 99.99 % major peak. As you increase sample volume to gain the sensitivity needed to measure a ppm peak, you also overload the column with the major component.

Note that the first peak retention time is the same in each case (400 s). The peak has become more asymmetric and the width increase is mostly on the downside of the peak. To overcome this extreme tailing problem, a ppm measurement usually requires the heartcut column switching system discussed in Chapter 11.

Partial peaks

Looking back at Figure 8.14a, it's easy to misidentify the peak as a **partial peak**, a peak missing a piece due to column switching. Seeing the sudden drop at the trailing edge of the peak, one might think column switching has removed a slice from the peak tail. Yet in this case, we know the peculiar peak shape is due solely to overlapping peaks.

When column switching removes a portion of a peak and only a part of the peak reaches the detector, the resulting partial peak will exhibit some characteristic features. A later chapter explores the advanced troubleshooting skills you'll need to diagnose those special shapes.

Table 8.1 summarizes the causes of common peak shapes and may assist with troubleshooting.

Table 8.1 Troubleshooting Peak Shapes.

Observed Shape	Proximate Cause	Root Cause	Mitigation
Normal skew	Transferring a spatially symmetrical peak from the column to the time domain of the chromatogram.	Inherent in all gas chromatographs.	Minimize sample volume. Use a more efficient column to keep peaks narrow.
	Poisson distribution occurs in column.	Very low plate number.	Use a more efficient column.
	Center of peak migrates faster than peak edges.	Nonlinear isotherm: solubility falls at higher peak concentrations.	Reduce sample volume to avoid excess analyte concentration in column and to minimize thermal effect.
		Heat of solution: front of peak is warmer than back of peak.	
	Second retention mechanism.	Some peak molecules adsorb from the front of peak and desorb later, adding to the back of peak.	Minimize adsorption sites in column and in flow path.
	Composite peak.	Precisely timed overlap of two or more chromatogram peaks having little resolution.	Modify columns to get adequate resolution. Measure the integrated sum of both peaks. Increase injector temperature.

(continued)

Table 8.1 (continued)

Observed Shape	Proximate Cause	Root Cause	Mitigation
Tailing peaks	Polar and nonpolar solute peaks are both exhibiting tailing.	A large liquid sample or one containing high boilers takes time to vaporize resulting in a slurred injection profile.	Reduce sample size. If using a sample splitter increase split vent flow.
		Dead legs in the analytical flow path. Dead legs affect the sharp early peaks more than the wide later peaks and large peaks more than small ones.	Ensure no dead legs exist in injector, detector, and tube connections. Increase carrier gas flow rate. For a mass-flow detector, increase flow by adding a makeup gas.
		Porous polymer surfaces in valve rotors or sliders, or tubing.	Employ less-porous materials, for example, PEEK.
		A slow accumulation of solids or oils in injector, connecting tube, or first column.	Improve filtration of sample fluid in the external sample conditioning system. Lower injector temperature to reduce the formation of polymer oils. Clean or replace affected parts.
		The columns may be contaminated.	Replace the column set.
		Overloading a PLOT column.	Reduce sample size.
	Only the polar solutes are tailing – the nonpolar peaks are unaffected. A small proportion of the polar peak molecules are migrating along the column much slower than the main peak does.	Adsorption and randomly delayed desorption of low-concentration polar analytes at active sites on contact surfaces.	If a packed column, consider using a treated support. If a capillary column, the liquid coating may not be fully covering the tube walls. Consider increasing the film thickness. Increase oven temperature. Shorten all the connection tubes. Consider silicon-treated tubing for column and connections.
		Using a nonpolar liquid phase incompatible with polar solutes.	Use a more polar liquid phase.
		Solute has a severely nonlinear isotherm on the chosen phase (probably an adsorbent solid column).	If possible, work at a higher column temperature. Use a different stationary phase.
		Harsh sample (acidic, basic, or reactive) may have created active sites in the flow path.	Employ columns, fittings, and tubing of resistant material.
	Only the nonpolar peaks are tailing (unusual).	Severe mismatch between solutes and liquid phase.	Use a less polar or nonpolar liquid phase.
	Only one peak is tailing.	A large offscale peak is tailing. Normal in ppm analysis due to very large concentration ratios.	Use a heartcut column system to remove most of the major component.
		A small coeluting peak is overlapping the back of a normal peak.	Use a different stationary phase. Measure the integrated sum of both peaks.

(continued)

Table 8.1 (continued)

Observed Shape	Proximate Cause	Root Cause	Mitigation
		The solute is decomposing or reacting with the carrier gas or liquid phase.	Reduce column temperature. Don't use hydrogen carrier gas. Use a different stationary phase.
Fronting peak	Front of peak migrates slightly faster than back of peak.	Nonlinear isotherm: solubility rises as concentration falls.	Reduce sample volume to avoid excess analyte concentration in column.
		Solid stationary phase with curved isotherm.	Find a better stationary phase.
	Column overload.	Excessive sample feed volume changes composition of liquid phase.	Reduce sample volume. Increase column diameter.
	Overlapping peaks.	A small coeluting peak is overlapping the front of a normal peak.	Use a different stationary phase. Measure the integrated sum of both peaks.
Triangular peak (no tail)	Steep front: migration rate rapidly increases with peak height.	Solid phase with strongly downward-curved isotherm. Reduced adsorption at higher concentrations.	Increase column temperature. Find a better stationary phase.
	Steep back: migration rate rapidly decreases with peak height.	Liquid phase with strongly upward-curved isotherm. Solubility of analyte increases as more analyte molecules dissolve.	Use a liquid phase of similar polarity as the analyte. Reduce the sample size.
Overlapping peaks	Partially separated peaks	Two peaks with insufficient resolution.	Optimize column temperature to put peaks in Power Region. Use a longer column. Use a different stationary phase.
	Onscale peak is wider than adjacent peaks from the same column.	Two or more merged peaks – separation factor is too low.	Measure the integrated sum of both peaks. Use a different stationary phase.
Partial peak		May be a single heavies peak from previous analysis cycle.	Adjust backflush settings to remove the heavies peak.
	Front is missing.	Column valve switching too late.	For diagnosis and time settings consult Chapters 10 and 11.
	Back is missing.	Column valve switching too early (or may be intentional).	
	Remnant peak.	Normal heartcut remnant peak (see Chapter).	No action required.
Split peak (bimodal)	Peak forms a valley at its apex.	Partial condensation of vaporized sample	Increase injector or column temperature.
	Looks like a plateau between two peaks	Analyte is reacting with carrier gas, liquid phase, or itself to form another chemical compound with a different retention factor.	Reduce the column temperature. Don't use hydrogen carrier gas.
Flat-top peak	Peak has a flat top. Signal processing routines looking for a slope change may fail to detect the peak.	Flat-top peak is actually two or more overlapping peaks of similar size. This sometimes occurs with backflushed peaks.	Use a different liquid phase or measure the sum of both peaks. Change Method settings to forced integration mode.

(continued)

Table 8.1 (continued)

Observed Shape	Proximate Cause	Root Cause	Mitigation
		Detector response has exceeded PGC signal limit.	If peak is to be measured, reduce sample size.
			Modify electronics to reduce detector sensitivity.
Jagged peaks	The narrow peaks on a chromatogram don't exhibit smooth curves.	Processor is sampling the peak signal too slowly. Inadequate data points to properly represent the peak.	Increase the sampling frequency. 40 Hz is often considered minimal and may be too slow for the sharp peaks from a capillary column.
		Data smoothing by simple averaging reduces the number of datapoints per peak.	Inhibit data smoothing routine.

Knowledge Gained

Practice

- A symmetrical Gaussian peak is the ideal shape of a chromatogram peak but not often seen in practice.
- Real chromatogram peaks are rarely symmetrical and some deviation from perfection is normal.
- Normal peaks rise to their apex faster than they return to the baseline.
- Tailing peaks exhibit a long exponential return to the baseline that may indicate a fault condition.
- Fronting or leading peaks rise to their apex slower than they return to the baseline.
- Normal peaks have more than 1.0 of measured skew, and fronting peaks have less than 1.0 skew.
- A skew of 0.8–1.2 (about ±20 %) is normal and usually acceptable for measurement.
- Theory shows the normal peak shape with a small positive skew is inevitable (see below).
- In addition, slow desorption from scarce active sites may contribute to normal peak skew.
- A partition or adsorption isotherm having limited linearity may skew the larger peaks.
- If the center concentration of a peak has a lower solubility, it travels faster than the peak edges.
- A continuously curved isotherm is common for solid stationary phases and guarantees skewed peaks.
- A curved isotherm may cause a triangular-shaped peak with an almost vertical rise or fall.
- A skewed peak may be the combination of two or more partially resolved peaks.
- When its concentration changes, a skewed peak is likely to exhibit a shift in retention time.
- An increase in concentration of a skewed peak may reduce its resolution from an adjacent peak.

- *A slow response of detector or electronics will make a fast peak asymmetric in the trailing direction.*

- *Fronting peaks are uncommon and may indicate column overload due to a large sample volume.*

- *All peaks tailing may be due to slow injection, deposits in the injector, or dead legs in connections.*

- *Polar peaks may tail on a nonpolar column or by adsorption on surfaces or on deposits in the injector.*

- *Moderately offscale peaks may seem wider than onscale peaks due to amplification of their base.*

- *In ppm analysis, the very large major peak forms a tail which can hide a small analyte peak.*

- *An asymmetric peak might be a composite of two or more overlapping peaks.*

- *Diagnosis of peak shape is an important skill for the PGC troubleshooter.*

Theory

- *For several reasons, a small degree of peak asymmetry is inevitable in chromatography.*

- *In theory, peak shape is a Poisson distribution, which is asymmetric at low plate number.*

- *The trailing half of a peak spends longer in the column and widens due to additional dispersion.*

- *The distortion due to additional dispersion is significant only with wide, low-efficiency peaks.*

- *The heat of sorption makes the front of a peak warmer and faster than the rear of the peak.*

- *The asymmetry caused by heat of sorption is more significant with larger peaks.*

- *Excessive sample volume may overload the column and produce a fronting peak.*

- *There is a different equation to calculate the efficiency of a column from an asymmetric peak.*

Did you get it?
Self-assessment quiz: SAQ 08
On practice

Consider the following mechanisms of retention:
- **A.** A linear partition isotherm.
- **B.** A linear partition isotherm that droops at high concentration.
- **C.** A linear partition isotherm that rises at high concentration.
- **D.** A linear partition isotherm coupled with slow desorption from scarce active sites.
- **E.** A continuously curved partition or adsorption isotherm.
- **F.** The injection of an excessively large sample volume.
    - **Q1.** Which of the above conditions is likely to form large or small perfectly symmetrical peaks?
    - **Q2.** Which of the above conditions may form a normal asymmetric peak?

Q3. Which one of the above conditions is most likely to form an asymmetric fronting peak?

Q4. Which one of the above conditions is most likely to form a tailing peak?

On theory

Consider the following mechanisms that may affect peak shape:

A. A Poisson distribution of molecules.
B. A Gaussian distribution of molecules.
C. An exponentially modified Gaussian distribution of molecules.
D. The transformation of an on-column peak into a chromatogram peak.
E. The heat of solution or adsorption.
F. Adsorption and desorption of analyte molecules at active sites.

Q5. Which of the above mechanisms always forms a perfectly symmetrical peak?

Q6. Which of the above mechanisms might form a normal asymmetric peak?

Q7. Which one of the above mechanisms might form a tailing peak?

References

Further reading

Although focused on capillary columns in laboratory practice, the GC troubleshooting book by Rood (2007) is a good guide.

For a more theoretical analysis of peak asymmetry, you might try the article by Hsu and Chen (1987).

Cited

Foley, J.P. (1987). Equations for chromatographic peak modeling and calculation of peak area, *Analytical Chemistry* **59**, 1984–1987.

Foley, J.P. and Dorsey, J.G. (1983). Equations for calculation of chromatographic figures of merit for ideal and skewed peaks. *Analytical Chemistry* **55**, 730–737.

Foley, J.P. and Dorsey, J.G. (1984). A review of the exponentially modified gaussian (EMG) function; evaluation and subsequent calculation of universal data. *Journal of Chromatographic Science* **22**, 40–46.

Giddings, J.C. (1965). *Dynamics of Chromatography: Principles and Theory*. New York, NY: Marcel Dekker.

Hsu, J.T. and Chen, T.L. (1987). Theoretical analysis of the asymmetry in chromatographic peaks. *Journal of Chromatography A* **404**, 1–9. https://doi.org/10.1016/S0021-9673(01)86831-X.

Rood, D. (2007). *The Troubleshooting and Maintenance Guide for Gas Chromatographers*. Weinheim, Germany: Wiley-VCH Verlag GmbH & Co. KGaA.

Scott, R.P.W. (2003). Extra-Column Dispersion, Book 10 Chrom-Ed Book Series by LIBRARY4SCIENCE, LLC. Accessed 2017/7/30 at: http://www.chromatography-online.org.

Scott, R.P.W. (2014). The Plate Theory and Extensions, Book 6 of Chrom-Ed Book Series by LIBRARY4SCIENCE, LLC. Accessed 2017/7/30 at: http://www.chromatography-online.org.

Waters, T. (2020). *Process Gas Chromatographs: Fundamentals, Design and Implementation*. Chichester, UK: John Wiley & Sons Ltd.

Table

8.1   Troubleshooting Peak Shapes.

Figures

8.1   Three Asymmetric Chromatogram Peaks.
(a) Normal Peak. (b) Tailing Peak. (c) Fronting Peak.
8.2   Measuring Peak Skew.
8.3   Typical Partition Isotherm.
8.4   Isotherms and Peak Shape.
(a) Normal Peak. (b) Ideal Peak. (c) Fronting Peak.
8.5   Typical Adsorption Peaks.
8.6   A Second Retention Mechanism.
8.7   The Poisson Distribution.
8.8   The EMG Peak Model.
8.9   Retention Shift with Concentration.
8.10 Effect of Peak Shape on Resolution.
(a) At Low Concentration. (b) At High Concentration.
8.11 Normal and Tailing Peaks.
(a) No Peaks Tail. (b) Some Peaks Tail. (c) Major Peak Tail.
8.12 Resolution of Normal Asymmetric Peaks.
(a) For Perfectly Symmetric Peaks. (a) For Normal Asymmetric Peaks.
8.13 Normal Peaks at Low Resolution.
8.14 Asymmetry due to Merged Peaks.
(a) Partial Separation at $R_s = 0.75$. (b) Single Merged Peak at $R_s = 0.50$.
8.15 Looking for Merged Peaks.
(a) Full Chromatogram. (b) Stretched display.
8.16 Effect of Peak Size.
(a) Small Peaks. (b) Large First Peak. (c) Tailing First Peak.

Symbols

Symbol	Variable	Unit
A_s	Asymmetry ratio.	None
a	Peak front half-width at 10 % of height.	s
b	Peak back half-width at 10 % of height.	s
N	Plate number.	None
t_R	Peak retention time.	s
τ	Standard deviation of chromatogram peak.	s
$w_{0.1}$	Peak width at 10 % of height.	s
σ	Standard deviation of on-column peak.	mm

Equations

8.1 $\quad A_s = \dfrac{b}{a}$ $\qquad$ Peak asymmetry ratio.

8.2 $\quad N = \dfrac{41.7(t_R/w_{0.1})^2}{b/a + 1.25}$ $\qquad$ Plate number of an asymmetric peak.

New technical terms

If you read the whole chapter, you should now know the meaning of these technical terms:

asymmetry ratio
bimodal peak
column overload
dead leg
exponentially modified Gaussian
leading peak

normal peak
partial peak
partition isotherm
Poisson distribution
second retention mechanism
self-sharpening peak

For more information, refer to the Glossary at the end of the book.

9

Columns in series

> *"For a general introduction to the practices and applications of multiple-column systems in PGCs, refer to our companion volume (Waters 2020, 157–181). Multicolumn systems all have one feature in common, columns in a series configuration always operate with different carrier pressures and velocities, and this affects their performance."*

The need for multiple columns

For many reasons, most process gas chromatographs use more than one column. These column arrangements attempt to satisfy more than one of these desired outcomes:

➢ To minimize the analysis **cycle time**.

➢ To remove all injected components before injecting another sample.

➢ To regroup and measure the aggregate concentration of a set of similar components.

➢ To obtain a resolution between components that no single column can achieve.

➢ To resolve and measure very small amounts of an analyte in the presence of a very large amount of something else.

➢ To protect a sensitive column or detector from damage by incompatible compounds in the injected sample.

To minimize analysis time

On an isothermal column, the retention time of solutes increases logarithmically with their molar masses (and with their boiling points) resulting in the so-called **general elution problem**; that is, *resolving the lighter peaks will always cause the heavier peaks to be strongly retained.*

While fast analysis may not be an important criterion for a research or laboratory chromatograph, it's always a high priority for a PGC used for process control. A multiple-column system working at constant temperature can give analytes a shortcut to a detector and unwanted components a quick exit to vent, thereby completing the desired analysis in less time than a temperature-programmed column could achieve. An isothermal PGC is also less costly to buy, consumes less energy, and is easier to maintain than a temperature-programmed PGC.

To do the housekeeping

The **housekeeping rule** of column system design requires a PGC to remove all injected components from the columns before injecting another sample. If not, the later components may interfere with a subsequent analysis. Even worse, they may accumulate in the columns and eventually reach the detector, upsetting the chromatogram baseline and invalidating the measurements. With solid-phase columns, such accumulation can even deactivate the stationary phase causing a complete loss of separation.

Temperature programming accelerates the **heavies** peaks but cannot guarantee to remove all components from the column. In contrast, we trust an isothermal **backflush column system** (described in Chapter 10) to remove all the later eluting components from the injected sample, whatever they are.

To regroup a set of components

Sometimes the user wants to measure the sum total of several components. The classic case is backflush-to-measure which regroups all the heavies into one peak for measurement. In other cases, the plant may need to measure groups of similar analytes, such as Total C_4 and Total C_5. We shall see that a **regrouping column system** always involves flow reversal in one of the columns.

To enhance the separation

Using two different stationary phases can have an enormous effect on analyte resolution. For instance, the first column may resolve Analyte A but not Analytes B and C. The second column may resolve Analyte C but not Analytes A and B. Working together, the two columns achieve what neither column could do alone: a fast resolution of all three components: A, B, and C.

The simple **distribution column system** described in Chapter 11 extends this principle to two or more second columns working in parallel. It does a partial separation on the first column, then distributes groups of analytes to different columns for final separation.

Even when one stationary phase is capable of separating all the analyte peaks, a second phase may be an advantage. Two carefully selected phases may provide optimal resolution between all the adjacent pairs of component peaks and thereby minimize the time for analysis.

To separate trace amounts of analyte

A chromatographic system has a limited dynamic range. The large sample volume and high detector sensitivity necessary to measure parts-per-million (ppm) of trace analytes results in severe column overload by the major component which precludes the separation of ppm analytes from the major tail. A **heartcut column system** (see Chapter 11) elegantly solves the problem. A short first column holds back the ppm peaks while most of the major component flows to vent. A switching device then allows the analytes and a much smaller amount of major component to enter a longer second column which performs the final separation.

To protect a sensitive column or detector

There might be compounds in a process sample that would damage a column or contaminate a detector if they came into contact with it. A classic example is the need to protect an active-solid column like molecular sieves from deactivation by water in the injected sample; another is to avoid contaminating a sensitive electron capture detector with halogen compounds. A two-column column system can isolate deleterious components and flush them to vent or steer them toward a compatible detector.

History of the technique

The early development of commercial gas chromatographs diverged, with process and laboratory instruments adapting to the needs of their respective users and to their different operating environments. From a column design perspective, they reached the apogee of their divergence in the 1980s. At that time process chromatographers started to adopt capillary columns while laboratory chromatographers rediscovered the advantages of dual-column systems.

Development of laboratory chromatographs

In the laboratory, the ability to separate and measure many analytes took precedence over analysis time. Often, laboratory GCs are set up for qualitative analysis – discovering what chemicals are present in the sample as well as how much of each. For this task, capillary columns were a godsend for laboratory analysis and won quick acceptance. With temperature programming, a single capillary column could quickly separate and identify scores of components, so the development of laboratory gas chromatographs focused on the special needs of capillary columns: low dead volumes, **split injection**, bonded stationary phases, temperature programs, and column ovens to reach ever-higher temperatures.

Until about 1990, most laboratory GCs employed a single capillary column, but the demand for separating and identifying even more components inexorably led to the adoption of capillary columns in series. Researchers anxious to put their mark on this "new" GC technique named it **two-dimensional gas chromatography** (Dimandja 2020). This new approach goes far beyond the simple act of combining two stationary phases. The first column is a powerful capillary to achieve a good separation of dozens of peaks. For samples containing scores or even hundreds of peaks, though, this is just a start. The second column is a short capillary with a liquid phase chosen to rapidly separate bunches of components not resolved by the first column. Of course, we are familiar with the idea. But when there are hundreds of peaks competing for a space on the chromatogram some of the peaks held back on the second column coelute with later peaks from the first column. The technique needed something to delay those later peaks.

The solution came with the invention of new methods of peak transfer between columns. The transfer device is generically known as a **modulator** and takes many forms, sometimes just a column valve or Deans switch. To get a programmable delay, more complex modulators use a low temperature trap to focus a bunch of peaks into a narrow band before reinjecting them into a second column (Liu and Phillips 1991).

A moment's thought reveals that the spacing between reinjections must be small so it's unlikely that the second column will be able to separate all the reinjected components. This is where it gets complicated. The modulator may reinject slugs of partially separated peaks into Column 2 that are only about one second wide. These slugs then separate and arrive at the detector as overlapping bands. The detector output is then a **correlogram** rather than a chromatogram and requires a complex mathematical procedure to deconvolute the mixed peak waveform.

The plethora of peaks generated by dual capillary columns raised the question of how to positively identify them all. The answer was to develop new detectors that use spectrometric methods to measure and identify the peaks. These new techniques fit neatly into the evolving classification of "hyphenated methods" in analytical chemistry and gave birth to the methodologies listed in Table 9.1.

Development of process chromatographs

Most process analyzers are based on established laboratory techniques adapted to suit the process environment. In contrast, the development of the process gas chromatograph occurred in parallel with its laboratory counterpart. Instrument manufacturers were not the first to design a process gas chromatograph. That recognition falls upon chemical and oil companies (notably Phillips Petroleum and Union Carbide) who seized upon the unique ability of a PGC to measure a few selected analytes quickly and accurately in a process stream containing many other components. The PGC rapidly advanced on its own independent trajectory, focused on the fast measurement of a few known analytes.

Table 9.1 A Lexicon of GC Methods.

1D GC	Single column gas chromatography
2D GC	Two-dimensional gas chromatography
C2DGC	Comprehensive 2-D gas chromatography
GC–GC	Selective 2-D gas chromatography (heartcut)
GCxGC	Comprehensive 2-D gas chromatography (all peaks)
GC-MS	Single column with mass spectrometric detector
GCxGC–MS	Dual column with mass spectrometric detector
GCxGC–VUV	Dual column with vacuum ultraviolet detector

This table is nonexclusive and is for general information only. Note that "heartcut" has a somewhat different meaning here than in PGC. In laboratory parlance, **heartcut methods** analyze only selected components, whereas **comprehensive methods** analyze them all.
Source: Adapted from Elsevier.

PGCs adopted multicolumn systems for all the reasons outlined above. They had backflush column systems soon after their 1956 introduction (Simmons and Snyder 1958). The heartcut column systems came a little later, after the flame ionization detector introduced by McWilliam and Dewar (1958) became commercially available. Isothermal PGCs equipped with multiple-column systems proved more adept at fast analysis and were less expensive and more reliable than those using temperature programming.

By design, most early PGCs could house only simple dual-column systems. Their key constraint was having a single detector, with no possibility of adding a second one. Column system designers had to be highly creative to use one detector to measure all the analytes. Unfortunately, these creative designs used complex multiple-column systems that were difficult to maintain, and hence unreliable. Relief eventually came from microprocessor technology, specifically the ability to process several detector signals simultaneously. This spurred the development of PGCs with multiple detectors and multiple column ovens, allowing a complex analysis to be broken down into a few simple backflush trains, each with its own detector.

PGC's also differ in their preference for packed columns over capillary columns. Even today, most PGCs use two or more packed columns. Back in 1964, McEwen demonstrated the use of capillary columns in a PGC, but their adoption has been slow. Simply installing a capillary column in place of a packed column is not enough. The whole PGC instrument must be adapted for the low flow rates used by capillary columns; the original valves and detectors were far too voluminous. The other drawback was the fragility of fused silica columns – many of the new capillary-column

Table 9.2 Comparing Laboratory and Process Gas Chromatographs.

	Laboratory GC	Process GC
Operating Environment:	Indoor, safe, and stable.	Outdoor, hazardous, and inconstant.
		Earlier Models
Separations:	Many known analytes.	A few known analytes.
Sample Inject:	By manual syringe.	By automatic valve.
Column Types:	Single packed column. Single capillary column (later).	Dual and multiple packed columns.
Column Switching:	Usually none.	Complex multiple-valve arrangements.
Detector(s)	Single detector, wide choice.	Single detector, limited choice. Several detectors (later).
Peak Identification:	By retention time.	By retention time.
Oven Temperature:	Wide range, high limit. Often programmed.	Narrow range, lower limit. Mostly isothermal.
Oven Type:	Single oven with internal fan.	Single oven; mostly air bath, but some using the thermal mass type (airless).
		Later Models
Measurements:	Vast numbers of unknown analytes.	Several known analytes.
Sample Inject:	Automated syringe and valve injection.	Multiple valve injections. Simple column trains.
Column Types:	Single or dual capillary columns. Two-dimensional chromatography.	Micropacked or capillary columns. Several parallel dual-column systems.
Column Switching:	Deans switch.	Complex analyses divided into several simple column systems, each with its own injector and column valve (or Deans switch) and detector.
Detector(s) Per PGC:	Supports few detectors, many types.	Supports many detectors, limited types.
Peak Identification:	By spectrometric detector (confirms).	By retention time (subject to error).
Oven Temperature:	Wide range, very high limit. Usually programmed temperature.	Narrow range, lower limit. Mostly isothermal.
Oven Type:	Single oven with internal fan.	Multiple ovens; many airless.

PGCs failed due to recurrent column breakage. When fitted with Silcosteel[1] (silicon-coated) metal capillary columns, a modern low-volume PGC offers a durable alternative. Yet it's still true that packed columns in series are often a better alternative for quickly measuring a few known analytes.

Table 9.2 summarizes the features of older and newer chromatographs.

The power of dissimilar columns

The real power of serial columns is their ability to adjust the net retention factor of each analyte to achieve optimum separation in minimum time. By eliminating unproductive time on the chromatogram when no peaks are emerging, series columns with different stationary phases can often

---
[1]*Silcosteel* is a trademark of Silco Tek Corporation.

resolve several analytes faster than any single column could achieve. When combined with column-switching techniques like backflush and heartcut, different columns in series can quickly resolve almost any limited set of designated analytes.

The separating power of dual stationary phases allows PGC applications chemists to limit their search to about one dozen phases they know well. Most column designers maintain a chart or database that gives them the retention factor of all the industrial chemicals they are likely to encounter on their chosen stationary phases. They may save this data as retention indices or more simply as retention times on standard columns, sometimes at two column temperatures.

Given this data, Figure 9.1 shows an easy way to choose the relative effect of each column. The chart has the retention times for each stationary phase on the left and right vertical axes. Then, the vertical line with the best spacing of analyte retention times indicates the optimum combination of columns. In selecting the best line, remember that later peaks will require a little more spacing than the early ones.

After determining the desired residence times on each phase, one must decide which phase shall be Columns 1 and 2, probably determined by the retention of components that need to be backflushed or heartcut – also clear from the diagram. Next, you'll need the total length of column required to get the desired resolution. While it's possible to calculate that, it's most often estimated from experience and confirmed by experiment.

It often takes several practical trials to find the optimum length of each column and this experimental work increases the cost of application engineering. Each trial also incurs the cost of new columns with different lengths or different liquid loadings. An option is to use a PGC that has multiple

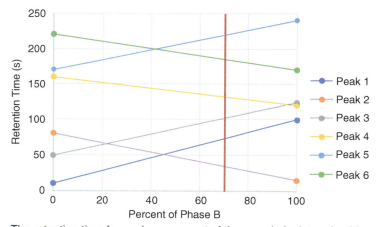

Figure 9.1 Dual Phase Selection Chart.

The retention time for each component of the sample is determined for stationary phase A and stationary phase B on standard columns. The values for Phase A are on the left vertical axis and for Phase B on the right vertical axis. In this example, the vertical red line indicates that a combination of 30 % Phase A and 70 % Phase B will provide the best separation of all six components.

column ovens operating at different temperatures. It may be cost-effective to fine-tune the contribution of each column by simply changing the temperature of a column.

The final task is to allow for the different carrier gas pressure in each column. When columns are in series each column works at a different pressure and this affects the retention times of the peaks. The next section explains the effect of pressure drop and how to adjust column lengths to compensate for it.

Pressures in series columns

A multiple-column system inevitably comprises two or more columns connected in series. In most process applications, the series columns have the same internal diameter and are running at the same carrier flow rate but they don't experience the same pressure. Therefore, their carrier gas velocities must be different.

This effect is present even in a single column, but we don't notice it. The mass flow of carrier gas is constant throughout the column but the gas expands as the pressure drops, causing its velocity to gradually increase. As the velocity increases, the rate of pressure loss also increases, causing a nonlinear pressure drop along the column. The carrier gas reaches its highest velocity and its highest rate of pressure loss at the column outlet.

From van Deemter (Equation 6.7), we know that column efficiency depends on carrier gas velocity, but velocity varies along the column, so different parts of a column work at different efficiencies. This is one reason that capillary columns are more efficient than packed columns; they exhibit smaller pressure drop so every part of the column is running closer to its optimum velocity.

Working with a single column, we observe only its average velocity and its average efficiency. But when that column is in two parts, the different velocities and efficiencies become apparent; the first column operates at a lower velocity than the second column.

> *When working with multiple columns, it's important to recognize the effects of low velocity on the first column. Peaks will elute more slowly from the first column than one would expect, judging by its length.*

The extent of the nonlinearity depends on the absolute pressure ratio between the column inlet and the column outlet. In practice, it's mainly an issue with packed columns. This is because:

➤ Packed columns are more common in serial configurations.

➤ Packed columns are more flow resistive than capillary columns and tend to run with higher pressure differentials.

The **SCI-FILE:** *On Pressure Drop* quantifies the effects of pressure drop in series columns. While it may be unnecessary to study the **SCI-FILE** in detail, note that the rest of this chapter will use the equations developed therein.

SCI-FILE: On Pressure Drop

Pressure loss in columns

The rate of pressure loss isn't linear throughout a column nor is the rate of velocity increase. They both reach maximum values at the column outlet. The extent of the nonlinearity depends on the ratio of the absolute pressures at the inlet and the outlet – a larger ratio results in a more severe effect.

The equations for calculating the pressure and velocity at specific points along a column use two ratios:

> P_r is the ratio of column inlet pressure (p_i) to column outlet pressure (p_o):

$$P_r = \frac{p_i}{p_o} \quad (9.1)$$

> P_{rz} is the ratio of the pressure (p_z) at a point (z) to the column outlet pressure (p_o):

$$P_{rz} = \frac{p_z}{p_o} \quad (9.2)$$

Equation 9.3 calculates the pressure ratio (P_{rz}) at a specified point (z) along the column:

$$P_{rz} = \sqrt{P_r^2 - \frac{Z}{L}(P_r^2 - 1)} \quad (9.3)$$

where Z is the distance from the column inlet to point (z) and L is the column length (Ettre and Hinshaw 1993, 51; Jennings et al. 1997, 120).

From Equation 9.3, Figure 9.2 plots gas pressure versus the distance along the column for various pressure ratios. As instantly evident, nonlinearity is more severe with large pressure differentials.

Average pressure in a column

Due to the nonlinear pressure drop, the **average pressure** in a column is higher than the average

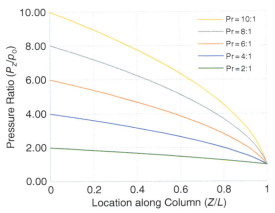

The rate of pressure loss along a column is nonlinear. Most of the pressure drop and the concurrent velocity increase occurs near the end of the column, particularly for high pressure ratios.

Figure 9.2 Nonlinear Pressure Drop.

of its inlet and outlet pressures. The average pressure ($\bar{p}$) is a function of outlet pressure (p_o):

$$\bar{p} = \frac{p_o}{j} \quad (9.4)$$

where $j < 1$.

The **gas compression factor** (j) is a function of the pressure ratio (P_r) applied to the column:

$$j = \frac{3}{2} \cdot \frac{(P_r^2 - 1)}{(P_r^3 - 1)} \quad (9.5)$$

Equation 9.5 is important when working with columns in series as each column has different inlet and outlet pressures.

Average velocity in a column

In Equation 2.2, we measured the average carrier velocity ($\bar{u}$) in a column by dividing the length (L) of the column by the retention time (t_M) of an unretained component:

$$\bar{u} = \frac{L}{t_M} \quad (9.6)$$

However, the average velocity is of limited use for columns in series; we need to know the average velocity in each column. Fortunately, there's an easy way to calculate the average velocity in a column from its average pressure.

Since the mass flow rate is constant throughout, the carrier gas velocity (u_z) at any point in a single column is inversely proportional to the carrier gas pressure (p_z) at that point. Expressed relative to the outlet pressure (p_o) and outlet velocity (u_o):

$$\frac{u_z}{u_o} = \frac{p_o}{p_z} \quad (9.7)$$

Velocity in series columns

Equation 9.7 is also true for multiple columns in series provided they have the same permeability; that is, the same internal diameter and packing density.[2] If so, the **average velocity** ($\bar{u}$) of the carrier gas in each column is inversely proportional to its average pressure ($\bar{p}$) in that column.

The average velocity in each column (designated by subscripts $_1$ and $_2$) follows from Equation 9.7 applied to each column. As a ratio, it reduces to:

$$\frac{\bar{u}_2}{\bar{u}_1} = \frac{\bar{p}_1}{\bar{p}_2} \quad (9.8)$$

Equation 9.8 is true for any column lengths.

Retention time in series columns

From Equation 9.6, the retention time (t_M) of an *unretained component* in a column of length (L) is:

$$t_M = \frac{L}{\bar{u}} \quad (9.9)$$

And from Equation 2.3, the retention time (t_R) of a *retained component* is:

$$t_R = t_M \cdot (k + 1) \quad (9.10)$$

where (k) is the retention factor of the retained component.

Combining Equations 9.9 and 9.10, the retention time of any component on any column is:

$$t_R = \frac{L}{\bar{u}} \cdot (k + 1) \quad (9.11)$$

Therefore, if two columns operate in series and have different stationary phases, and t_{R1}/t_{R2} is the ratio of retention times for a particular component on Columns 1 and 2, then:

$$\frac{t_{R1}}{t_{R2}} = \frac{L_1}{L_2} \cdot \frac{\bar{u}_2}{\bar{u}_1} \cdot \frac{(k_1 + 1)}{(k_2 + 1)} \quad (9.12)$$

Equation 9.12 is the general rule for the retention of a designated peak in series columns.

If the permeability of the two columns is the same, we can combine Equations 9.8 and 9.12 to give a similar expression based on the average pressure in each column:

$$\frac{t_{R1}}{t_{R2}} = \frac{L_1}{L_2} \cdot \frac{\bar{p}_1}{\bar{p}_2} \cdot \frac{(k_1 + 1)}{(k_2 + 1)} \quad (9.13)$$

Retention time in series columns

When both columns have the same stationary phase, the retention factor of the peak is the same, therefore:

$$k_2 = k_1$$

Then, from Equation 9.13:

$$\frac{t_{R1}}{t_{R2}} = \frac{L_1}{L_2} \cdot \frac{\bar{p}_1}{\bar{p}_2} \quad (9.14)$$

And if the two columns are also the same length:

$$L_1 = L_2$$

$$\therefore \quad \frac{t_{R1}}{t_{R2}} = \frac{\bar{p}_1}{\bar{p}_2} \quad (9.15)$$

Equation 9.15 reveals that the retention time in each column is proportional to its average carrier

---

[2]This may not be true for packed columns or for particle-coated capillaries due to variations in packing or coating geometry.

pressure. Therefore, if the columns were identical, Column 1 would retain each peak longer than Column 2. This would be undesirable in most separations, so the first column is usually shorter than the second one. The main text gives some examples of column length calculations.

Pressure effect on solubility

By combining Equations 9.8 and 9.15, we discover that peak retention times in identical serial columns differ only by their average carrier velocity:

$$\frac{t_{R1}}{t_{R2}} = \frac{\bar{u}_2}{\bar{u}_1} \quad (9.16)$$

It may come as a surprise that the carrier pressure in a column doesn't have a direct effect on the peak retention time.

The distribution constant (K_c) is the fundamental physical law governing the solubility of a gas in a liquid, and hence, gas chromatographic behavior. At equilibrium, the ratio of solute concentration in the stationary phase $[A]_S$ to solute concentration in the mobile phase $[A]_M$ is constant. Equation 4.4 defines the constant:

$$K_c = \frac{[A]_S}{[A]_M}$$

Note that the ratio of solute *concentrations* stays constant, not the ratio of solute masses. Although a change in carrier gas pressure automatically affects the partial pressure of a solute vapor, it does not change its concentration. Therefore, it has no effect on solute solubility. Since the distribution constant does not change, the gradual drop of carrier pressure in a column doesn't affect retention times directly – but the associated increase in carrier gas velocity certainly does.

Example calculations

These examples use equations from the **SCI-FILE** above. Since most of these equations use ratios of pressures, they can accept any units of measure, but they must be expressed as **absolute pressures**. In this book, we use pressure units of kilopascal (kPa) or bar (bar), where 100 kPa = 1 bar. To emphasize the use of absolute pressure, we use the symbol **bara** for absolute pressure and the symbol **barg** for gauge pressure.

Pressure in series columns

Let's look at two identical columns connected in series with an inlet pressure of 5 bara and an outlet pressure of 1 bara. Simple averaging predicts a junction pressure of 3 bara but Equation 9.3 gives a higher value; using $P_r = 5$ and $Z/L = 0.5$, the relative pressure (P_{rz}) at the junction is:

$$P_{rz} = \sqrt{25 - 0.5(25 - 1)}$$

$$P_{rz} = 3.61$$

Then, from Equation 9.2:

$$p_z = P_{rz} \cdot p_o = 3.61 \text{ bara}$$

From this, we know that Column 1 has an inlet pressure (p_i) of 5 bara and an outlet pressure (p_o) of 3.61 bara. From Equation 9.1, its pressure ratio (P_{r1}) is:

$$P_{r1} = \frac{p_i}{p_o} = 1.39$$

From Equation 9.5, its gas compressibility factor (j_1) is:

$$j_1 = \frac{3}{2} \cdot \frac{1.39^2 - 1}{1.39^3 - 1} = 0.831$$

Then, from Equation 9.4, its average pressure ($\bar{p}_1$) is:

$$\bar{p}_1 = \frac{p_o}{j_1} = 4.34 \text{ bara}$$

Now consider Column 2. Its inlet pressure is 3.61 bara and its outlet pressure is 1.00 bara, therefore:

$$P_{r2} = \frac{p_i}{p_o} = 3.61$$

$$j_2 = \frac{3}{2} \cdot \frac{3.61^2 - 1}{3.61^3 - 1} = 0.392$$

$$\bar{p}_2 = \frac{p_o}{j_2} = 2.55 \text{ bara}$$

Figure 9.3 illustrates these data. Although the columns are identical, they are working at different pressures. On average, the carrier pressure in Column 1 is 1.7 times higher than in Column 2:

$$\frac{\bar{p}_1}{\bar{p}_2} = \frac{4.34}{2.55} = 1.7$$

For two identical columns connected in series, Table 9.3 shows their junction pressure and the ratio of their average pressures ($\bar{p}_1/\bar{p}_2$) for various inlet pressures. The table assumes the second column vents to 1 bara, as most second columns connect to a detector at atmospheric pressure.

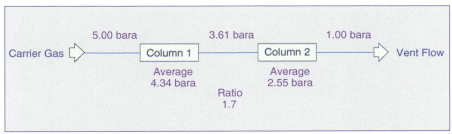

Showing the pressure distribution in two identical columns connected in series with an inlet pressure of 5 bara and an outlet pressure of 1 bara.

Figure 9.3 Pressure Distribution in Series Columns.

Table 9.3 Average Pressures in Identical Series Columns.

Inlet p_i Bara	Column 1 $\bar{p}_1$ Bara	Junction p_z Bara	Column 2 $\bar{p}_2$ Bara	Ratio $\bar{p}_1/\bar{p}_2$
10	8.64	7.11	4.82	1.79
8	6.92	5.70	3.90	1.77
6	5.20	4.30	2.99	1.74
4	3.49	2.92	2.11	1.65
3	2.64	2.24	1.70	1.55
2	1.80	1.58	1.31	1.37

Junction and average column pressures for two identical columns in series, assuming outlet pressure is 1.00 bara. These ratios apply to two equal columns of any length. Notice that the ratio of average pressures is about the same for inlet pressures above 4 bara.

Velocity in equal columns

In our current example, the two columns are equal in length. With a carrier gas inlet pressure of 5 bara, the ratio of average pressures is 1.7. From Equation 9.8, it follows that the average carrier velocity ($\bar{u}_2$) in Column 2 is 1.7 times faster than the average velocity ($\bar{u}_1$) in Column 1:

$$\bar{u}_2 = \frac{\bar{p}_1}{\bar{p}_2} \cdot \bar{u}_1 = 1.7\,\bar{u}_1$$

Retention time in equal columns

For column system design or troubleshooting, we need to predict the time a component peak will spend in each column. For a peak on two identical columns, any difference in retention is due only to their average carrier gas velocities, which in turn depends on their average pressures. Equation 9.15 confirms that for equal columns, the retention time of a peak in Column 1 is 1.7 times longer than its retention time in Column 2:

$$t_{R1} = \frac{\bar{p}_1}{\bar{p}_2} \cdot t_{R2} = 1.7 \cdot t_{R2}$$

For example, if a component peak passes through both columns with an overall retention time of 270 s, we calculate its retention (t_R) in each column as:

$$t_{R1} + t_{R2} = 270 \text{ s}$$

$$t_{R1} = 1.7\, t_{R2}$$

$$t_{R2} = 100 \text{ s}$$

$$t_{R1} = 170 \text{ s}$$

Note the much longer time in Column 1.

Table 9.4 Pressures and Retention Times for Columns in Series.

Length L_2/L_1 Ratio	Col 1 $\bar{p}_1$ Bara	Junction p_z Bara	Col 2 $\bar{p}_2$ Bara	Ratio $\bar{p}_1/\bar{p}_2$	Time in Col 1
1	4.34	3.61	2.55	1.70	63 %
1.6	4.51	3.92	2.78	1.62	50 %
2	4.58	4.12	2.88	1.59	44 %
3	4.69	4.36	3.03	1.55	34 %
5	4.79	4.58	3.17	1.51	23 %
7	4.85	4.69	3.24	1.49	18 %
9	4.88	4.75	3.29	1.48	16 %

Showing the junction and average pressures for two similar columns of different length, assuming the inlet pressure is 5.00 bara and the outlet pressure is 1.00 bara. The **Time** column shows the percentage of its retention time that a peak spends in Column 1 when both columns have identical packing or coating.

Choosing column lengths

A long retention time on the first column is rarely appropriate in column systems. Usually, Column 1 does a quick partial separation, then passes a group of components to another column for final separation. Therefore, Column 1 is usually short – sometimes only one-tenth of the length of column 2. For two columns with the same packing or coating and various length ratios, Table 9.4 shows the percentage of its overall retention time that a component spends in Column 1. With a pressure ratio of 5.0, Column 2 must be 1.6 times longer than Column 1 just to achieve an equal retention time.

It's possible to build a simple computer program to calculate column lengths (see Annino and Villalobos (1992, 339–360) but PGC column designers usually rely on the trial-and-error method based on their experience of peak retention in serial columns.

The adjustment of PGC settings for timed events is always empirical. Uncertainties in column construction and operation have an unpredictable effect on retention times. The packing or coating of serial columns might not be exactly uniform, their endplugs may not exhibit the same pressure drop, and the two columns might not be at exactly the same temperature. Due to these random inconsistencies, the adjustment of column valve timing is always a practical task requiring skill and understanding. Even so, the calculations presented here may help you to visualize what's happening inside your columns and to improve your column design and troubleshooting skills.

Column systems

Multiple-column systems in PGCs comprise two or more columns and one or more devices to direct different component peaks to different columns.

Traditional switching devices are mechanical, including diaphragm, rotary, and slide valves using pneumatic or solenoid actuators.[3]

PGCs using capillary columns may use the pressure balance technique introduced by Deans (1965, 1968) and now enhanced by electronic pressure regulation.

PGCs can now host several detectors. This allows an applications engineer to break a complex application into several simple column systems each working independently to measure only a few of the analytes, a rapid-analysis technique known as **parallel chromatography**. The individual column systems each have a sample injector, two or more columns, a column switch, and a detector. They are variously known as "applets" or "column trains" and may be intermingled in a common oven or encapsulated for easy maintenance. Parallel chromatography is a welcome simplification to column system design.

The next two chapters describe the functions performed by column switching and some key points in their design and usage. They fall into two categories:

> Those that reverse the flow direction in a column, discussed in Chapter 10.

> Those that maintain the flow direction in a column, discussed in Chapter 11.

A complete column system will often combine functions. For instance, a system using heartcut should also include a backflush to protect it from strongly retained components.

Knowledge Gained

- *Multiple-column systems are effective in PGCs to separate analytes in minimum analysis time.*

- *Backflush removes with certainty all injected compounds from columns before the next injection.*

- *Regrouping a class of compounds into a single peak enables a single aggregate measurement.*

- *Distribution of peaks on different stationary phases achieves separations impossible on one phase.*

- *Heartcut can resolve low-ppm impurities in 99.99 % pure samples, not possible with a single column.*

- *Column switching can protect a column or detector from damaging components in the sample.*

- *Process and laboratory GCs evolved to suit the disparate needs of their users.*

- *Laboratory GCs favored a capillary column with temperature ramp, good for many unknown peaks.*

---

[3]For an introduction to chromatographic valves, see Waters (2020, 157–181).

- ❖ Some now use multiple columns in arrangements called two-dimensional gas chromatography.
- ❖ Process GCs favored many isothermal packed columns, good for measuring a few specified peaks.
- ❖ Some now use capillary columns and temperature-programmed column ovens.
- ❖ Dual columns with different stationary phases can separate a few analytes from almost any sample.
- ❖ Peak resolution on two phases may be optimized simply by adjusting the column temperatures.
- ❖ Columns in series run at different pressures and different carrier velocities.
- ❖ The pressure drop along a column is nonlinear, mostly occurring near the column exit.
- ❖ Equation 9.3 gives the pressure at any point in the column, and at the junction of two series columns.
- ❖ From this, one can calculate the inlet and outlet pressure for each column.
- ❖ The gas compression j-factor relates the average pressure to the column inlet and outlet pressures.
- ❖ The average velocity in any column is inversely proportional to its average pressure.
- ❖ The peak residence time in a column is directly proportional to its average pressure.
- ❖ The first column is at high pressure and peaks move slowly, so it's often shorter than the second column.
- ❖ Packed columns have more pressure drop than capillaries and suffer more velocity change.
- ❖ A PGC may deploy several independent applets, each measuring a few of the desired analytes.
- ❖ These parallel systems each have a simple column arrangement with dedicated injector and detector.

Did you get it?

Self-assessment quiz: SAQ 09

Q1. Which of these is NOT an advantage gained by running columns in series?
- **A.** removing all injected compounds from the columns before injecting another sample.
- **B.** analyzing different components on independent parallel systems, each having its own sample injector, column valve, columns, and detector.
- **C.** achieving a complete separation of analytes faster than is possible by any single stationary phase.
- **D.** reliably measuring less than 100 ppm of an impurity in a 99.99 % pure process fluid.
- **E.** completing the analysis of several designated components faster than any single column could achieve.

Q2. What is the housekeeping rule in process gas chromatography and what kind of column system can satisfy it?

Q3. What is the general elution problem in gas chromatography and what are the two options for overcoming it?

Q4. *Why have process gas chromatographs been slow to adopt the laboratory practice of using mainly capillary columns?*

Q5. *What are two disadvantages of using identical columns in series that have the same packing or coating?*

Q6. *Two columns having the same internal diameter and packing operate in series. The second column is twice the length of the first column. Given that Column 1 inlet pressure is 3.0 barg and a Column 2 outlet pressure is zero barg:*
  A. *What is the pressure at the junction of the two columns?*
  B. *What is the average pressure in Column 1?*
  C. *What is the average pressure in Column 2?*
  D. *What is the average pressure in both columns combined?*

Q7. *In Question Q6, if an analyte peak has a 100 s retention time on Column 1, what will be its retention time on both columns combined?*

References

Further reading

You might like the paper by Villalobos and Annino (1992), one of the few published papers on serial columns in PGCs.

For a good historical review on the development of the process gas chromatograph, read Bostic and Clemons (2009).

If you would like more detail on the use of multiple-column systems in both laboratory and process gas chromatographs, read the article by Mahler et al. (1995).

Finally, if you're interested in the new laboratory GC techniques, you'll enjoy the detailed anthology edited by Snow (2020).

Cited

Annino, R. and Villalobos, R. (1992). *Process Gas Chromatography*. Instrument Society of America.

Bostic, S.M. and Clemons, J.M. (2009). Online gas chromatography: Celebrating 50 years. *Process and Control Engineering*, 12 February 2009, 1–6. Accessed on 2024/08/03 at: https://pacetoday.com.au/50-years-online-gas-chromatograph/

Deans, D.R. (1965). An improved technique for back-flushing gas chromatographic columns. *Journal of Chromatography A* 18, 477–481.

Deans, D.R. (1968). A new technique for heart cutting in gas chromatography. *Chromatographia* 1, 18–22. https://doi.org/10.1007/BF02259005.

Dimandja, J.-M.D. (2020). Introduction and historical background: the "inside" story of comprehensive two-dimensional gas chromatography. In *Basic Multidimensional Gas Chromatography* (ed. N.H Snow), 1–40, London, UK: Academic Press.

Ettre, L.S. and Hinshaw, J.V. (1993). *Basic Relationships in Gas Chromatography*. Cleveland, OH: Advanstar Communications.

Jennings, W., Mittlefehldt, E., and Stremple, P. (1997). *Analytical Gas Chromatography*, Second edition. San Diego, CA: Academic Press

Liu, Z. and Phillips, J.B. (1991). Comprehensive two-dimensional gas chromatography using an on-column thermal modulator interface. *Journal of Chromatographic Science* 29 (6), 227–231.

Mahler, H., Maurer, T., and Mueller, F. (1995). Multi-column systems in gas chromatography. In: *Chromatography in the Petroleum Industry*, Journal of Chromatography Library Volume 56 (ed. E.R. Adlard), 231–268. Amsterdam, Netherlands: Elsevier.

McWilliam, I.G. and Dewar, R.A. (1958). Flame ionization detector for gas chromatography. In *Gas Chromatography 1958* (ed. D.H. Desty), 142–152. London, UK: Butterworths Publications Ltd.

Simmons, M.C. and Snyder, L.R. (1958). Two stage gas-liquid chromatography. *Analytical Chemistry* 30 (1), 32–55.

Snow, N.H. ed. (2020). *Basic Multidimensional Gas Chromatography*, Volume 12 in the Separation Science and Technology series. London, UK: Academic Press.

Villalobos, R. and Annino, R. (1989). Series coupled capillary columns in process gas chromatography – theory and practice. Paper 89-0311. In *Proceedings of the 1989 Calgary Symposium* 57–67. Research Triangle Park, NC: International Society of Automation.

Waters, T. (2020). *Process Gas Chromatographs: Fundamentals, Design and Implementation*. Chichester, UK: John Wiley and Sons, Ltd.

Tables

9.1 A Lexicon of GC Methods.
9.2 Comparing Laboratory and Process Gas Chromatographs.
9.3 Average Pressures in Identical Series Columns.
9.4 Pressures and Retention Times for Columns in Series.

Figures

9.1 Dual Phase Selection Chart.
9.2 Nonlinear Pressure Drop.
9.3 Pressure Distribution in Series Columns.

Symbols

Symbol	Variable	Unit
$[A]_M$	Solute concentration in the mobile phase	µg/L
$[A]_S$	Solute concentration in the stationary phase	µg/L
$[A]_M$	Solute concentration in the mobile phase	µg/L
j	Gas compression factor	none
k	Retention factor	none
K_c	Distribution constant	none
L	Length of column	m
p_i	Carrier gas pressure at column inlet	bara
p_o	Carrier gas pressure at column outlet	bara
$\bar{p}$	Average carrier gas pressure in column	bara
p_z	Carrier gas pressure at point (z) along column	bara
P_r	Ratio of carrier pressures at column inlet and outlet	none

Subscripts

Sub_1 Pertaining to Column 1 in a series column system
Sub_2 Pertaining to Column 2 in a series column system

Equations

9.1 $\quad P_r = \dfrac{p_i}{p_o}$ — Column inlet/outlet pressure ratio.

9.2 $\quad P_{rz} = \dfrac{p_z}{p_o}$ — Ratio of pressure at point (z) to column outlet pressure.

9.3 $\quad P_{rz} = \sqrt{P_r^2 - \dfrac{Z}{L}(P_r^2 - 1)}$ — Pressure ratio at a specified point (z) along the column.

9.4 $\quad \bar{p} = \dfrac{p_o}{j}$ — Average pressure in a column.

9.5 $\quad j = \dfrac{3}{2} \cdot \dfrac{(P_r^2 - 1)}{(P_r^3 - 1)}$ — Gas compression factor.

9.6 $\quad \bar{u} = \dfrac{L}{t_M}$ — Average carrier gas velocity in column.

9.7 $\quad \dfrac{u_z}{u_o} = \dfrac{p_o}{p_z}$ — Carrier gas velocity ratio at a specified point (z) along the column.

9.8 $\quad \dfrac{\bar{u}_2}{\bar{u}_1} = \dfrac{\bar{p}_1}{\bar{p}_2}$ — Average carrier velocities are inversely proportional to average pressure.

9.9 $\quad t_M = \dfrac{L}{\bar{u}}$ — Retention time of an unretained component.

9.10 $\quad t_R = t_M \cdot (k + 1)$ — Retention time of a retained component.

9.11 $\quad t_R = \dfrac{L}{\bar{u}} \cdot (k + 1)$ — Retention time of any component on any column.

9.12 $\quad \dfrac{t_{R1}}{t_{R2}} = \dfrac{L_1}{L_2} \cdot \dfrac{\bar{u}_2}{\bar{u}_1} \cdot \dfrac{(k_1 + 1)}{(k_2 + 1)}$ — General expression for relative retention times on series columns.

9.13 $\quad \dfrac{t_{R1}}{t_{R2}} = \dfrac{L_1}{L_2} \cdot \dfrac{\bar{p}_1}{\bar{p}_2} \cdot \dfrac{(k_1 + 1)}{(k_2 + 1)}$ — Retention times on series columns having the same permeability.

9.14 $\quad \dfrac{t_{R1}}{t_{R2}} = \dfrac{L_1}{L_2} \cdot \dfrac{\bar{p}_1}{\bar{p}_2}$ — Retention on the same phase depends on the column lengths and average pressures.

9.15 $\quad \dfrac{t_{R1}}{t_{R2}} = \dfrac{\bar{p}_1}{\bar{p}_2}$ — Retention on identical series columns is proportional to their average pressures.

9.16 $\quad \dfrac{t_{R1}}{t_{R2}} = \dfrac{\bar{u}_2}{\bar{u}_1}$ — Retention on identical series columns is inversely proportional to average velocity.

Technical terms

If you read the whole chapter, you should now know the meaning of these technical terms:

absolute pressure
average pressure
average velocity
backflush column system
bara and **barg**
comprehensive method[a]
correlogram
cycle time
distribution column system
gas compression factor

general elution problem
heartcut method[a]
heavies
housekeeping rule
modulator
parallel chromatography
regrouping column system
split injection
two-dimensional GC[a]

[a] As used in Laboratory GCs.

For more information, refer to the Glossary at the end of the book.

10

Backflush systems

"The backflush function is so useful in process chromatography that it's present in nearly all PGC column systems. Backflushing keeps the first column clean and protects the second column from damaging chemicals in the process sample, while also delivering the fastest analysis."

Introduction

Backflush functions

Backflush is the most common function of serial column systems and features in nearly all process gas chromatographs (PGCs), so it's important to thoroughly understand it. The primary reason backflush is so popular in PGCs is the need for a *housekeeping function*. Backflushing will easily and quickly remove all injected substances from the first column before injecting another sample.

Another reason for a backflush system is to provide a **guard column** to prevent reactive components from reaching the main analytical column. An example is the necessary use of a Porapak T (or equivalent) guard column to prevent water from reaching and deactivating a molecular sieve column.

Reversing the flow in a column can also regroup selected peaks for measurement or further separation by a subsequent column.

The main varieties of the backflush technique are:

➤ Backflush to vent.

➤ Backflush to measure.

➤ Backflush to regroup.

In addition, the backflush column contributes to the final separation of analyte peaks. The second column completes the separation and sometimes contains a different stationary phase. This adds to the power of a backflush

---

Process Gas Chromatography: Advanced Design and Troubleshooting, First Edition. Tony Waters.
© 2025 John Wiley & Sons Ltd. Published 2025 by John Wiley & Sons Ltd.

system; two different columns in series can often accomplish separations that no single column could achieve within the analysis time constraint.

Backflush to vent

The simplest backflush system uses two columns in series coupled to a 6-port, 8-port, or 10-port valve. Figure 10.1 shows the popular 10-port valve system that combines sample injection and backflush functions into one valve. This single-valve system is always applicable for gas samples, but most liquid samples need a separate sample injection valve as discussed below.

From Figure 10.1, it's evident that actuation of the 10-port valve injects the gas sample and places the two columns in series. The valve stays in the inject position while the analyte peaks (for example, A and B) pass through Column 1 and enter Column 2. At that time, the valve is deactuated to reverse the carrier flow in Column 1 while sustaining an identical flow in Column 2. From that moment onwards, the column system saves time by doing two jobs at once; Column 2 completes the resolution of the analyte peaks while Column 1 flushes all the heavies (for example, C and D) to an atmospheric vent.

The simple explanation of backflush is straightforward. Although every component remaining in Column 1 traveled a different distance along the column, they all traveled for exactly the same time. Assuming, for the moment, that the backward flow rate equals the forward flow rate, each of these components takes the same time to return to the starting point, regardless of how strongly the liquid phase retains them. Thus, all the

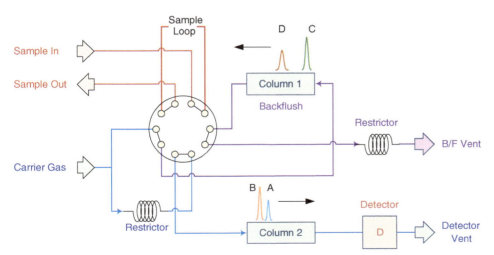

Showing the popular 10-port diaphragm valve in the deactuated state; other 10-port valves have an equivalent function. The valve has two positions. In the rest condition shown, the system is backflushing Column 1 and refilling the sample loop. Valve actuation connects the two columns in series and diverts the carrier gas through the sample loop to inject the sample into Column 1. The small peaks show the typical position of sample components just before and just after valve deactuation.

Figure 10.1 Ten-port Backflush Column System.

backflushed peaks elute from the column at exactly the same time and emerge as a single composite peak.

> *Be sure to understand why the heavies peaks all come out of Column 1 at the same time, regardless of their retention time in Column 1.*

After backflushing Column 1 for a time equal to the forward-flow time, the center of the combined heavies peak reaches the front end of the column, so half the peak is still in the column. A little more time (or a higher backflush flow rate) is necessary to completely remove the heavies.

We shall see that this **regrouping function** is not perfect. Even so, the backflush technique is a wonderful asset to a PGC because it removes all the injected substances, *even if we don't know what they are.*

In Figure 10.1, note the use of flow restrictors to control column flow rates. The first restrictor has the same flow resistance as Column 1 and ensures that the detector flow is the same in both valve positions. The second restrictor sets the backflush flow rate, discussed in detail below. These flow restrictors may be small needle valves or capillary tubing cut to the appropriate length. The latter method is cumbersome but may ultimately be more reliable. In modern PGCs, electronic pressure regulators (EPRs) may replace some of these passive flow restrictors.

> *The valve arrangements in Figures 10.1 and 10.3 illustrate an important principle of column system design: upon power failure, they put the columns into backflush mode and don't inject any more samples. Always ensure that loss of power or actuating gas leaves the column system in safe mode.*[1]

Backflush to measure

The backflush-to-vent mode discards the heavies peak, venting it to atmosphere. The alternative backflush-to-detector mode allows the PGC to regroup the later components and measure them as a composite heavies peak, as shown in Figure 10.2.

There are two ways to do this.

In an older PGC that has only one detector, the backflush vent flow tees into the flow from Column 2 and both flows mix together before entering the detector. In this arrangement, the backflush peak must arrive at the detector when no other peaks are eluting from Column 2. Often, it's the first peak on the chromatogram.

A modern PGC will dedicate a second detector to separately measure the heavies peak in the backflush vent flow, as shown in Figure 10.3. The detection of the heavies is then independent of the main analysis and peak timing is no longer an issue; both detectors can measure peaks concurrently.

---

[1] The exception is a diaphragm valve actuated solely by the pressure of an actuating gas, with no spring return. Loss of actuating gas pressure may leave all ports open to each other, potentially damaging the columns.

Introduction

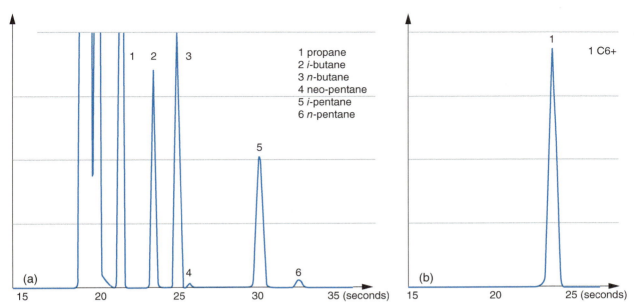

On the chromatogram in Figure 10.2a, this micro-PGC measures paraffin impurities in natural gas within about 35 s. Concurrently, the analyzer backflushes the heavies from the first column into a second detector, regrouping them into a single C_6+ peak. These chromatograms illustrate the rapid analysis achievable with capillary columns.

Figure 10.2 PGC Chromatogram using Backflush to Measure.
Source: Qmicro B.V. Reproduced with permission.

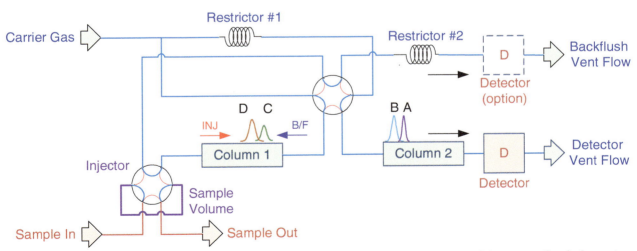

Showing two six-port diaphragm or rotary valves in the deactuated state. In the rest condition (blue connections), the system is backflushing Column 1 and refilling the sample loop. Actuation of the column valve (red connections) connects the two columns in series. A brief actuation of the injector valve then injects the sample into Column 1. When the desired analytes are in Column 2, deactivation of the column valve backflushes all the heavies from Column 1. It's easier for a second detector to measure the heavies peak; otherwise, it may tee into a sole detector. The six-port injector shown is suitable for gas sample injection; liquid samples may use four-port rotary or plunger valves.

Figure 10.3 Six-port Backflush Column System.

The calibration of a composite heavies peak cannot be exact because it comprises multiple substances – some unknown – that have differing detector sensitivities. Calibration is often based on a single substance in the calibration sample that is commonly present in the process sample, such as n-pentane. The measurement is then known as "total heavies as n-pentane." While slightly less accurate than measuring a single peak, this measurement is often useful and it can become essential when normalizing all measurements to sum to 100 %.

Pressure balance switching

Valveless column switching is an attractive option for a PGC. The Live Tee is a device that can switch capillary or micropacked columns without a mechanical valve and with virtually no extracolumn volume. It's an adaptation of the *Deans Switch*, which controls carrier gas flow by applying external pressures to a special column coupling (Deans 1965, 1968).

To follow the flow paths in the Live Tee, refer to Figure 10.4. When attached, the two columns are almost butt jointed and a narrow-bore capillary tube enters the end of each column forming slim annular passages between the outer diameter of the capillary and the inner diameter of the column ends. Carrier gas must then flow from Column 1 into Column 2 via the capillary connector, wherein it incurs a small pressure drop.

The Live Tee has carrier gas inlets and vents on both its upstream (P_m-) and downstream (P_m+) sides. The upstream vent is known as the Cut Vent. Chapter 11 will discuss its function in more detail. It usually vents via a restrictor to the atmosphere or to a detector. The downstream vent is known as the Purge Vent and normally connects via another restrictor to the main

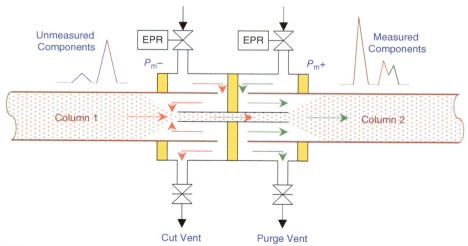

Figure 10.4 Live Tee in Pass-through Mode.

This mode allows the measured peaks into Column 2 prior to backflushing Column 1. The electronic pressure regulators (EPR) are set so that P_m- is higher than P_m+, causing carrier gas from P_m- to flow through the connecting capillary, taking the flow from Column 1 with it (red arrows). See the main text for a full explanation.

detector to act as makeup or fuel gas. Some applications have these vent connections reversed.

A Live Tee works by adjusting the Column 1 inlet pressure (not shown) and the P_m- and P_m+ pressures applied to the tee. In the pass-through mode shown in Figure 10.4, carrier gas flowing from Column 1 to Column 2 via the capillary tube creates a slight pressure drop. An EPR then sets the pressure on the P_m- inlet slightly higher than the outlet pressure of the upstream column. This creates a carrier gas flow into the inlet which purges the annular space between the inner column wall and the capillary connector, adding a small increment to the carrier flow going into Column 2. Some of the added gas exits via the cut vent restrictor.

In this mode, carrier gas flows through Column 1 in the normal way and analyte peaks pass from Column 1 into Column 2.

Meanwhile, the downstream EPR sets the gas pressure on the P_m+ inlet slightly higher than the outlet of the capillary connector, causing carrier gas to flow through the annular space between the column and capillary and adding another small increment of flow to the main column. The small flows introduced to the column at both sides of the tee eliminate unswept dead volumes to prevent any broadening of peaks as they pass through.

To backflush Column 1, a valve or EPR (not shown) drops the column inlet pressure low enough to allow flow of carrier gas from the P_m- side of the live tee backward through Column 1 and out the backflush vent as shown in Figure 10.5. If the PGC uses split injection, the splitter vent will usually double as a backflush vent, but in the case of direct injection, a pneumatic valve can open to induce backflush flow (Siemens 2008, 2-89–2-94).

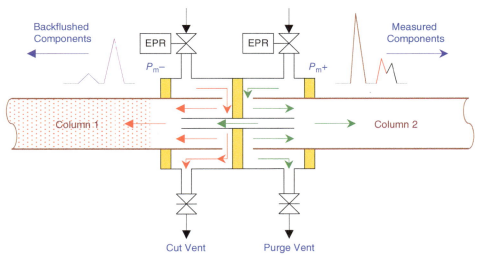

This mode backflushes Column 1 and continues the constant flow though Column 2. To initiate backflush, a valve (not shown) vents the Column 1 head pressure via a restrictor or detector, reversing the flow in Column 1. The reversed differential pressure causes carrier from P_m+ to flow through the connecting capillary to backflush Column 1 (green arrow). See the main text for a full explanation.

Figure 10.5 Live Tee in Backflush Mode.

While highly effective and reliable when operating correctly, the Live Tee setup procedure is more complex than a valved system and may be beyond the capabilities of some analyzer technicians. In addition, it relies on the performance of the EPRs to maintain constant pressure under varying flow conditions. Note that a Live Tee backflushes the first column at a much lower pressure than its forward flow. We shall see this is a good way to remove all the heavies from the column, but it's not so effective at regrouping the heavies peaks for composite measurement.

In a PGC, the Live Tee typically provides both backflush and heartcut functions. When the application needs only backflush a simplified version called a "Back-T" is adequate. This version doesn't have a capillary tube between the columns and has a single EPR inlet and a single vent. The EPR is set slightly higher than the exit pressure of Column 1 to ensure a small inlet flow. Since the EPR then maintains a constant pressure at the column junction, venting the Column 1 inlet pressure to atmosphere is sufficient to reverse its flow direction, thereby backflushing the column to an atmospheric vent (Siemens 2008, 4-127–4-139).

Setting the backflush time

We use a trial-and-error procedure to find the optimum backflush event time. The goal is to ensure the last analyte peak from Column 1 is fully into Column 2 before initiating backflush. Adjust the timing while analyzing a constant sample that contains the analyte most strongly retained by Column 1. Use the process sample if possible. If using a separate test sample, it should have a composition similar to the process sample; analytes blended in nitrogen are unlikely to have the same retention times on the first column.

Many samples contain a high concentration of one substance, called the **major component**. Typically, the tail of this component is not completely backflushed and a small background amount remains in the columns. It takes a few analysis cycles for the retained amount to settle and form a steady-state background level, adding to the ever-present column bleed. This background level may affect peak retention times, so don't adjust any event time settings until the PGC completes several analysis cycles.

Some systems include an **intercolumn detector**, strangely known as an "ITC", to monitor the effluent from Column 1. In column systems using valve switching, the peaks exiting the first column will pass through this thermal conductivity detector and enter Column 2 unchanged. In systems using Live Tee switching, the gas from the purge vent may flow into the main detector to provide the same function. The ITC chromatogram may be helpful for setting the backflush timing. However, the partial separation of components on Column 1 can sometimes make it difficult to interpret the ITC signal.

When checking the valve timing, temporarily increase the backflush event time by two seconds and confirm the area of the last analyte peak (or its measured concentration) does not increase. When measuring peak area, peak height is an unreliable indicator of valve timing because the backflush might remove a large chunk of the peak tail with no effect on the peak height. If the later backflush event has no effect on peak area the timing is correct. Don't forget to reset the event time.

Stuttering backflush

Backflush can also assist in a difficult separation. A short-duration backflush delays a few analyte peaks, briefly moving them backward in Column 1, and then forward again into the next column. Of course, the backflush valve must be separate from the injector valve and has more work to do, potentially reducing its reliability. This **stuttering technique** allows some time for the peaks in Column 2 to move ahead of those briefly delayed in Column 1. Once all the analyte peaks have eluted from Column 1, the analyzer performs a complete backflush-to-vent or backflush-to-detector routine.

Parallel chromatography

A modern PGC will often employ *parallel chromatography* to analyze a complex sample. Several independent modules operate in parallel, each designed to measure a few of the analytes. Typically, each self-sufficient module comprises a single 10-port valve, two columns, and a detector, and is set up to measure a few of the analytes and backflush all later peaks. Effectively, each module is a separate gas chromatograph tasked with a simple analysis. These individual modules are easier to design and simpler to maintain than trying to measure all the analytes from one sample injection. PGC manufacturers can also save their custom engineering costs by creating standard modules for common applications.

Regrouping systems

A three-column system, as in Figure 10.6, can employ backflush of the middle column to group classes of analytes together for composite measurement (Nuss 1963; Villalobos and Turner 1963).

To measure total C_4 and total C_5, for instance, the system uses nonpolar columns to keep all the C_5 peaks separate from all the C_4 peaks. Early peaks, such as C_3 and lighter, pass through Column 2 and enter Column 3 before the column valve actuates. When all C_4 peaks are in Column 2 and the C_5 peaks are still in Column 1, the column valve actuates, reversing the flow in Column 2 and loading the C_4 peaks into Column 3 in reverse order. Meanwhile, the C_5 peaks enter the other end of Column 2. When the valve deactivates, these C_5 peaks enter Column 3 in reverse order. Column 3 negates the separation done by Column 1, bringing each batch of peaks back together again.

258 Backflush systems

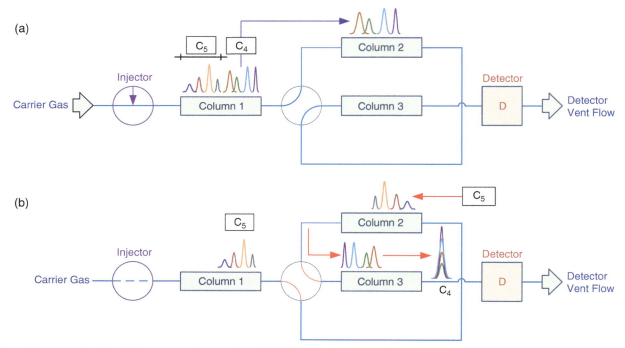

This system uses backflush to regroup classes of analytes together for composite measurement.
In Figure 10.6a, the first class (e.g. the C_4 peaks) enters Column 2 which is long enough to hold them all, while the second class (e.g. the C_5 peaks) stays in Column 1.
In Figure 10.6b, actuation of the column valve backflushes the C_4 peaks from Column 2 into Column 3 in reverse order. Meanwhile, the C_5 peaks enter the other end of Column 2.
Then, deactuation of the valve backflushes the C_5 peaks into Column 3, again in reverse order. Column 3 is exactly long enough to undo the separation caused by Column 1 regrouping each class together for measurement.

Figure 10.6 A Regrouping Column System. (a) Components to be regrouped are loaded into Column 2. (b) Components in Column 2 are backflushed into Column 3 for regrouping.

Although not shown in Figure 10.6, a regrouping system will usually backflush Column 1 after the last analyte enters Column 2.

To determine the length of Column 3, calculate the average pressure in each column and adjust the length of Column 3 to achieve retention times equal to Column 1. From Chapter 9, we know that Column 3 should be at least 1.6 times longer than Column 1. Column 2 should be as short as possible to ensure effective regrouping but long enough to contain each set of analyte peaks coming out of Column 1. It might need a higher liquid loading or a different liquid phase altogether.

Regrouping column systems rely on critical column lengths to work successfully and may not be reliable in the long term. They're also susceptible to setup errors that end up measuring the wrong peaks. A modern PGC is more likely to employ a simple column system to measure the individual components and to sum them electronically, even if not fully resolved.

Backflush theory and practice

Limitations

The backflush technique yields important benefits in process chromatography, so the theory and practice of backflushing developed rapidly alongside the first commercial PGCs (Villalobos et al. 1961; McEwan 1964).

Simple backflush theory predicts an exact regrouping of peaks after flow reversal, but regrouping is never perfect. It's instructive to discover why.

Consider a backflush system with the reversed carrier flow in Column 1 set equal to its forward flow; that is, the Column 1 flow rate is the same in both valve positions. We have seen that each heavies peak should then migrate backward through Column 1 at the same speed as it went in. All the heavies peaks would then arrive at the front end of the column at exactly the same time, *regardless of their various retention times* on that column. If it worked like that, regrouping would recombine all the heavies peaks remaining in Column 1 as a single composite peak.

In practice, backflush doesn't regroup the heavies precisely. Strangely, the more strongly retained peaks elute first, followed by those less strongly held; in Figure 10.3, Peak D beats Peak C to the end of the column. This effect is also visible in the real chromatogram in Figure 10.2. The failure to regroup is at first surprising. The explanation comes down to pressure variations within the columns:

> - Peak C travels almost the entire length of the first column thereby experiencing an average carrier gas pressure and velocity about equal to those for the whole column. When the flow direction reverses, Peak C again travels the length of the column, so its forward and reverse retention times are about equal – as expected.

> - Peak D enters the column at full carrier gas pressure and exits at lower pressure and higher velocity. Therefore, its reverse retention time is less than its forward retention time and it elutes slightly before Peak C.

When backflushing to vent, the premature elution of the heavier peaks is a good feature as it guarantees the removal of all injected components. If we ensure the removal of Peak C, the backflush is certain to remove all other heavies – whatever they are.

To guarantee the removal of heavies, most chromatographers set the backflush flow from 50 to 100 % higher than the forward flow. Unfortunately, this extra flow in the backflush column reduces its exit pressure and allows a strongly retained peak (like Peak D) to elute even faster than one less strongly retained (like Peak C), aggravating the regrouping problem. Instead of increasing the backflush flow rate, it's better to design the columns to ensure the backflow period is longer than the forward-flow period. Given more time, backflushing guarantees the removal of all heavies without additional flow rate. Ultimately, it's necessary for the net flow in

Column 1 to be backward, and the designer can accomplish this with more time or more flow.

> *The net carrier flow direction in a backflush column must be backward, so any traces of heavy components move inexorably toward the backflush vent.*

If a PGC starts the backflush after the half-time point, it becomes necessary to increase the backflush flow rate to compensate for the shortage of time. We shall see that this practice may be indicative of a poor column design or simply due to a bad habit. It wastes carrier gas and increases the pressure pulse felt by the column each time the valve switches.

Another pressure effect is the sudden inrush of gas that occurs at the instant of flow reversal. The backflush valve reverses the column flow, applying full carrier pressure to the tail end of Column 1. As the pressure gradient reverses, there is a sudden inflow of gas at one end of the column and a sudden outflow at the other end. In our example, Peaks C and D are near the ends of the first column, so they would both get a short velocity boost due to these transient increases of carrier flow. This is another potential cause of ineffective regrouping.

Column design and operation

In general, the column designer has three options:

➤ Use the same packing or coating in both columns.

➤ Use a nonpolar Column 1 and a polar Column 2.

➤ Use a polar Column 1 and a nonpolar Column 2.

Using the same stationary phase

When both columns contain the same packing or coating, the backflush function has only one task: to remove the heavies from Column 1. The designer chooses a stationary phase and decides the total column length necessary to separate the analyte peaks from each other and from any other components that might be present. Generally, this is possible only when measuring a few light components – the separation of heavier analytes becomes difficult on a single liquid phase because the number of dissimilar molecules increases rapidly with molar weight.

The designer then apportions part of the total length to each column. Generally, the two columns must not be equal in length, yet we often see backflush systems with two identical columns. This is poor design. With equal columns, we know from Table 9.4 a peak will spend 63 % of the analysis time in Column 1 and only 37 % in Column 2. If the last analyte peak fully elutes from Column 1 in 126 s, it will zoom through Column 2 and fully elute in 74 s for an analysis time of 200 s. Meanwhile, an equal-flow backflush will take another 126 s for a total of 252 s plus a little extra time to fully elute the composite heavies peak. The analysis is complete in 200 s but the cycle time must extend to about 270 s, an increase of 35 %.

Now we see a reason for chromatographers to increase the backflush flow rate: it might be to compensate for poor design. Let's see how we might do better.

Given a total column length adequate for good analyte resolution, how should we divide that length between the two columns? We already know that the first column should be shorter than the second column, so let's consider a length ratio of 1.6. From Table 9.4, we see that peak retention times on each column will then be equal, in our example 100 s each.

> Note that the table is based on a column head pressure of 5 bara, so you might have to recalculate a column length ratio for your actual inlet pressure.

Now, the last analyte fully elutes from Column 1 in 100 s and is completely clear of Column 2 in an additional 100 s, for the same 200 s analysis time. Backflush also elutes the heavies in 100 s, synchronizing the analysis and cycle times. However, to completely clear the column, the backflush will need a little extra time or perhaps a little higher flow rate.

To avoid having a higher backflush flow consider a higher length ratio, perhaps 2.0 would be enough? Look again at Table 9.4. When the second column is twice the length of the first column, a peak that fully exits both columns in 200 s will now exit the first column in 88 s (44 % of the cycle time) and the second column in 112 s. The backflush now has 112 s to flush out the heavies, 24 s longer than the forward flow. This is a good design – it saves carrier gas and is gentle with the columns.

The final constraint is the need for Column 1 to separate the last analyte from the following unmeasured component. If Column 1 is too short, it may not have enough resolution to separate and backflush the following component and part of that component will enter Column 2. If so, it will separate on Column 2 and appear at the end of the chromatogram as an unmeasured partial peak. Some chromatographers like to see this, as the size of the partial peak gives a very good indication that the backflush timing is correct or not. But it does increase the cycle time.

The other approach is to make Column 1 long enough to fully separate the first backflushed peak from the last analyte peak. If that results in excessive analysis time, there is justification for increasing the backflush flow rate.

Using different stationary phases

A common arrangement is a nonpolar first column followed by a polar second column. For example, the nonpolar column might retain and then backflush the heavy hydrocarbons allowing the polar column to separate some alcohol peaks from the remaining light hydrocarbons. In effect, the backflush clears a space on the chromatogram for the alcohol peaks.

The opposite arrangement of a polar column followed by a nonpolar one is less common, but useful for backflushing the polar components before separating the nonpolar components. For example, the polar column might retain and then backflush the aromatics, clearing a space for the nonpolar column to separate the nonaromatic peaks in boiling point order.

A worked example

Calculated backflush event timing

For backflush to work as desired, the column valve must initiate the backflush at exactly the right time. This **event time** is one of the **Method** settings you program into the PGC controller. In actual practice, the optimum backflush time is set by trial and error, but it's easier to troubleshoot if you can visualize what's happening inside the columns. To gain that understanding, we offer a worked example of calculations one can perform to discover the optimum backflush event time and analysis cycle time. By working through the math, we will discover why the heavies peaks do not exactly regroup when backflushed.

Please realize this mathematical approach is only to enhance your understanding; it's not intended as an exact procedure for column design or maintenance.

Consider a backflush column system like the one in Figure 10.3. Let's start with these assumptions:

- Both columns use the same packing or coating.
- The second column is twice as long as the first column.
- The carrier gas pressure at Column 1 inlet is 4.00 bara.
- The carrier gas pressure at Column 2 outlet is 1.00 bara.
- Peaks A and B are analyte peaks and the overall retention time of Peak B is 180 s.
- Peaks C and D are backflushed from Column 1 and when backflush starts, Peak D is half the distance down the column as Peak C.

In a backflush system, Column 1 must fully separate all the slow-eluting heavies peaks from the desired analyte peaks. Figure 10.3 illustrates this key requirement by arbitrarily showing Peaks A and B continuing to the detector while Peaks C and D flow backward to the backflush vent. The resolution created by Column 1 between the last analyte peak (B) and the first heavies peak (C) is critical.

Valley point

Ideally, the PGC should initiate backflush when all the molecules of Peak B have entered the second column and no molecules of Peak C have left the first column – in other words, right at the **valley point** between Peak B and Peak C.

Valley points move through a column in the same way that peaks move. To grasp that concept, it's helpful to imagine that a marker peak exists between Peaks B and C, and the marker peak has a retention time coincident with the valley point. Let's call it Peak V. Then, the ideal time to initiate backflush is the retention time for Peak V on Column 1.

To ensure complete separation of B and C molecules, Column 1 should deliver an adequate resolution between them. As shown in Figure 10.7, a resolution of $R_s = 2$ is enough to completely separate similar-sized peaks. In practice, more resolution is desirable, and it may be necessary if the peaks differ in size. You can adapt these calculations to suit any desired resolution.

Working from the desired resolution, estimate the location of Peak B and Peak C at the instant of backflush initiation. Recall from Equation 2.5 that resolution (R_s) is a function of peak separation (S) and base width (w_b), both measured in distance units:

$$R_s = \frac{S}{w_b}$$

To get a resolution of $R_s = 2$, the separation between Peaks B and C must be twice the average peak width. Therefore, when the valley point reaches the

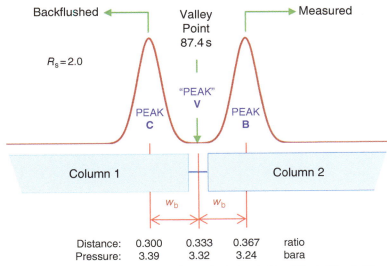

Figure 10.7 Idealized Backflush Event Setting.

Shows the idealistic location of peaks at the instant of backflush. The fictitious "Peak V" is the valley point between the last analyte peak (B) to exit Column 1 and the next peak (C), which is backflushed. From the desired analysis time and the average column pressures, we estimate the backflush event time. In this example, the calculated event time is 87.4 s. In reality, it would then be necessary to optimize the setting by experiment.

end of Column 1, it has traveled one-third of the total column length (L) and the center of Peak B must be one base width ahead of it. That puts the center of Peak B at a distance (Z_B) from the inlet of Column 1:

$$Z_B = 0.33\,L + w_b$$

At the same instant, the center of Peak C must be one base width behind the valley point, at a distance (Z_C) from the inlet of Column 1:

$$Z_B = 0.33\,L - w_b$$

These deductions assume the base widths of Peaks B and C are about the same. The width of the marker Peak V would also be the same. Find the base width of Peak V in distance units by rearranging Equation 2.26 to express base width as a function of its plate number (N) and retention distance (Z_V):

$$w_b = \frac{4 \cdot Z_V}{\sqrt{N}}$$

For a short, packed column a conservative estimate of plate number is 1600, giving a Peak V base width (w_b) of $Z_V/10$, or 10 % of its retention distance. If using more efficient columns, adjust this number to suit. On this basis, the base width of our imaginary Peak V is:

$$w_b = 0.033\,L$$

And the locations of Peak B and Peak C are:

$$Z_B = 0.367\,L$$

$$Z_C = 0.300\,L$$

Figure 10.7 illustrates these peak locations. This data is the basis for calculating the average carrier gas pressure and velocity experienced by each peak. The next section estimates the backflush event time simply by calculating the retention time of Peak V on Column 1.

Timed events

In any backflush system, the PGC processor must issue two timed commands, **start backflush** and **start next analysis** (that is, *inject next sample*). When the columns have the same packing or coating, both of these events depend on the timing of the valley point, "Peak V":

- ➤ The *start backflush* event should occur when the valley point reaches the end of the first column.
- ➤ The *start next analysis* event should occur when the valley point reaches the end of the second column.

Ideally, the analysis ends when the valley point reaches the detector. Of course, it's not a valley anymore because no sample molecules can elute after this time – they're all backflushed. That's why it's safe to inject the next sample.

The following calculations use equations from the **SCI-FILE** *On Pressure Drop* in Chapter 9.

To evaluate the valley point timing, start by calculating the carrier gas pressure ratio at the junction of Column 1 and Column 2. In this example, the column inlet pressure (p_i) is 4.00 bara and the vent pressure (p_o) is 1.00 bara giving an overall pressure ratio (P_r) of 4.00.

By Equation 9.3, the junction pressure ratio (P_{rz}) is:

$$P_{rz} = \sqrt{16 - 0.333 \times (16 - 1)}$$

$$P_{rz} = 3.32$$

From Equation 9.2, the actual junction pressure is:

$$p_z = P_{rz} \cdot p_o = 3.32 \text{ bara}$$

This is what we now know:

> Column 1 has an inlet pressure of 4.00 bara and an outlet pressure of 3.32 bara, for a pressure ratio $P_r = 1.21$.

> Column 2 has an inlet pressure of 3.32 bara and an outlet pressure of 1.00 bara, for a pressure ratio $P_r = 3.32$.

> The columns together have an inlet pressure of 4.00 bara and an outlet pressure of 1.00 bara, for a pressure ratio $P_r = 4.00$.

Now calculate the average pressure in each column. First, use Equation 9.5 to work out the compressibility factor (j) for each column individually and for both columns together:

$$j_1 = \frac{3}{2} \cdot \frac{1.21^2 - 1}{1.21^3 - 1} = 0.904$$

$$j_2 = \frac{3}{2} \cdot \frac{3.32^2 - 1}{3.32^3 - 1} = 0.423$$

$$j_{1+2} = \frac{3}{2} \cdot \frac{4.00^2 - 1}{4.00^3 - 1} = 0.357$$

Then, from Equation 9.4, calculate the average carrier gas pressure ($\bar{p}$) in each case:

$$\bar{p}_1 = \frac{3.32}{0.904} = 3.67 \text{ bara}$$

$$\bar{p}_2 = \frac{1.00}{0.423} = 2.37 \text{ bara}$$

$$\bar{p}_{1+2} = \frac{1.00}{0.357} = 2.80 \text{ bara}$$

From these average pressures, calculate the retention time of Peak V on each column. For columns with the same packing or coating, Equation 9.13 simplifies to:

$$\frac{t_{R1}}{t_{R1+2}} = \frac{L_1}{L_{1+2}} \cdot \frac{\bar{p}_1}{\bar{p}_{1+2}}$$

In this example, the desired cycle time is 200 s, which defines the retention time (t_{R1+2}) of Peak V on both columns. Therefore, the retention time (t_{R1}) of Peak V on Column 1:

$$t_{R1} = 200 \times \frac{1}{3} \times \frac{3.67 \text{ bara}}{2.80 \text{ bara}} = 87.4 \text{ s}$$

And the retention times of Peak V on Column 2 is:

$$t_{R2} = 200 - 87.4 = 112.6 \text{ s}$$

Table 10.1 summarizes these data. The backflush event time is 87.4 s, or 43.7 % of the cycle time. That leaves 112.6 s for backflushing the heavies from Column 1, about 29 % longer than their time going in. This is good column design.

Table 10.1 Calculation Data for Backflush Event Time.

| | | | Forward Flow | | | | |
| | Serial Pressure Effects | | Peak V | | | Peak B | |
Variable	Equation	Column 1	Column 2	Both Columns	Traveled Column	Remaining Column	Unit
Peak migration distance	(Z/L)	0.333	0.667	1.00	0.367	0.633	ratio
Inlet pressure (p_i)	9.3	4.00	3.32	**4.00**	4.00	3.24	bara
Outlet pressure (p_o)	9.3	3.32	1.00	**1.00**	3.24	1.00	bara
Pressure ratio (P_r)	9.1	1.21	3.32	4.00	1.23	3.24	ratio
Compression factor (j)	9.5	0.904	0.423	0.357	0.892	0.432	ratio
Average pressure ($\bar{p}$)	9.4	3.67	2.37	2.80	3.63	2.32	bara
Retention time (t_R)	9.13	87.4	112.6	**200.0**	87.4	96.3	s
Backflush start and finish times		87.4		200.0	Retention time = 183.7		s

This data illustrates the calculation of backflush event time, as described in the text. Starting data (**bold**) includes the carrier gas inlet and outlet pressures, desired analysis time, and the relative lengths of the two columns. Both columns have the same internal diameter and the same packing or coating. The "Peak V" data is for the valley point between the measured Peak B and the backflushed Peak C.

Peak B retention time

The calculation of peak location and movement is much the same as outlined above. Follow this procedure:

- Estimate the location of the peak at a particular time of interest.
- Calculate the pressure at that location, using Equation 9.3.
- Calculate the pressure ratios from column inlet to peak location and from peak location to column outlet.
- Calculate the average pressure experienced by the peak when migrating to that location.
- Calculate the pressure ratio and average pressure in both columns combined.
- From the retention distance yet to travel and the average carrier gas pressures, estimate the retention time of the peak.

The first step is to estimate the location of the peak at a particular time of interest. In this example, we know from Figure 10.7 that:

- Peak B has migrated 0.367 of the column length when the backflush initiates at 87.4 s into the cycle.
- Therefore, Peak B has 0.633 of the column length yet to go.

Calculate the carrier gas pressure at the Peak B location using Equation 9.3. Then use Equation 9.2 to find the pressure ratios across the column lengths the peak has already traveled and has yet to travel.

Here are the numbers for Peak B:

- Pressure at the start of column is 4.00 bara.
- Pressure at the current peak location is 3.24 bara.
- Therefore, the pressure ratio already traveled is 1.23.
- Pressure at the end of column is 1.00 bara.
- Therefore, the pressure ratio yet to travel is 3.24.

Apply Equations 9.4 and 9.5 to get the average pressures experienced by the peak. Table 10.2 lists the calculated values. Here are the key results:

- Average pressure Peak B has traveled is 3.63 bara.
- Average pressure Peak B is yet to travel is 2.32 bara.

Finally, estimate the time it takes for Peak B to reach the detector by applying the ratios of average pressures and column lengths, per Equation 9.13. Here's how it works out:

- The peak has already traveled for 87.4 s.

Backflush systems

Table 10.2 Calculation Data for Backflushed Peaks.

Pressure Effects		Column 1					
		In Forward Flow			In Backward Flow		
Variable	Eqn	Peak B	Peak C	Peak D	Peak C	Peak D	Unit
Peak migration distance (Z/L)		0.367[a]	0.300[a]	0.150	0.300	0.150	ratio
Starting pressure (p_i)	9.3		**4.00**		3.94	3.64	bara
Ending pressure (p_o)	9.3	**3.24**	3.39	3.71	**3.32**		bara
Pressure ratio (P_r)	9.1	**1.23**	1.18	1.08	1.19	1.10	ratio
Compression factor (j)	9.5	**0.892**	0.916	0.962	0.913	0.953	ratio
Average pressure ($\bar{p}$)	9.4	**3.63**	3.70	3.86	3.64	3.48	bara
Retention time (t_R)	9.13		**87.4**		85.8	78.8	s
Backflushed peaks separation						7.0	s
Total retention time		**183.7**			173.2	166.2	s

[a] From Figure 10.7.

*Illustrating the calculation of peak movement in forward and backflush modes, as described in the text. Data in **bold** are from Table 10.1. Both columns have the same internal diameter and the same packing or coating. The data show that Peak D reaches the end of the backflush column seven seconds before Peak C, causing an imperfect regrouping of the backflushed peaks.*

> The remaining travel distance is longer by the ratio 0.633/0.367.
> The remaining average pressure is lower by the ratio 2.32/3.63.

Apply Equation 9.13 to find the retention time of Peak B on Column 2. Because it's the same peak on the same liquid phase $k_2 = k_1$. So the only factors affecting this retention time are the column length ratio and the average pressure ratio:

$$t_{R2} = t_{R1} \cdot \frac{L_2}{L_1} \cdot \frac{\bar{p}_2}{\bar{p}_1}$$

$$t_{R2} = 87.4 \times \frac{0.633\,L}{0.367\,L} \times \frac{2.32\,\text{bara}}{3.63\,\text{bara}} = 96.3\,\text{s}$$

Thus,

> The additional peak travel time is 96.3 s.
> The overall retention time of Peak B is 183.7 s.

The center of Peak B arrives at the detector 16.3 s before the end of the analysis. This is perfect column design! In practice, the cycle time may be extended to allow for other functions before the next sample injection.

Regrouping effectiveness

To calculate the time necessary to flush out the backflushed peaks, follow the same procedure as used above. Start by estimating where the peaks are at the instant of backflush. In this example, Figure 10.7 shows Peak C has reached 0.300 of the column length and it's given that Peak D is at half that distance.

Next, use Equation 9.3 to calculate the pressure at those locations. Then compute the pressure ratios and the average pressures each peak experiences over its distance traveled in the forward direction. Of course, you already know the time of travel – it's the time before the backflush event, 87.4 s.

Now consider the backward travel. During backflush, the column has the same starting and ending pressures, but in reverse. At the instant of backflush, you know the location of Peak C and Peak D. Again, use Equation 9.3 to find the pressure at those peak locations. Then calculate their pressure ratios and average pressures during backflush.

Each peak travels the same distance forward and backward. To get their travel times, simply multiply the forward travel time of each peak (87.4 s) by its ratio of average pressures in backward and forward travel.

These calculations are not difficult but are tedious to do manually. It's best to set up a simple spreadsheet to do the job. Check your answers with the data given in Table 10.2. You should find that Peak D reaches the backflush detector about seven seconds before Peak C, causing partial separation of the backflushed peaks:

➢ The overall retention time of Peak C is 173.2 s.
➢ The overall retention time of Peak D is 166.2 s.

Regrouping column

A PGC measuring total heavies might include a short additional column before the detector to fully regroup them. In the present example, for instance, it would be easy to calculate the add-on column length necessary to retain Peak D exactly seven seconds longer than it retains Peak C. While effective for a few known components, this practice is risky because a component strongly retained in Column 1 would take a long time to transit the extra column. It's better to improve regrouping by minimizing the pressure drop across Column 1. Make it as short as possible and set the backflush flow equal to the forward flow. In a difficult application, it might be effective to increase the permeability of Column 1 by using a coarser mesh size or a larger internal diameter.

Knowledge Gained

❖ *In PGC, it's necessary to remove all injected sample components before injecting another sample.*

❖ *Backflush does the housekeeping and is the most common function of a multiple column system.*

❖ *Backflush removes all injected components from the columns even if we don't know what they are.*

❖ *The PGC may vent backflushed components or regroup them together for measurement.*

- *The backflush system uses two columns that may contain the same or different stationary phases.*
- *The PGC may use a mechanical valve or pressure balance switching to reverse the flow in Column 1.*
- *The column valve can be a diaphragm, rotary, or slide type, typically with 6 or 10 ports.*
- *All heavies in Column 1 travel forward for the same time, so they take the same time to return.*
- *Thus, all the heavies peaks regroup and elute together at about the same time.*
- *Peaks that travel only a short distance in Column 1 are backflushed at a lower pressure, higher velocity.*
- *Thus, the strongly held peaks elute slightly before the less-retained peaks.*
- *An additional column may regroup specified B/F peaks but it will strongly retain any later peaks.*
- *The backflush start time is set by trial and error, to ensure no analyte peak is reduced in size.*
- *To ensure the backflush is not cutting an analyte peak, peak area is a better indication than height.*
- *An intercolumn detector may assist in the setting of backflush start time.*
- *The PGC may measure the composite regrouped backflush peak as a "total heavies" peak.*
- *A composite regrouped peak cannot be accurately calibrated but may still be a useful measurement.*
- *A brief backflush may be a delaying tactic to give time for earlier peaks to separate on Column 2.*
- *In parallel GC, several independent backflush modules may each measure a few of the analytes.*
- *Backflushing can regroup a bunch of analyte peaks into one composite peak for measurement.*
- *Most PGCs have the backflush flow rate higher than the forward flow rate, which may not be necessary.*
- *Careful selection of column lengths may permit the backflush flow to equal the forward flow.*
- *To remove every trace of heavies, the average flow in a backflush column must always be backward.*
- *Thus, to ensure complete heavies removal, the backflush must use extra flow, extra time, or both.*
- *Some PGCs use the same packing or coating in both columns, others use different stationary phases.*
- *If the same packing, the second column should be about twice the length of the first column.*
- *Backflush starts at the valley point between the last analyte peak and the first heavies peak.*
- *Valley points move through columns in the same way as peaks – they also have retention times.*
- *If both columns have the same packing, the analysis ends when the B/F valley point reaches the detector.*
- *We can calculate peak retention times in each column using the equations developed in Chapter 9.*

Did you get it?

Self-assessment quiz: SAQ 10

Q1. Why is the backflush function used in almost all PGCs?
Select all the correct answers:
   A. To remove all injected components from the column system before injecting another sample.
   B. To enable the measurement of very low concentrations of one or more analytes in a sample of the process fluid.
   C. To measure the total heavies concentration in a sample of the process fluid.
   D. To minimize the analysis time by removing unmeasured components.
   E. To measure different analytes in two identical samples of a process fluid analyzed concurrently.

Q2. Explain why the backflushed heavies peaks don't all exit the column at the same time.

Q3. What is an intercolumn detector and how is it used?

Q4. Why can't we accurately calibrate a backflushed heavies peak?

Q5. What two methods are available to ensure the net flow in the first column is backward?

Q6. State three different options for the types of column in a two-column backflush system.

Q7. Explain why two identical columns are unlikely to be a good arrangement for a backflush column system.

References

Further reading

The classic book on process chromatography (Annino and Villalobos 1992) gives a good account of backflush as practiced by PGCs in the late 1980s.

Cited

Annino, R. and Villalobos, R. (1992). *Process Gas Chromatography: Fundamentals and Applications*. Research Triangle Park, NC: Instrument Society of America.

Deans, D.R. (1965). An improved technique for back-flushing gas chromatographic columns. *Journal of Chromatography A* **18**, No. 1965, 477–481. https://doi.org/10.1016/S0021-9673(01)80403-9.

Deans, D.R. (1968). A new technique for heart cutting in gas chromatography. *Chromatographia* **1**, (1968) 18–22.

McEwen, D.J. (1964). Backflushing and two-stage operation of capillary columns in gas chromatography. *Analytical Chemistry* **32**, No. 2, 279–282.

Nuss, G.R. (1963). Applications of group-type analyses in multiple column configurations and in trace analysis. In: *Gas Chromatography* (ed. L. Fowler), 119–125. New York, NY: Academic Press.

Siemens (2008). *Maintenance Manual, MAXUM Edition II Process Gas Chromatograph*. 6/2008 Edition 2000596-001. Houston, TX: Siemens Energy & Automation, Inc.

Villalobos, R., Brace, R.O., and Johns, T. (1961). The role of column backflushing in gas chromatography. In: *Gas Chromatography – 1959 ISA Symposium* (ed. H.J. Noebels, R.F. Wall, and N. Brenner), 39–54. New York, NY: Academic Press.

Villalobos, R. and Turner, G.S. (1963). The role of column backflushing in gas chromatography: II. Multiple column systems. In: *Gas Chromatography* (ed. L. Fowler), 105–118. New York, NY: Academic Press.

Tables

10.1   Calculation Data for Backflush Event Time.
10.2   Calculation Data for Backflushed Peaks.

Figures

10.1   Ten-port Backflush Column System.
10.2   PGC Chromatogram using Backflush to Measure.
10.3   Six-port Backflush Column System.
10.4   Live Tee in Pass-through Mode.
10.5   Live Tee in Backflush Mode.
10.6   A Regrouping Column System.
        (a) Components to be regrouped are loaded into Column 2. (b) Components in Column 2 are backflushed into Column 3 for regrouping.
10.7   Idealized Backflush Event Setting.

Symbols

Symbol	Variable	Unit
j	Gas compression factor	none
k	Retention factor	none
L	Total length of column	mm
N	Plate number	none
p_i	Carrier gas pressure at column inlet	bara
p_o	Carrier gas pressure at column outlet	bara
$\bar{p}$	Average carrier gas pressure in column	bara
p_z	Carrier gas pressure at point (z) along column	bara
P_m-	Pressure upstream of Live Tee	barg
P_m+	Pressure downstream of Live Tee	barg
P_r	Ratio of carrier pressures at column inlet and outlet	none
P_{rz}	Ratio of carrier pressures at point (z) and outlet	none
R_s	Resolution of adjacent peaks	none
S	Spatial separation between adjacent peaks	mm
t_R	Retention time	s
w_b	Spatial base width of peak	mm
Z	Distance to point (z) from column inlet	mm

Subscripts

Sub$_1$ Pertaining to Column 1 in a series column system
Sub$_2$ Pertaining to Column 2 in a series column system
Sub$_{1+2}$ Pertaining to both columns in a series column system

Sub$_B$ Pertaining to Peak B
Sub$_C$ Pertaining to Peak C
Sub$_V$ Pertaining to Peak V

New technical terms

These are the technical terms rendered in bold font in this chapter. You should now know the meaning of these terms.

electronic pressure regulator	**Method**
event time	**regrouping function**
guard column	**start backflush**
intercolumn detector	**start next analysis**
Live Tee	**stuttering technique**
major component	**valley point**

For more information, refer to the Glossary at the end of the book.

11

Heartcut systems

> *"Although originally coined in PGC parlance for a column system that could measure low concentrations of impurities in process fluids, laboratory analysts now use "heartcut" to describe any multiple column system that separates and measures a limited number of components in the injected sample. Recognizing this trend, this chapter looks at several ways to quickly separate a few analytes from a process fluid."*

Origin and development

Early days

Early PGCs with chunky thermal conductivity detectors (TCD) had no problem measuring peaks larger than 1,000 parts-per-million (ppm), but the introduction of the flame ionization detector (FID) changed all that (McWilliam and Dewar 1958). The FID was an instant success as it allowed the PGC to measure a few ppm of hydrocarbons. Several other high-sensitivity detectors soon followed for the ppm-level measurement of selected analytes, such as the flame photometric detector (FPD) for sulfur and the electron capture detector (ECD) for halogen compounds.

The new detectors not only opened a new application area for the PGC – **trace analysis** – but also brought some new challenges as their high sensitivity revealed baseline problems not previously observed. Chief among these is the difficulty of measuring a very small peak that follows a very large one.

To measure a low concentration, a large sample volume is necessary and this is likely to exceed the *sample capacity* of the column. As the large sample volume enters the column it occupies a significant space, becoming a wide peak. The high concentration of major component then saturates the stationary phase and aggravates the adsorption problem, causing extreme tailing. Figure 11.1a illustrates the enormity of the problem, even when measuring

---

Process Gas Chromatography: Advanced Design and Troubleshooting, First Edition. Tony Waters.
© 2025 John Wiley & Sons Ltd. Published 2025 by John Wiley & Sons Ltd.

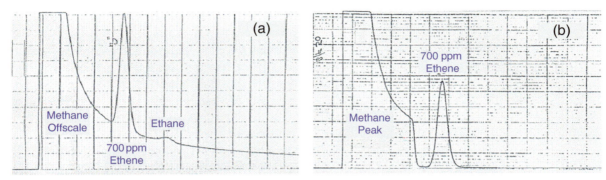

These older chromatograms illustrate the problem well. On the left, Figure (a) shows the effect of allowing all the peaks into the second column. The ethene and ethane peaks are riding on the methane tail. On the right, Figure (b) shows the effect of backflushing the methane tail. Column 2 has pulled the ethene peak away from the methane peak, yielding a baseline separation. To measure a 700 ppm ethene peak, backflush is an adequate solution, but it's not good enough for measuring less than 100 ppm. If you look closely there is still about 1 % of baseline interference, and that would be a problem at higher sensitivities.

Figure 11.1 Large Peak Tailing Problem.
The chromatogram in (a) is without heartcut, whereas the one in (b) is after heartcut.
Source: Author's collection.

700 ppm. A modern PGC could easily measure these ethene and ethane peaks using electronic baseline correction routines, but the tail of the major peak would still be a problem. The tails from subsequent sample injections would accumulate and eventually drive the baseline offscale, rendering the analysis impractical.

At lower concentrations, the problem becomes worse. When attempting to measure less than 100 ppm, the baseline after the major peak is offscale for the whole time the analyte peaks come out, obscuring their presence on the chromatogram and making an analysis possible only with a heartcut system. In the petrochemicals industry, for example, heartcut makes it possible for a PGC to measure a few ppm of the carbon oxides and acetylene in polymer-grade ethylene. Similarly, heartcut enables the measurement of a few ppm of propyne and propadiene in polymer-grade propylene. These industrial applications combine backflush and heartcut to achieve otherwise impossible analyses in just a few minutes.

Heartcut functions

The heartcut column system enables a PGC to measure concentrations less than 100 ppm. It's not difficult to measure such low concentrations – an FID can easily measure 1 ppm of a hydrocarbon. The difficult part is separating the 1 ppm from 99.9999 % of something else.

Early PGC applications engineers developed the heartcut technique to separate these low-concentration analytes for measurement (McEwen 1964). The etymology of the word "heartcut" incorporates the notion of surgically removing a small peak from an encompassing large one, and this remains its meaning in PGC today. As we shall see, the heartcut action doesn't literally "cut out a peak," it samples a time slice of the gas emerging from the first column and we strive to ensure that slice contains all the analyte molecules.

In laboratory usage, however, *heartcut* has come to mean the isolation of a selected group of components for further separation and analysis by two-dimensional gas chromatography – in contrast with the term *comprehensive*, meaning the complete analysis of all components in the sample (Dimandja 2020). This definition of heartcut abandons the concept of a vast size differential between the concentrations of analyte and major component. Yet one common characteristic of heartcut systems remains; they all maintain a singular direction of carrier gas flow, a feature that distinguishes all heartcut systems from all backflush systems.

The heartcut technique might not be necessary for peaks preceding the major component. For low-ppm peaks eluting before the major, you can measure the desired analytes and then backflush the major component to vent. Although the backflush might not totally remove the tailing molecules, the net reverse flow in the backflushed column should keep the major tail out of the second column.

For analytes at higher ppm concentrations, backflush can be effective even for a peak that elutes after the major component. Figure 11.1b illustrates how backflush can remove part of the major tail, thereby enabling a baseline separation of the small peak that follows.

Heartcut column system

PGCs use the heartcut column system to measure trace amounts of impurities in process fluids. It's a powerful technique, sometimes capable of separating peaks that are a million-to-one in size disparity, yet it uses the simple column system shown in Figure 11.2. Described later is a valveless system using the *Live Tee* device.

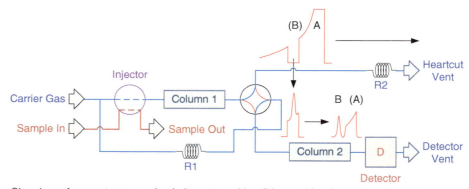

Figure 11.2 Heartcut Column System. Showing a four port rotary valve in its rest position (blue path); other valves have equivalent function. After sample injection, most of the major component (A) flows to vent. Just before an analyte peak is due to exit Column 1, a short-duration 90° valve rotation (red path) allows that peak to enter Column 2, together with a small slice from the unseparated tail of the major component. Once the analyte is fully into Column 2 the valve reverts to its rest state (blue path) and Column 2 separates the small analyte peak (B) from the remnant of major component (A).

In contrast to the backflush systems discussed in Chapter 10, a heartcut column system experiences no change in flow direction or flow rate when the column valve operates. In Figure 11.2, the flow restrictor R1 (which might instead be an electronic pressure controller) ensures the carrier gas flow rate from the detector vent is identical in both valve positions. Flow restrictor R2 then ensures the very same flow rate from the heartcut vent.

After sample injection, the column valve remains in its rest condition to allow the major component peak to quickly pass through the first column and flow to vent via flow restrictor R2. This column (sometimes called the **cutter column**) retains the analyte peak but cannot completely separate it from the major tail. The valve actuates when the analyte peak is about to leave Column 1, allowing that peak and a slice of the major tail to enter the second column. As soon as the analyte peak is fully into Column 2, the column valve returns to its rest state directing the remaining major tail and any other components to flow out of the heartcut vent. In effect, the heartcut action injects a sample into the second column that contains carrier gas plus only two components – the whole analyte peak and a small portion of the major component. It's usually easy to separate two components, so Column 2 tends to be reliable.

Pressure balance column switching

An option to using mechanical valves for column switching is the *Live Tee*. As discussed in Chapter 10, the Live Tee is a device that uses carrier gas pressure to switch capillary or micropacked columns without a mechanical valve and with virtually no extracolumn volume (Deans 1968).

As shown in Figure 11.3, the two columns are almost butt jointed and a narrow-bore capillary tube enters the end of each column forming slim

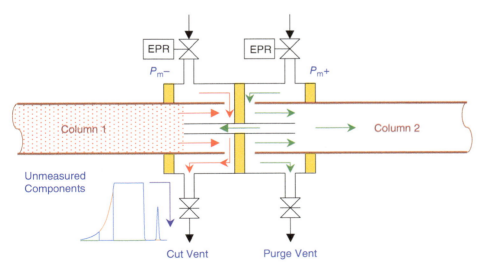

This mode diverts the effluent flow from Column 1 to the Cut Vent and continues the constant flow through Column 2. The carrier gas pressure on P_m+ is higher than on P_m-, reversing the flow through the connecting capillary. No components enter Column 2 until the device switches to the heartcut mode shown in Figure 11.4.

Figure 11.3 Live Tee in Vent Mode.

annular passages between the outer diameter of the capillary tube and the inner diameter of the column ends. Carrier gas can flow between the columns only via the capillary connector, wherein it incurs a small pressure drop.

The carrier gas inlet and vent ports on both upstream (P_m-) and downstream (P_m+) sides of the Live Tee allow it to control the flow direction in the columns. When operating in vent mode (Figure 11.3), the pressure applied to P_m- is slightly lower than the pressure on P_m+, reversing the flow through the capillary tube and diverting the entire flow from Column 1 into the upstream Cut Vent, from where it flows to atmosphere via a flow-control restrictor. The downstream Purge Vent flow normally goes via another restrictor to the detector as makeup gas or fuel gas.

To execute a cut (Figure 11.4), the upstream electronic pressure regulator applies a pressure at P_m- slightly higher than the pressure at P_m+, driving the flow from Column 1 through the capillary tube and into Column 2.

After completing its heartcut functions, the Live Tee can also backflush the first column as described in Chapter 10. As noted therein, the setup procedure for a Live Tee is more complex than for a valved system and may require special training for maintenance technicians.

Column system design

Application engineers work from experience and empirical trials when designing columns for heartcut systems. Calculation is ineffective for deciding column lengths but it's good to remember that the first column runs

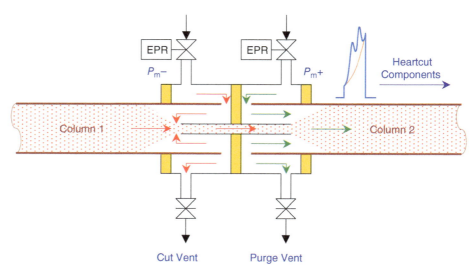

Figure 11.4 Live Tee in Heartcut Mode. This mode directs the measured peaks and a small remnant of the major component into Column 2. The electronic pressure regulators (EPR) are set so that P_m- is higher than P_m+, causing carrier gas from P_m- to flow through the connecting capillary, taking the flow from Column 1 with it (red arrows).

about half as fast as the second one. Thus, the first column will be short, probably less than one-fifth the length of the second one.

Making the cuts

Modes of operation

There are three ways to operate a heartcut column system:

- Make a single cut to capture and measure only one analyte.
- Make a single cut to capture and measure two or more analytes.
- Make two or more cuts to capture and measure multiple analytes.

These three modes of operation are progressively more difficult to design and maintain. While the column configuration is the same in each case, the columns need to separate the increasing flock of analyte peaks and remnants captured by the more complex heartcut actions.

A single cut for one analyte

The real chromatogram in Figure 11.5 shows the power of heartcut when measuring a single analyte. Look at the flat baseline on both sides of the

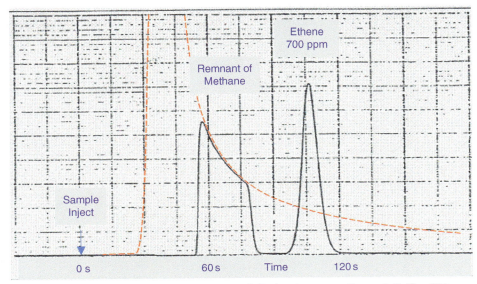

This older chromatogram illustrates heartcut well. Both columns are Porapak Q. The 700 ppm ethene peak enters Column 2 together with a small portion of the methane tail. Column 2 easily separates the ethene peak for measurement, with a nice flat baseline on both sides of the peak. We see from the size and shape of the remnant that Column 1 has separated the ethene peak well away from the major methane peak and only a small amount of methane tail has entered Column 2. It's interesting that the top surface of the remnant retains the original curvature of the methane peak tail, superimposed here in red.

Figure 11.5 Single-cut Heartcut Chromatogram.
Source: Author's collection.

peak. Of course, a 700 ppm peak is easy to measure, yet this simple technique can quickly separate almost any analyte – at any concentration – in any process sample. We shall see that it gets more difficult when multiple measurements are necessary.

> *When configured to make just one measurement, a PGC can quickly separate and measure almost any analyte – at any concentration – in any process sample.*

The schematic diagram in Figure 11.6 shows the elegance of a single cut. This is the simplest and most reliable heartcut system. Make sure you fully understand its exceptional performance before delving into the complexity of multiple cuts.

After sample injection, the first column retains the desired analyte, partially separating it from the major component which mostly flows to vent. To capture the analyte peak, the column valve turns ON a moment before it starts to exit the first column. The valve remains ON just long enough for all the analyte to flow into the second column. Then, as soon as the analyte peak has completely entered the second column, the column valve turns OFF.

The chromatogram displays only two peaks: one is the analyte and has a typical peak shape and the other is a peculiar-shaped artifact. The artifact is a slice of the major component tail that entered the second column while the valve was open – along with the analyte peak. Some users like to call it a

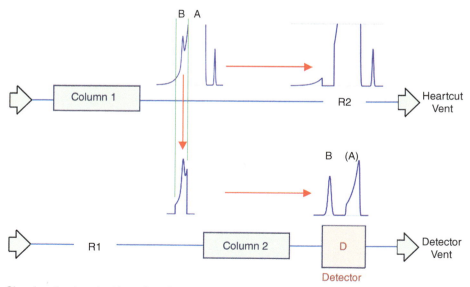

Figure 11.6 Schematic of a Single Cut. Showing the heartcutting of a single analyte (B) into the second column, together with a small slice of the major component. Since the remnant peak (A) is similar in width to the analyte peak (B), the separation of A and B is easily accomplished by the second column.

ramp as it often has a triangular shape. We prefer the more general term, **remnant peak**.

A modern PGC chromatogram will also feature **marker spikes** on the baseline added by the processor to indicate the event times for sample injection, column switching, and analyte peak gating. Nothing else should appear on the chromatogram as all the other components discharge to vent.

A single cut for two analytes

When a PGC must measure two or more ppm analytes, there are two possible heartcut solutions. Here, we review the method of catching two components with one valve action – it's rarely possible to catch more than two. In the next section, we consider multiple valve actions to capture several analytes.

To capture two analytes with one cut it's necessary for the first column to:

➤ Separate both analyte peaks from the major and its tail.

➤ Not separate the two analyte peaks from each other, so we can capture them both in a single cut.

Immediately, we notice the extra complexity of measuring two components. What kind of column would separate the two analytes from the major but not from each other?

That depends on the nature of the analytes. It might be easy if there are no other components to worry about and the two analytes each have the same number of carbon atoms, like measuring propane and propene in ethane, for instance. For that analysis, a nonpolar column would hold both C_3 peaks back from the ethane peak without separating them from each other. Then, a single heartcut would deliver both C_3 peaks to a porous polymer column for final separation. The chromatogram would show three peaks, one remnant, and two analytes. Not all applications are that easy, though. If you are measuring higher carbon numbers, for instance, the choice of columns is more difficult due to the large number of possible components (realistically, about 15 C_4 hydrocarbon peaks are possible). A single cut might capture fragments of several peaks.

Unfortunately, even when we select the first column to get minimal separation between the two analytes, there is always some separation. This immediately makes the heartcut more difficult, as depicted in Figure 11.7. A partial separation between analyte peaks means the valve has to be ON longer and this makes the remnant peak wider. The second column then has the more difficult task of separating the two analyte peaks from a wider remnant. In most cases the remnant will appear first, followed by both analyte peaks, but occasionally the second column will move one analyte peak in front of the remnant and the other behind it.

For some types of analytes, a single cut can never work. If we are measuring ppm levels of two chemical homologs, for instance, we know they will separate on any column by the *doubling rule*, so a single cut would be impossibly wide. This kind of analysis always needs multiple heartcuts.

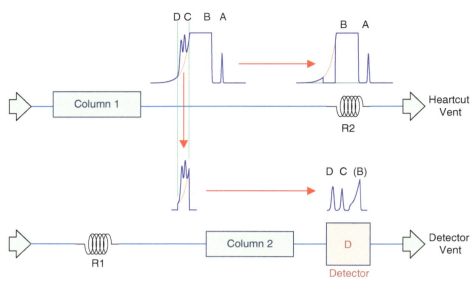

Figure 11.7 Single Cut for Two Analytes.

Showing the single cut of two analytes (C and D) into the second column, together with a slice of the major component. Since the remnant peak (B) is wider than the analyte peaks (C and D), it's more difficult for the second column to separate the two analytes from the remnant.

Multiple cuts

A single heartcut can rarely capture more than two analytes. Three or more analytes are likely to separate on the first column, creating a remnant peak that is too wide to separate from all the analyte peaks. To solve this problem, we use a different tactic; instead of trying to keep the analyte peaks together on the first column, we deliberately separate them and then do multiple cuts, one for each analyte – shown schematically in Figure 11.8.

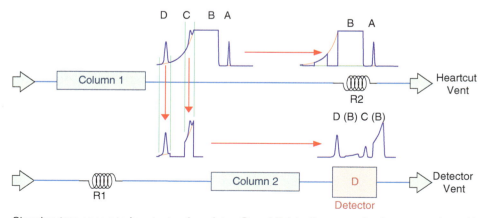

Figure 11.8 Multiple Cuts for Two Analytes.

Showing two separate heartcuts of analytes C and D into the second column, together with two small slices of the major component. The first column determines the spacing of the two remnant peaks (B) and since they are both made up of identical molecules the second column cannot increase their separation. The second column must be the exact length necessary to fit the analyte peaks between the remnants.

When the PGC performs a separate heartcut for each analyte, each cut loads a different analyte peak into the second column, together with a gradually declining remnant of major tail. The second column separates the analyte peaks from the remnant peaks but it cannot increase the separation between remnants because all remnants contain the same molecules:

> *Each of the remnants entering the second column comprises the same kind of molecules, so that column cannot increase the separation between them.*

Once the remnants enter the second column, their widths and the time between them are fixed by the heartcut timing and cannot be changed. Therefore, the first column must create a gap between the analyte peaks wide enough to contain one of the analyte peaks. Actually, the baseline time between remnants *decreases* as they pass through the second column due to the rounding of their square sides. And the analyte peaks get wider. Therefore, the time between the two heartcuts must be greater than their duration. That amount of separation might be difficult to achieve and once accomplished for the first two analytes, subsequent analytes can be very late indeed.

As an additional challenge, the design of the second column is more difficult for multiple heartcuts. The second column must be exactly the right length to drop each peak neatly into the fixed gap between remnants. Usually, an analyte peak lands in the gap just after its associated remnant – but not always. Occasionally, the second column will rapidly elute an analyte *before* its remnant, or strongly retain an analyte to elute *much later* on the chromatogram, after two or more remnants have passed.

Avoiding complexity

Clearly, multiple heartcuts are more difficult to design, setup, and maintain – and are unlikely to be reliable. For the highest reliability, keep in mind the familiar yet eternally true mantra: **KISS** (keep it simple, stupid!).

When a PGC measures ppm concentrations, analytical difficulty seems to follow a square law. While we can't rigorously prove it to be true, this *law* serves to illustrate our argument:

- A PGC can easily measure one analyte in any sample.
- Measuring two analytes is four times harder.
- Measuring three analytes is nine times harder.
- Four is sixteen times harder and unlikely to work.
- Seven is forty-nine times harder and guaranteed to fail.

> *At a Mid-West refinery, we found a PGC making seven heartcuts on each sample injection. The chromatogram contained seven analyte peaks and seven remnant peaks, all mixed up together. The site personnel could not make it work. Nothing was gained.*

From experience, we strongly recommend you dedicate each heartcut system to measuring only one or two analytes. Resist the ever-present temptation to do more. Highly skilled application engineers can make a heartcut system do just about anything you ask for, but your maintenance people will not be able to keep it working on site.

When measuring low concentrations:

➢ Request only measurements vital to process operations.

➢ Be satisfied with a few reliable measurements.

➢ Never ask for an additional measurement that would be "nice to know".

Adding a backflush

Once heartcutting is complete, it's best to backflush the first column. Although backflush adds a valve function, it doesn't increase measurement complexity and is a sensible precaution. Without backflush, a heavy component from a previous injection might coelute with the analyte and enter the second column. Such recurring incursions of heavy peak remnants, no matter how small, will eventually disrupt the baseline and spoil the analysis.

> *The saving gained by running without backflush is not worth the risk. Never run an isothermal PGC without backflushing the first column.*

Never run without backflush (Gokeler 2011). Applications engineers sometimes omit the backflush in an attempt to save money or to simplify maintenance. Their justification is that the first column is very short (sometimes only half a meter) and the heartcut valve action is very early in the analysis cycle. Therefore, the first column has a long forward flush to vent, which should remove any late components before the next injection. That might be true for some exceptionally pure process fluids, most of the time. But it risks the occasional process upset that will take the analyzer out of service for a long time. And without expert troubleshooting that analyzer may never be the same again. For more information on backflush systems refer to Chapter 10.

Setting the valve timing

The heartcut valve ON and OFF times must be precisely set. When measuring low-ppm levels of the analyte, setting these times is always a delicate trial-and-error procedure. For ppm analysis, an intercolumn detector is unlikely to be helpful because the excessive tail of the major peak will make it impossible to see the tiny analyte peak.

The setup procedure

Whenever possible, use a test sample known to contain a constant amount of each analyte within the upper half of the measurement range. Settings made with a lower concentration may not accommodate higher analyte concentrations in the process sample.

The main constituents of your test sample should closely match the composition of the process fluid. Don't use a test sample with a radically different composition (such as one with nitrogen as its major component) because it will change the analyte peak retention times on the first column, invalidating your valve time settings. When a high concentration of major component saturates the stationary phase, the solubility of the analytes decreases and their retention times diminish.

Another issue to watch is the frequency of sample injections. You need to set the heartcut times when conditions within the first column are stable. The first few sample injections setup a background level of major component in the stationary phase of the first column. Analyte retention times won't become constant until that background level is constant, so run the PGC through several automatic cycles before you set the valve timing and then allow the Method to continue uninterrupted while you make the adjustments. If your column system also backflushes the first column, ensure the backflush occurs just after the last heartcut on every cycle while you are setting the heartcut times. Don't adjust the backflush after completing the heartcut timing.

Adjust the column valve ON and OFF times to get the maximum analyte peak and the minimum remnant peak on the chromatogram. To finalize each setting make small adjustments that reduce the valve ON duration. Stop when the analyte peak gets smaller and revert to the previous setting. Then add a safety margin that increases the valve ON duration by one or two seconds while ensuring that the larger remnant peak thus caused doesn't spoil the resolution of the analyte peak. Technicians sometimes use the analyte peak height to indicate the correct setting, but the integrated peak area is a better confirmation that none of the analyte is missing.

Once your heartcut is properly set and working well, note the duration of valve actuation. Since the cut duration is a function of the peak width, it's unlikely to change. For future performance checks, move the entire cut one second forward and backward in time to confirm that the whole peak (or group of peaks) is cut into Column 2.

Setting the heartcut times is a blind procedure which can make it a difficult one. There is always a risk of catching the wrong peak:

> *When setting up a PGG heartcut take time to confirm that you have captured the correct peak. The PGC measures what you tell it to measure and measuring the wrong peak is by far the worst error it can make!*

To confirm that your heartcut is set correctly, spike your test sample with a little – a very little – of the analyte. The chromatogram peak should increase, confirming that your heartcut will measure the analyte. Recall that this outcome doesn't confirm the identity of the peak captured from a process sample. For that to be true, the first column must separate the analyte from all other components in the process fluid. This is a reasonable assumption only if the composition of the sample is well known and a trusted application engineer designed the column.

A simple way to spike a sample is to run the pure analyte through a short length of rubber or polymer tubing. Then disconnect the tubing, blow out the analyte with compressed air or nitrogen, and insert the tubing into the line connecting your test sample to the PGC sample inlet. The doctored tubing will contaminate the test sample with your analyte, perhaps too much at first, but the level will decline as the test sample purges out the tubing.

When better facilities are available, collect a process sample in a gas bladder and analyze it with the PGC until constant results are obtained. Then, using a syringe, inject a calculated amount of one analyte into the bladder. After mixing, an increased PGC response confirms that you are capturing the correct component. Continue to test other analytes in the same way.

Setting multiple cuts

Although the carrier gas flow rates in a heartcut column system should be equal in both valve positions, their settings will not be perfect. And even if they are, the injection of a large sample will affect the flow rate in the first column. Thus, each heartcut may cause a small change in the retention time of subsequent peaks and it's necessary to set multiple heartcuts in retention-time order. After the first cut is working well, keep it running while setting up the second cut. Then progressively add any further cuts. Do this even if some early analytes are not present in your current test sample.

General rules of procedure

The general rules are:

- Never adjust the valve timing on a single sample injection.
- Always set the heartcut valve times while the PGC is running continuous automatic cycles.
- Allow several automatic analysis cycles to run before making an adjustment.
- Never setup a later cut before setting up an earlier one.
- Always check the subsequent cuts after adjusting an earlier cut.
- Never adjust the backflush timing after the heartcut timing is set.

Diagnosis

Precise observation of peak shape and a logical interpretation of the cause are essential prerequisites to successful troubleshooting. A single heartcut generates a chromatogram like Figure 11.9 with only two peaks, yet many

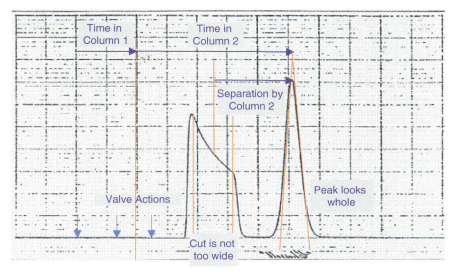

This figure illustrates information deduced from a heartcut chromatogram. The observed separation is due solely to Column 2 and is adequate. The remnant is small and narrow, indicating good separation by Column 1 and proper valve timing. Note the half-peak shapes on either side of the remnant are sharper than a true peak would be because the remnant travels only through the second column. Triangulation reveals no evidence that column valve switching has cut into the sides of the analyte peak, but it's best to check this by observing the integrated area of the peak.

Figure 11.9 Deductions from the Chromatogram.

clues are visible that tell you what the columns are doing. Take the time to fully understand this simple chromatogram and you'll be well-prepared to diagnose more complex heartcut techniques.

Look closely.

Observing the remnant peak

Take a moment to think about the shape of the remnant peak. Recall that its sides are vertical when it enters Column 2 because pure carrier gas enters the second column before and after valve actuation. Then, as the remnant travels through the second column its leading and trailing edges form standard peak shapes. If you look closely at Figure 11.9, you can see that both edges of the remnant look like half a peak, but compared with the analyte peak they are a little sharper than expected for a peak at that location on the chromatogram. This is because they passed through only one column. Also, you should notice that the leading edge of the remnant is sharper than its trailing edge. This is because it spent less time in the column. The unnaturally sharp sides are one way to distinguish a remnant peak from a bunch of component peaks.

All the molecules that make up the remnant peak are identical, so they migrate along the column at the same speed. Therefore, discounting the spreading of its sharp sides, the remnant doesn't get any wider as it passes through the second column. Its core width on the chromatogram is a good indicator of the valve ON duration. If the remnant core is much wider than the analyte peak, the valve stays ON too long.

The chromatogram in Figure 11.9 also shows how the top of a remnant retains the shape of the major tail. It's unusual to see this because remnant peaks are usually much larger and go offscale. When the chromatogram signal is highly sensitive, a large remnant can look like a square wave, flying offscale for a while and then rapidly returning to the baseline. But if the chromatogram sensitivity is low enough, the remnant may show a slice of the major tail dropping rapidly. Then, the remnant appears on the chromatogram as an almost triangular peak. Seeing these triangular remnants led chromatographers to call them *ramps*.

The shape of the remnant is a good indicator of how well the first column is separating the analyte peak from the major component. Recall that the ethene peak and methane remnant enter Column 2 at the same time. Therefore, the low curved top of the methane remnant in Figure 11.9 is where the ethene peak was at the beginning of Column 2. We see that Column 1 was powerful enough to pull the ethene peak well down the methane tail. In most heartcuts, the first column doesn't achieve such a large separation and more of the major tail enters the second column, forming a large remnant peak. Conversely, some heartcuts can form very small remnants that look like peaks. This can happen when capturing a later peak well separated from the major, and even when capturing a peak that elutes *before* the major. An early cut can catch a low-level remnant from the tail of the previous injection. This tiny remnant might look like a rogue peak on your chromatogram. If you backflush the first column it should go away.

Observing the analyte peak

The analyte peak enters Column 2 while the column valve is ON, so the average of the valve ON and OFF times is a good estimate of analyte retention time on Column 1. To trim your estimate, you can even make a small allowance for the expected skewness of the peak. The chromatogram shows the total retention time of the analyte peak, so it's easy to estimate its retention time in Column 2.

When the timing of the column valve is set correctly, its ON duration exactly contains the analyte peak. The ON duration is therefore an indicator of the full width of the analyte peak as it enters the second column. The analyte peak will spread as it travels through the second column and appear a little wider on the chromatogram, yet its width should still be similar to the remnant width. If the remnant is much wider, the valve timing is wrong.

When the valve timing is correct, we can deduce that:

> If the separation is good but the remnant is too large, Column 1 has failed. This is the most likely occurrence, due to the large injections of process fluid into Column 1.

> If the remnant peak is a reasonable size and shape, yet the analyte separation is inadequate, Column 2 is at fault.

> If the analyte peak is wider than the remnant peak. Column 2 is at fault. It must be causing the additional width.

It's possible to cut the back or front of an analyte peak without affecting its height. Look closely at the shape of the analyte peak to assess whether any part of it is missing. This can be difficult to see; expanding the horizontal axis and drawing triangulation lines might help. To completely avoid this concern set the valve timing by monitoring peak area rather than peak height.

Similar column systems

Today's analytical lexicon extends *heartcut* terminology to any chromatographic system that doesn't measure all components of the sample. While we exclude backflush systems from that category, some simple PGC column systems do fit into the definition. Here, we consider:

➤ The trap-and-hold column system.

➤ The distribution column system.

➤ Parallel chromatography.

Trap-and-hold column system

This unique technique – now rarely used – is also known as trap-and-release or trap-and-bypass and allows a complex analysis using a single detector.

The trap-and-hold column system immobilizes selected peaks while other peaks separate and clear the columns. It then releases the trapped peaks for further separation and measurement. Refer to Figure 11.10. When the column valve operates, Column 2 traps selected analyte peaks while

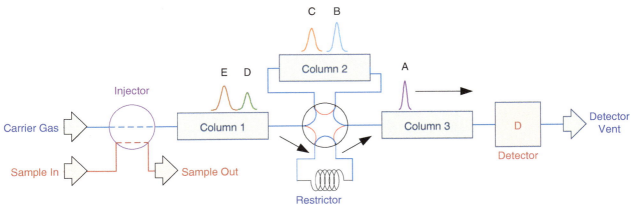

The six-port rotary or diaphragm valve in its rest state (blue paths) puts all three columns in series. Other valves have similar function. The flow resistance of the restrictor is equivalent to Column 2. A valve rotation of 60° (red paths) seals Column 2, trapping selected peaks (for example, B and C), and connecting Column 1 to Column 3 to allow later peaks (for example, D and E) to bypass Column 2. When the column valve returns to its rest state (blue) the trapped peaks enter Column 3 for further separation before reaching the detector.

Figure 11.10 Trap-and-Hold Column System.

some other peaks pass from Column 1 into Column 3. When released, the trapped peaks enter Column 3 for final separation.

A good feature of the trap-and-hold system is the conservation of carrier gas. It uses only half the carrier gas as a conventional heartcut system.

Trap-and-hold can sometimes act as a heartcut and at other times just to delay a few components while the other peaks are clearing Column 3. Most complete systems also include a backflush for Column 1 that starts as soon as all of the analyte peaks have exited the first column.

While the trap-and-hold technique was useful for expanding the application of older PGCs having limited detector capability, it has some disadvantages:

- The trapped component peaks gradually become wider due to longitudinal diffusion of their molecules, thereby reducing the resolution of adjacent peaks.
- The trapped peaks take longer to elute, increasing the analysis time.
- Switching the trap column in and out of circuit may cause pressure and flow upsets that disturb the detector baseline.

Although parallel column systems have generally superseded trap-and-hold, you may encounter some older systems in the field still using the technique.

Distribution column system

One of the original PGC column arrangement is the distribution column system. The simplest version of this is the dual column system shown in Figure 11.11. The idea is to divide the sample into small groups of components and then distribute each group to a different second column for final separation. The first column partially separates the injected sample. The column valve then directs each group of components to an appropriate second column. To optimize the resolution of the analytes, the secondary columns usually employ different stationary phases and might be different lengths. Serial restrictors equalize their flow rates. Secondary columns may have dedicated detectors, otherwise their effluents merge and flow into one detector. The system might also include a backflush of the first column to remove unwanted heavies.

Parallel chromatography

Microprocessors can now monitor several detectors at once, making it possible for PGCs to house two or more complete chromatographic systems each with a dedicated sample injector, column system, and detector. Each of these systems can independently measure a selected few of the desired analytes using a simple column system. This parallel chromatography simplifies application engineering and maintenance and allows manufacturers to design standard column systems for common applications, reducing cost.

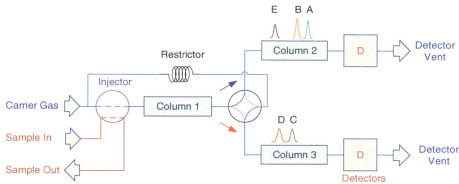

The four-port rotary or diaphragm valve in its rest state (blue paths) allows partially separated components from Column 1 to pass into Column 2 for final separation (for example, A and B). A valve rotation of 90° (red paths) sends selected peaks (for example, C and D) to Column 3. Then, moving the valve back to the rest position may allow later analytes (for example, E) into Column 2. After all analyte peaks are out, Column 1 ideally would be backflushed to remove unwanted heavies (not shown). Other valves have similar function. A flow restrictor or electronic pressure regulator ensures the detector flow is the same in both valve positions. Another flow restrictor may be in series with one of the secondary columns to equalize the carrier gas flow rates in both columns.

Figure 11.11 Distribution Column System.

The real power

We have seen that column-switching techniques can reduce analysis time, do the housekeeping, and separate extremely small peaks for measurement. Yet valve configurations are just the tools we use; the real power of a column system is in the columns themselves.

All isothermal column systems have one feature in common; they employ two or more columns in series. If the serial columns contain the same stationary phase, their performance is easy to predict. All peaks will have the same retention time and width as on a single column of the same length, so system design is straightforward. The designer determines the total length of column needed to separate the analyte peaks, then apportions that length between the serial columns.

But to release the real power of a serial column system one of the columns must employ a different stationary phase. Then, the first column will partly separate the desired analytes from other components, and a second column will use a different stationary phase to complete their separation.

Though hundreds of stationary phases are available, most application engineers limit their choice to about one dozen, which they get to know well. Even so, over 100 stationary phase combinations are possible in a column system having two columns. This is the true power of the serial column system.

Knowledge Gained

On trace analysis

- *The high sensitivity of some detectors enables a PGC to measure low-ppm concentrations.*
- *Low-ppm measurements need a large sample volume that overloads the column.*
- *Highly sensitivity detectors reveal the long tail of molecules following a major component.*
- *Without heartcut, it may not be possible to measure small peaks eluting after the major.*
- *Backflushing the major tail may be adequate to measure a following high-ppm analyte.*
- *Removing the major tail before and after a small peak allows measurement of low-ppm analytes.*
- *In laboratory usage, heartcut means any system that doesn't measure all the components.*

On heartcut systems

- *The heartcut column system is simple, comprising just two columns and one column valve.*
- *Some PGCs use a "Live Tee" instead of a valve, especially with small-bore columns.*
- *The valve directs selected analyte peaks from Column 1 into Column 2 and all other peaks to vent.*
- *Column 2 separates analyte peaks from the small slice of major that enters with the analytes.*
- *With heartcut, a PGC can measure almost any analyte, at any concentration, in any process fluid.*
- *The simplest and most reliable heartcut makes one cut for a single analyte.*
- *A single cut generates just one remnant peak and one analyte peak.*
- *In a simple heartcut, the separation visible on the chromatogram is entirely due to Column 2.*
- *The shape of the remnant peak is entirely due to the analyte separation by Column 1.*
- *It's more difficult to design a system to capture and separate two analytes with one cut.*
- *It's most difficult to capture multiple analytes with multiple cuts, so the PGC is likely to fail.*
- *PGC complexity and risk of failure increases with the square of the number of analytes.*
- *Measure only the analytes needed for process control; never add "nice-to-know" measurements.*
- *PGCs with a heartcut column system should also have a backflush function.*

On valve timing

- *Always adjust the timing of the first cut before adjusting the second and subsequent cuts.*
- *Check the heartcut timing with a test sample similar in composition to the process fluid.*
- *Always run several automatic analyses before adjusting the valve timing.*
- *If using backflush, set it to occur on every cycle while setting the heartcut timing.*

- ❖ *Spike the test sample to confirm the heartcut is selecting and measuring the correct component.*

- ❖ *Measuring the wrong peak is the worst error a PGC can make!*

On remnants

- ❖ *A remnant has vertical sides when entering Column 2, which soften as it transits that column.*

- ❖ *The width of the remnant is the same as the duration of the cut that formed it.*

- ❖ *The core width of a remnant doesn't increase in Column 2, but its sides spread into half peaks.*

- ❖ *If correctly adjusted, the cut duration is slightly wider than the full width of the analyte peak.*

- ❖ *If the remnant peak is much wider than the analyte peak, the valve timing is wrong.*

- ❖ *The top of a remnant follows the downslope of the major tail and is sometimes called a ramp.*

- ❖ *The height of a remnant indicates how well Column 1 separates the analyte from the major.*

- ❖ *The center of a cut indicates the retention time of that analyte on Column 1.*

- ❖ *The time between a remnant and its analyte peaks is their separation on Column 2.*

- ❖ *Column 1 must separate analytes enough to make wide gaps between multiple remnants.*

- ❖ *Column 2 cannot increase the separation of remnants as they all move at the same speed.*

- ❖ *If the remnant peak is the right size but not separated from an analyte, Column 2 has failed.*

- ❖ *If the separation is good but the remnant is too large, Column 1 has failed.*

On similar systems

- ❖ *Stopping the flow in a column can trap-and-hold selected peaks for later separation.*

- ❖ *A trap-and-hold column can delay certain peaks or perform a heartcut function.*

- ❖ *A dual column distributes selected peaks to two or more second columns for final separation.*

- ❖ *The secondary columns each have their own detector or the flows merge into one detector.*

- ❖ *Parallel chromatography uses two or more complete GC systems to measure all the analytes.*

- ❖ *Each of the parallel systems uses a simple design to measure a few of the desired analytes.*

- ❖ *Standardized parallel systems are simple to design, maintain, and troubleshoot.*

- ❖ *The real power of a column system comes from using different stationary phases for fast analysis.*

Did you get it?

Self-assessment quiz: SAQ 11

Q1. Which one of the following statements best describes the basic function of a heartcut system as used in a PGC?
 A. To adequately isolate for measurement one or more low-ppm analytes that elute after a high-percentage component.
 B. To adequately isolate for measurement a low-ppm analyte that coelutes at the same time as another low-ppm analyte.
 C. To adequately isolate for measurement a low-ppm analyte by combining the effects of two different stationary phases.
 D. To adequately isolate for measurement a low-ppm analyte by backflushing the tail of the major component.

Q2. Consider a simple PGC heartcut system:
 A. At minimum, how many columns does it need?
 B. Which column would typically be the longest?
 C. At minimum, how many column valves does it need?
 D. At minimum, how many detectors does it need?
 E. At minimum, how many atmospheric vents does it have?
 F. How many different carrier gas flow rates does it use?

Q3. Consider the chromatogram from a PGC heartcut system and assume all desired analytes are present in the injected sample and are well separated by the columns:
 A. For a single cut for one analyte, how many peaks and remnants are present on the chromatogram?
 B. For a single cut for two analytes, how many peaks and remnants are present on the chromatogram?
 C. For two cuts for two analytes, how many peaks and remnants are present on the chromatogram?

Q4. Imagine you want to check the timing of a single heartcut to ensure that none of the analyte peak is missing. Which of the adjustments listed below would be an appropriate temporary change to the valve timing?
 A. Set the valve ON command one second earlier.
 B. Set the valve ON command one second later.
 C. Set the valve OFF command one second earlier.
 D. Set the valve OFF command one second later.

Q5. Assess the relative difficulty of designing a PGC column system to use each type of heartcut listed below:
 A. One cut for one analyte.
 B. One cut for two analytes.
 C. Two cuts for two analytes.

Q6. List the reasons why it's more difficult for a heartcut system to make multiple cuts than to make a single cut.

Q7. Graduate Question: Look at the chromatogram in Figure 11.5 and assume the column valve goes ON at 33 seconds and OFF at 52 seconds. Then, in the listed order, estimate the retention time of:
  A. Ethene in the first column.
  B. Ethene in the second column.
  C. Methane in the second column.
  D. Methane in the first column.

References

Further reading

The book chapter by Mahler et al. (1995) summarizes their experience of multiple column applications in Siemens PGCs.

Gokeler (2011) gives a good summary of the advantages and disadvantages of traditional and modern PGC column arrangements.

Cited

Deans D.R. (1968). A new technique for heart cutting in gas chromatography. *Chromatographia* **1**, No. 1–2, 18–22. https://doi.org/10.1007/BF02259005.

Dimandja, J.-M.D. (2020). Introduction and historical background: the "inside" story of comprehensive two-dimensional gas chromatography. In: *Basic Multidimensional Gas Chromatography* (ed. N.H. Snow), 1–40. London, UK: Academic Press.

Gokeler, U. (2011). Column switching simplification in process gas chromatography. Paper S01.2 in: *Proceedings of the ISA 56th Analysis Division Symposium 2011, League City, TX*. Research Triangle Park, NC: International Society of Automation.

Mahler, H., Maurer, T., and Mueller, F. (1995). Multi-column systems in gas chromatography. In: *Chromatography in the Petroleum Industry* (ed. E.R. Adlard). Journal of Chromatography Library Volume **56**, 231–268. Amsterdam, Netherlands: Elsevier.

McWilliam, I.G. and Dewar, R. A. (1958). Flame ionization detector for gas chromatography. In: *Gas Chromatography 1958* (ed. D. H. Desty). London, UK: Butterworths Publications Ltd.

McEwen, D.J. (1964). Backflushing and two-stage operation of capillary columns in gas chromatography. *Analytical Chemistry* **36**, No. 2, 279–282.

Figures

11.1 Large Peak Tailing Problem.
11.2 Heartcut Column System.
11.3 Live Tee in Vent Mode.
11.4 Live Tee in Heartcut Mode.
11.5 Single-cut Heartcut Chromatogram.
11.6 Schematic of a Single Cut.

11.7 Single Cut for Two Analytes.
11.8 Multiple Cuts for Two Analytes.
11.9 Deductions from the Chromatogram.
11.10 Trap-and-Hold Column System.
11.11 Distribution Column System.

Symbols

Symbol	Variable	Unit
P_m-	Pressure upstream of Live Tee	barg
P_m+	Pressure downstream of Live Tee	barg

New technical terms

If you read the whole chapter, you should now know the meaning of these technical terms:

cutter column **ramp**
dual column system **remnant peak**
KISS **trace analysis**
marker spikes **trap-and-hold column system**

For more information, refer to the Glossary at the end of the book.

12

PGC troubleshooting

"Don't rush into troubleshooting until you're sure a problem exists. Then, determine whether the problem is an instrument problem affecting the baseline or a chromatographic problem affecting the measurements. If you don't know which, you may spend a long time trying to solve a non-existent problem."

Is there a problem?

The usual suspects

It's not unusual for Operations to suspect a measurement and they might have good reasons for doing so, based on their knowledge of plant operation at that time. Yet there are likely to be several other possibilities. Perhaps the measurement is correct and its unexpected value indicates a process problem. Or perhaps the sample arriving at the PGC is not representative of the process; it's well known that most problems with PGCs are due to an inadequate sampling system. Or perhaps the PGC really is the culprit.

Often a maintenance technician will instinctively respond to a call for service by recalibrating the PGC. Indeed, the operators may specifically ask for a calibration, believing it to be the solution to all measurement problems. But there's at least one problem with this unthinking response:

➤ Either: recalibrating a sound PGC that doesn't need calibrating has just introduced more variability into perfectly good measurement data.

➤ Or: recalibrating a failing PGC has just masked a problem for a while, hiding an incipient failure and creating more work for later.

Even a well-executed calibration procedure doesn't prove the analyzer is working well. For instance, there might be an error in the stated composition of the calibration fluid or the PGC might not be integrating the peaks properly. Calibration may not detect such underlying errors. If calibration can't be trusted, how does anyone know whether a PGC is generating accurate data – today, tomorrow, or any day? Without a **validation program** they don't know and never will. No one knows.

To avoid this intolerable situation, you need a way to prove the measurement is correct. It would be a mistake to make any changes to a PGC without knowing for certain that there's something wrong with it. To gain that confidence, you need a validation program.

Why validation?

A robust validation program will rapidly detect analyzer system failures, provide proof of analyzer accuracy, improve analyzer reliability, and ultimately build trust in the analyzer. A multitude of things can go wrong with a PGC system; some occur suddenly, some gradually. Without a way to detect failures, Operations could receive bad analyzer data over an extended period with potentially costly or even dire consequences. That's why regular PGC validation is a *nonnegotiable necessity*. The validation data will convince you, Operations, and other interested parties that the PGC is operating correctly or not. And your validation program can go beyond simply reporting malfunctions, it can employ statistical quality control tools to enhance PGC accuracy, precision, and reliability.

An effective PGC validation program is a nonnegotiable necessity.

Note that a validation program does not fix analyzer failures, it merely illuminates them. Unless a validation program is accompanied by a quick action to investigate and repair detected faults, it will do nothing to improve PGC performance and credibility. Acting on the information that validation provides and implementing repairs that eliminate the **root cause** of failure will prevent recurrence, increase availability, and enhance trust.

Choose your champion

Implementing a robust validation program can be a significant undertaking in any process plant, depending on the chosen validation strategy. The critical first step is for the responsible manager to designate a responsible *Champion* charged with managing the project implementation and accountable for its success. Your champion should also be a skilled politician, able to satisfy the diverse needs and wishes of the many interested parties in Operations, Maintenance, Laboratory, and Process Automation.

Benefits

A key benefit of a validation program is having constantly updated performance data when the PGC is working properly. The user always knows how well the PGC is currently performing each measurement. The performance data might include:

- Precision of measurement.
- Accuracy of measurement.
- Repeatability of peak retention time.
- Detector noise levels.
- Overall availability or reliability.
- Mean time between failures and mean time to repair.

Having logged this base-level information, the PGC can advise when its performance is faltering, allowing the user to plan needed repairs before an expected failure occurs. Signs of future failure could include:

- Calibration drift predicting time to next calibration.
- Retention time drift predicting column failure.
- Peak gradually moving out of its integration window.
- Gradual decline of repeatability.
- Increasing detector noise level.
- Incipient failure needing attention.
- Imminent failure needing urgent attention.
- Catastrophic failure disrupting other work priorities.

Validation strategies

The purpose of a validation program is to increase our confidence in the analyzer measurements and to provide hard evidence to support that trust.

> *To have confidence in our analyzers we must prove to ourselves and anyone else who questions the data that the measurements are valid and worthy of our trust.*

A validation program can take many forms, some simple and inexpensive and others comprehensive and more costly. The appropriate form will depend on the value of the measurement to process operations.

A detailed discussion of validation techniques is beyond the scope of this book. For a deeper understanding, consult the sources of information cited in *Further Reading* at the end of this chapter. What follows is a brief overview of the key validation practices commonly employed in a robust PGC validation program.

Watching process data

An essential maintenance practice is to routinely examine the trend of recent PGC measurements, as illustrated in Figure 12.1. Regular inspection of the process data history will familiarize an analyzer specialist on what is "normal" for their process. In most processes, the analyte concentrations change gradually. The sudden onset of noise, spikes, or steps in the data may indicate the analyzer is in trouble. Knowing this, perhaps you can preempt that call from Operations by fixing the problem before it has a detrimental effect on their process.

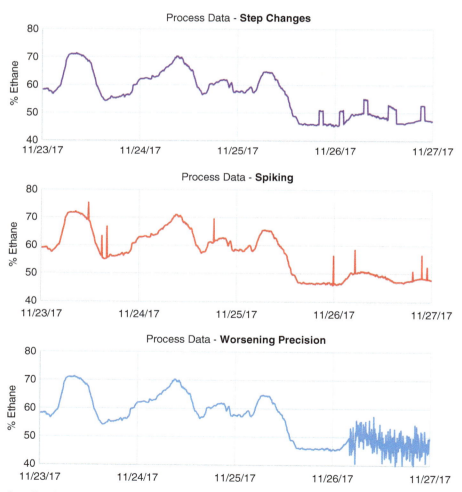

Figure 12.1 Diagnosing Process Data Trends.
Source: Courtesy of Peak Performance Analytical Consulting.

An effective and inexpensive procedure to add to your validation program is a routine visual check of trends in the process measurement data from the PGC. Some fault symptoms may be clearly visible. For instance:

Top: Steps and plateaus have recently appeared in the process data.
Middle: Spikes are endemic in the process data.
Bottom: Severe noise has recently appeared in the process data.

Alarming the calibration factors

A PGC shouldn't need frequent calibration. If it does, the root cause should be investigated and clearly understood. Yet, even with lengthy calibration intervals, much value is gained by logging the peak calibration factors and setting alarms to detect significant deviation. Any drift in sensitivity or peak retention time, or deterioration of a calibration sample will show as a calibration change. Set an alarm from the analyzer if a calibration factor changes by more than (say) 5 %.

Statistical quality control

This classical validation procedure requires the PGC to analyze at regular intervals a **validation fluid** containing the analytes and known to have a constant composition similar to the process fluid. The results are plotted on a **control chart** to make unexpected variations clearly visible. Setting up the control chart involves some statistics but most modern PGCs can do the math for you. The math detects trends and lets you know when an analyzer needs attention *before it fails!* Maintenance is then proactive and unnecessary work is eliminated.

> *With statistical performance validation, routine calibration is no longer necessary – in fact, unnecessary calibrations are detrimental, they just increase the measurement variance.*

To track measurement precision and predict future trends, a validation fluid need not be certified. A large cylinder filled with process fluid will often suffice. It's good practice to ask the process laboratory to analyze this sample several times and record the average results. This agreed composition will help with the future diagnosis of any difference between the laboratory and process analyzer measurements.

When a PGC has to demonstrate product or environmental quality, measurement accuracy becomes the prime concern. Measurement accuracy is more difficult to validate because it requires the validation fluid composition to be traceable to an **agreed reference value**.

For some analytes, validation may not be possible because their concentration in a stored sample is unstable. Yet you might be able to validate other components separated by that PGC and thereby capture some of the benefits.

The frequency of validation is an important consideration and for each PGC it will depend on the perceived value of accurate measurements. As always, value has a cost. Frequent validation will ensure the best data quality at the highest cost. A lower frequency of validation is less costly but risks substandard measurements going undetected for a longer time.

Statistical validation uses performance data gathered over a past time period. It takes about 15 validations to establish this historical data. For non-critical measurements, the optimum validation frequency is daily. The system will then take 15 days to build an adequate historical database from which it can make predictions. Often this span will increase to 20 or 30 days and then remain constant, with each new datum replacing the oldest one. From this database, performance errors will be detected within one day. This is a recognized *best practice* in many quality assurance (QA) laboratories; they validate their analyzers by analyzing a constant QA sample at the beginning of each day.

The basic control chart validates only the precision of the measurements. To also validate their accuracy the analyzer must be calibrated to an agreed reference value (ARV), which might be the analysis of a calibration sample certified by its manufacturer or the analysis of a validation fluid attested by the plant laboratory. It's important for all interested parties to agree on the validity of the reference value.

For a process analyzer, automated validation is essential. The validation fluid is connected to the PGC as another stream and programmed to run at the appropriate time. A modern PGC will collect the data, do the stats, and output alarm notifications when it needs attention.

Some arguments for control charting are:

> It provides automatic daily assurance that the analyzer is working well.
> It determines the statistical precision of the measurement.
> It quantifies confidence in the measurements.
> It anticipates failures before they happen.
> It eliminates unnecessary calibrations and reduces the usage of an expensive **calibration sample**.

Some arguments against control charting are:

> The need to provide a quantity of validation fluid large enough to last for many months of validations.
> The need to take an analyzer offline for one or two analyses per day. However, if the analysis is that critical so is the need for validation!
> The cost and complexity to set up and maintain the Analyzer Management System that collects and processes the data may hinder its adoption and utilization.

Alarming the validation fluid analysis

This is a simplified version of the control charting technique just described. It's the simplest, cheapest, and least labor-intensive strategy to build confidence in the analyzers and alert Maintenance and Operations to occasional failure. The PGC measures a validation fluid each day (or other period) but doesn't perform a statistical analysis. Instead, a simple alarm warns when a validation result deviates by more than (say) $\pm 5\%$ from its nominal value. The alarm levels should be set to the best compromise between too many nuisance alarms and failing to detect important deviations. This technique detects failures but cannot predict them. It catches faulty injection, retention time change, detector malfunction, inadequate peak integration, calibration error, or validation fluid deterioration.

Comparing PGC and laboratory analyses

Casual comparison of PGC results with the Plant Laboratory analysis is common and should lead to an investigation to find the cause for any mismatch – it usually comes down to differences in sampling techniques.

Many process plants perceive or designate the laboratory analysis as the most trusted measurement and use it to evaluate PGC performance. But even the most careful comparison adds the variance of the laboratory analysis to the variance of the PGC results. Sadly, this strategy is also prone to generating false alarms mostly due to different tapping points, asynchronous sample grabs, dissimilar conditioning techniques, or perhaps the failure of the laboratory method itself.

An automated and more scientific approach employing statistical principles is available but rarely encountered. With the continuing integration of plant data systems, the automated comparison of PGC and laboratory results may become commonplace. Already, some plants have assigned a single manager responsible for all analytic functions, both laboratory and process. The day may come when the main function of the Process Laboratory is to validate the process analyzers.

Comparing redundant PGC analyzers

The most effective way to validate is to duplicate the measurement by another analyzer. In critical applications, it's not unusual to see three oxygen analyzers measuring the same process stream. In most plants, PGCs are too expensive for that. However, it may now be cost-effective to duplicate a measurement or an entire analysis using a second column train in the same PGC. Although susceptible to common-mode errors affecting both column trains, duplicate measurements give an immediate validation of the selected measurements in real time. Statistical comparison of the duplicate results yields a high level of confidence in the validity of the measurements.

Comparing PGC with another analyzer type

This is a similar technique to duplicating analyzers, but rarely plausible as PGCs measure analytes that other techniques can't measure.

Troubleshooting

Two kinds of problem

As noted earlier, troubleshooting always starts with the sampling system. Waters (2013) has described in detail the many techniques of sampling system design and optimization. PGC troubleshooting work should not start until the sampling system has been exonerated: it must be known without doubt that the sample reaching the analyzer is fully representative of the process composition and is compliant with relevant analyzer specifications.

Troubleshooting is like detective work; you look for evidence and make logical deductions to reach tentative conclusions. Evidence for the root cause of PGC failure is usually visible on a chromatogram, and the first task of the expert troubleshooter is to distinguish between two radically different kinds of symptom:

> **Instrumentation symptoms** are indications that something is wrong with the detection system, preventing it from producing a flat and smooth **baseline**.

> **Chromatographic symptoms** are indications that something is wrong with the separation of components and the measurement of analytes.

You need to know which of these problems you're dealing with.

Here, we show how to regain a flat and smooth baseline by diagnosing and fixing *instrumentation problems*. To avoid the risk of a wrong diagnosis, always test and confirm the baseline is truly flat and smooth before continuing with the next step. If fixing the baseline also eliminates the problem, exit the procedure with the problem solved.

The next chapter introduces a unique method for diagnosing *chromatographic problems*. It explains what the symptoms indicate and why they are there. This complete understanding of what's happening in the PGC is a necessary prelude to diagnosis and repair. It finds the root cause of the problem and indicates some ways to eliminate it.

Our recommended procedures don't include indiscriminately changing the PGC settings or randomly swapping the hardware devices. Guesswork isn't a reliable approach to PGC troubleshooting. A keen observation of symptoms and some clear thinking are more likely to yield success.

Evaluating the baseline

What is a baseline?

An ideal PGC detector would respond only to analyte molecules and have zero response to anything else. Then, the baseline would be true zero and any signal from the detector would accurately represent the number of molecules present. But real detectors respond to the presence of other molecules and to variations in their operating conditions such as temperature and pressure. These additional responses are still present and contributing to its output signal while the detector is responding to an analyte peak. Since no detector can measure its own baseline while measuring a peak, the PGC processor must predict the baseline level under each analyte peak.

> *The baseline under a peak is the estimated trajectory of the detector signal if no peak were present.*

Some chromatography software is able to predict a curved baseline under a peak, but that option is not generally available. PGC data processors assume the baseline is a straight line, either horizontal or sloping. When using the horizontal setting the processor measures the detector signal at a convenient time when no peaks are present, and then assumes it doesn't change. If expecting a sloped line, it measures the detector signal just before and just after the peak and uses the average of the two as the best estimate of the underlying baseline.

Why is it important?   The PGC baseline measurement is more critical than a typical instrument zero because the peak integration process magnifies any error. Consider the area of a simple triangular peak: base times half height. A 1 % error in estimating base level would add or subtract a rectangular area equal to 2 % of the triangular area. It's worse for a chromatogram peak because we can't measure the baseline level that close to the peak.

From Figure 2.5, we know the triangulated base width of a Gaussian peak is four times its standard deviation (σ) and includes about 95.5 % of its molecules. An integration period of twice the base width (8σ) would include 99.99 % of the analyte molecules and is wide enough to measure the baseline before and after the peak with negligible error from the peak itself. The effect of a baseline error equal to 1 % of the peak height is then given by:

> Area (A_p) of Gaussian peak of height (h): $A_p = h\sigma/0.40$.
> Area (A_r) of rectangle 8σ wide and $0.01h$: $A_r = 0.08h\sigma$.
> Percentage error in peak area estimate is: $0.08 \times 0.4 = 3.2\%$

This calculation illustrates the importance of getting a flat and smooth baseline. When using automatic peak detection, any additional **baseline noise** necessitates a higher threshold setting for the start and end of integration, potentially omitting a large rectangular area from the bottom of the peak. Forced integration provides a better area estimate because it averages out the noise (Waters 2020, 284–285), yet it still relies on an accurate estimate of the baseline level. Either way, the quality of the analytical measurement is highly reliant on the quality of the baseline.

What's the problem?   Baseline troubleshooting starts with an observation – the baseline is not flat and smooth:

> A flat baseline is a chromatogram zero that follows a straight line. Ideally, it should be exactly horizontal but a slight slope is tolerable provided it is smooth.
> A smooth baseline is a chromatogram zero essentially devoid of the rapid jitter usually characterized as noise or spiking. However, noise is always present; it must be low enough to have an insignificant effect on peak detection and measurement.

Any deviation from flat and smooth is an anomaly that could interfere with the accurate measurement of analyte concentration. Our troubleshooter must diagnose the root cause of the anomaly and then eliminate it.

Chromatogram troubleshooting is challenging because three things happen at once. In addition to measuring the analyte peaks eluting from the column, the detector is responding to its baseline conditions and to carrier gas variations caused by column switching. It makes sense to separate these effects by ensuring the baseline is flat and smooth before attempting to diagnose a chromatography problem.

Checking the baseline

Baseline problems may be visible on a full chromatogram but it's best to evaluate them without the added complexity of sample injection and column switching.

Ensure your column and detector ovens are at the correct temperature and the carrier gas flow rates are properly set. Power-up the detector. Then, run your PGC in standby mode with the Method paused and no programmed actions occurring. Set the baseline at about 10 % of scale on the display device, or higher if it's dropping rapidly.

Observe the baseline. If the PGC has been out of service, it may take several hours or days for the chromatogram signal to stabilize. This is due to the long elution times of substances absorbed by the columns during downtime. If the contamination is severe, protect the detector by shutting off its power and, if necessary, by disconnecting it from the column effluent. Continue the carrier gas flow for a day or two before checking again. Be patient; there's not much you can do but wait. It might be possible to accelerate the process by increasing the oven temperature, but only if you're certain that a higher temperature won't damage the columns.

If facilities are available, consider a plan to remove the columns and recondition them in a laboratory oven following the manufacturer's recommendations.

If the baseline is improving, wait for it to become flat and smooth. A slight downslope is normal after column contamination; it's okay to continue troubleshooting while it levels out. If the baseline is not improving, a fault exists and the troubleshooting task is to find the cause and fix it.

Prime suspect – the detector

The detector is the source of the chromatogram signal, so when the baseline is not flat and smooth, it's natural to focus on the detector. PGCs use two kinds of detector, those that respond to the *concentration* of analyte molecules in the carrier gas and those that respond to the *number* of analyte molecules in the carrier gas. You need to know their different behaviors.

Concentration-sensing detectors

These detectors output a signal proportional to the instantaneous concentration (g/mL) of analyte molecules currently present in the carrier gas. The common example is the thermal conductivity detector (TCD).

A TCD will respond to any foreign molecules in the carrier gas and its baseline signal is sensitive to variation in carrier flow rate and to pressure waves. It also responds to variation in detector block temperature or detector vent pressure, and the reference element doesn't fully compensate for these effects. Having a low source impedance gives the TCD some immunity from induced signal noise and spiking.

Mass flow-sensing detectors

These detectors output a signal proportional to the instantaneous rate of arrival (g/s) of analyte molecules in the detector. As such, they are immune to the addition of pure makeup gas, making them ideally suitable for use with low-flow capillary columns. The most common example is the flame ionization detector (FID).

The FID baseline signal is highly sensitive to hydrocarbon molecules in the carrier gas but is not affected by other molecules nor by variation in temperature, vent pressure, or carrier gas flow rate. However, it will output spurious signals from any hydrocarbon impurities in the carrier gas, fuel gas, or combustion air. If the flame jet gets dirty it can release particles into the flame causing random spikes on the baseline. A high source impedance renders the FID susceptible to radio frequency interference and spiking.

Another thought

The detector may not be causing the problem.

It's easy to assume you know the cause and then spend a long time trying to justify it. You might get lucky, but it's not an efficient procedure. Troubleshooting starts with a clear symptom of malfunction; it should then proceed logically from a knowledge of the possibilities. If you don't know what's possible, you may miss the root cause.

A logical approach

To avoid the chance of missing something, start by dividing the potential causes into binary categories that cannot exclude any cause, even those presently unknown. The detector is certainly a good place to start. There are two mutually exclusive possibilities: *something is wrong* with the detector or *nothing is wrong* with the detector. No other option is tenable. We could easily design experiments to test for those two options but for the moment it's sufficient just to think about them. In each case, there are two mutually exclusive consequences:

If something is wrong with the detector:

➤ The observed baseline includes features caused by the detector responding to variations in its external working environment, or:

➤ The observed baseline includes features caused by an internal detector fault.

If nothing is wrong with the detector:

➤ The observed baseline is the correct detector response to molecules present in the carrier gas, or:

➤ The observed baseline has features introduced during signal processing that are not present in the original detector output signal.

That logic provides four comprehensive and mutually exclusive reasons for the baseline anomaly. The root cause must be hiding in one of these. While further binary divisions may be possible, these four categories may be enough to highlight the possible causes of a baseline anomaly. Let's try to identify them.

The detector is working but responding to external influences

In this scenario, the detector is working well but is responding to changes in its external working environment. These environmental influences normally should not be present. Concentration detectors are much more sensitive to their working environment than mass flow detectors.

As a practical test, disconnect the detector vent outlet at the PGC from the vent piping, allowing it to vent freely to atmosphere.

If the anomaly disappears:

➤ Pressure variations in the vent piping are causing the anomaly.

➤ Find the cause and fix it.

If not, consider these other possible causes:

➤ The detector temperature is changing – which should initiate an alarm.

➤ Air flow or drafts are affecting the detector, possibly from a fan or oven purge.

➤ The flow rate of carrier or fuel gas is changing, possibly due to column ageing or the mechanical failure of a valve or regulator.

➤ The detector is subject to excessive vibration.

The detector is not working correctly

In this case, the baseline anomaly is coming from the detector itself and is due to a local support inadequacy or a detector hardware failure.

If due to a local support inadequacy:

- Is the flow rate of carrier, makeup, or fuel gas excessive or inadequate?
- Is the detector excitation voltage abnormal or unstable?

If due to an internal fault:

- Are there deposits of solid or liquid material inside the detector flow paths?
- For a TCD, is a sensor element aged or damaged?
- For a flame detector, is the flame alight? Is the flame too small, too large, or unstable? Is the collector electrode correctly aligned with the jet? Is liquid water collecting in the detector (more likely with an FPD)?
- For an ionization detector, is the applied voltage within acceptable limits?
- Is ambient light leaking into a photon sensor or has the sensor failed?
- Is there a partial short circuit or preamplifier fault?

Clean, repair, or replace the detector.

The detector is responding to molecules

According to this notion, there's nothing wrong with the detector. The baseline anomaly is due to the detector correctly responding to molecules in the carrier gas. If so, where did the molecules come from?

The suspect molecules came from outside the PGC:

- The molecules are part of the injected sample that escaped backflush or heartcut.
- The injector valve is leaking sample molecules into the carrier gas stream.
- The molecules come from a contaminated gas cylinder, pressure regulator, or gas supply line. Perhaps the gas purifiers need reconditioning or replacing.

The suspect molecules came from inside the PGC:

- The molecules are column bleed from an unstable liquid phase, perhaps due to temperature change.
- The molecules are a product of the high-temperature oxidation of the liquid phase by air diffusing into the carrier gas.
- The molecules are coming from solid or liquid deposits anywhere in the carrier gas path.
- The contamination is due to desorption of molecules previously adsorbed on tubing or stationary phase.

If the anomaly is a correct detector response to foreign molecules in the carrier or fuel gases, the detector is not at fault; it's indicating a fault elsewhere in the gas flow path.

The detector is not generating the anomaly

This option postulates that the baseline anomaly is not coming from the detector at all. The detector output signal is flat and smooth, but it's then modified by electronic interference. For a practical test, connect a display device directly to the analog output of the PGC (if available). For troubleshooting it's best to see the real detector output signal, not one reconstructed from digital data. If the anomaly disappears, the problem is most likely electronic and caused by the data processing system, not by the PGC itself.

If the baseline symptom is still present, suspect an electronic problem within the PGC rather than a signal processing issue:

- Insufficient shielding of signal wires or preamplifier, allowing radio-frequency interference.
- Poor grounding.
- Dirty, corroded, or loose electrical connections.
- Inadequate insulation, partial short circuits.
- Dirty electric power.
- Electronic component instability or failure.
- Display device instability or failure.

Working from the symptoms

Armed with a clear understanding of the possible malfunctions, you are ready to diagnose the root cause of the anomaly. If the baseline is improving, wait. If the baseline is not improving, distinguish between these four instrument symptoms:

- **Baseline Drift:** continuous upward or downward change in the detector signal, often exponentially approaching a flat baseline, as in Figure 12.2a.
- **Baseline Wander:** a slow, random, and cumulative variation in the detector signal, as in Figure 12.2b.
- **Baseline Cycling:** a smooth waveform impressed upon an otherwise constant detector signal, as in Figure 12.2c.
- **Excessive Baseline Noise:** erratic short-term variations in an otherwise constant detector signal, as in Figure 12.2d. Noise may also include regular or random spikes.

Prime suspect – the detector 311

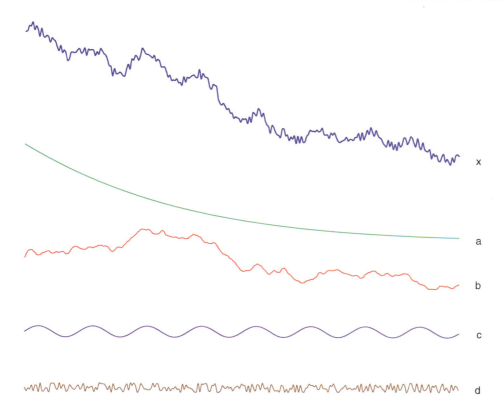

Four types of detector instability may contribute to baseline noise. The first diagnostic task is to recognize which of these are present in the suspect baseline:
x: Observed (blue curve) is the actual baseline signal output from the PGC.
a: Drift (green curve) is a gradual change in the baseline level, usually downward.
b: Wander (red curve) comprises random small changes in the baseline level.
c: Cycling (purple curve) is a regular and smooth oscillation of the baseline level.
d: Noise (brown curve) is short-term random variation in an otherwise flat baseline.

The observed baseline (x) is the cumulation of the four detector instabilities. It's easy to see the contribution of the gradual downward drift (a). The fast-acting noise component (d) is also clearly visible. Without seeing the individual curves, though, it would be more difficult to distinguish the presence of both wander (b) and cycling (c).

Figure 12.2 Four Patterns of Baseline Disturbance.

Each kind of disturbance forms a distinct pattern on the baseline that is easy to recognize when none of the others are present. When several are active at the same time it's more difficult, as the patterns can combine to yield a noisy, drifting baseline. Figure 12.2x shows a baseline affected by all four of the basic patterns. The appearance of a real baseline critically depends on the amplitude of each individual contribution.

When present, each of the four contributing patterns indicates a different fault condition. Since it might not be easy to identify all the patterns present when combined in the baseline, the best approach is to identify the dominant pattern, find its source, and try to correct it. Reducing or removing that source is certain to improve the baseline signal. Then, other patterns may emerge that require a different resolution.

Diagnose the instrumentation symptoms and repair the underlying malfunctions *before* attempting to troubleshoot a chromatographic symptom.

Diagnosis

Diagnosing baseline drift

Drift is a gradual change in the detector baseline signal, as in Figure 12.3. The main causes of drift are:

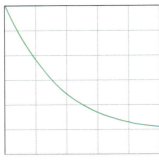

Figure 12.3 Baseline Drift.

> The slow elution of solvent traces remaining in a new column recently installed. The detector response will be very high at first and then decay exponentially to a flat baseline. There may be several long bumps in the decay curve due to minor components present in the original solvent.

> Multiple real peaks eluting from a column that has not been used for a while. When a column is in contact with the atmosphere it absorbs a surprising array of chemical compounds. It can take many days to purge all these foreign substances from the column. The baseline may slowly go up and down as discrete components elute from the column.

> The slow conditioning of a new detector or a gradual change in the detector environment, including temperature, pressure, and carrier or fuel gas flow rates.

> Contamination of the sample injector, detector, carrier gas, or detector fuel gases. A newly installed carrier gas cylinder may contain impurities that affect the detector. It's a good practice for the process laboratory to analyze the gas from each cylinder before connecting it to the PGC.

In programmed temperature operation, an upward drift is common due to the increased column bleed at higher temperatures. Column bleed is the contamination of the carrier gas by molecules from the liquid phase in the column. For packed columns, the contamination is most likely due to an increase in liquid phase vapor pressure, resulting in a gradual upward drift as the column temperature rises. For capillary columns, it's more likely due to a polymeric liquid phase breaking down to form more volatile fragments. This process starts to occur at a higher column temperature, resulting in a distinct rise in the baseline later in the analysis cycle. Some laboratory GCs attempt to eliminate this predicable drift by storing a blank baseline (one with no sample injection) and subtracting it from each live analysis chromatogram.

Drift is rarely due to pressure changes in the detector as those are usually more erratic. Yet a slow change in barometric pressure, if uncompensated, could result in an upward or downward drift in the baseline.

A little drift is common, and the PGC data processing techniques compensate for it by running autozero routines and acquiring baseline measurements before and after each measured peak. Table 12.1 gives guidance for troubleshooting a more severely drifting baseline.

Table 12.1 Diagnosing a Drifting Baseline.

Observation		Indication
Baseline level has changed or is gradually changing, possibly moving in discrete steps. *Newly installed gas cylinder?*	Yes	A higher baseline may indicate contamination of the carrier gas or detector fuel gas with something the detector responds to. The baseline may go up in steps as different carrier gas contaminants get through the columns: • A newly installed gas cylinder is a prime suspect. • A cylinder containing the wrong gas was installed as carrier gas. A lower baseline means the new gas is purer than the old gas.
	No	• Replace or repair any conditioning device in the carrier gas or fuel gas supply lines.
Baseline level is very high and is decreasing exponentially toward its normal level. *Newly installed column?* *Newly installed detector?*	Yes	A new column contains traces of the solvent used in its manufacture. The column manufacturer may recommend conditioning at a higher temperature: • Wait for baseline to become adequately flat. If you're certain the column can withstand a higher temperature, increase the oven temperature to accelerate the process. • Be patient: it might take several days. To shorten the time, condition the new column in a laboratory oven before installation. A new detector may also take some time to stabilize.
	No	The column feeding the detector contains one or more large, highly retained components that have escaped backflush or heartcut. Check the column valve is working and is correctly timed: • Wait. It may take hours or days for the column to recover. The contamination can also come from the environment. See the next step below.
Baseline is moving up and down in random smooth humps, each very much wider than a normal peak. *Was the PGC recently shut down without carrier gas flow?* *Was a new carrier gas regulator recently installed?*	Yes	Multiple large peaks are eluting from the column and are not separating from each other: • When the PGC is not in use, always maintain a low carrier gas flow. When the columns have no gas flow, they absorb many different chemicals from the environment. A new carrier gas regulator may contain traces of oil or solvent that is contaminating the carrier gas. The level of contamination can depend on ambient temperature: • Use only non-venting stainless-steel regulators for carrier and detector gases, not those with polymeric diaphragms. Wait for the columns to clear. Be patient: it might take several days.
	No	The humps may be unresolved heavies peaks from a prior sample injection. Check the backflush operation and timing. If the backflush is working but the problem persists, it might be

(continued)

Table 12.1 (continued)

Observation		Indication
		due to the injector valve continuously leaking sample fluid into the carrier gas:
		• Purge the sample injector with carrier gas to remove all traces of sample fluid. Then run the Method for several cycles, injecting carrier gas samples. If the humps gradually disappear, repair or replace the leaking valve.
		• If the baseline doesn't recover, it might be due to a severely contaminated column or detector:
		• Replace the columns and try again.
		• To test the detector, bypass the columns with a capillary tube, allowing pure carrier gas to flow into the detector.
Baseline rises toward the end of analysis. *Programmed temperature?*	Yes	Increased column bleed at higher temperature:
		• This is probably normal for the analysis. If intolerable, it may be necessary to reduce the final column temperature and accept a longer analysis.
	No	It's unlikely, but the baseline rise might be due to many partially separated heavy peaks:
		• If so, check the backflush operation and timing.
Baseline gradually and continually rises or falls during each analysis. *Programmed temperature?*	Yes	In programmed temperature operation, the carrier flow rate into the detector will decrease slightly as the column temperature rises:
		• To overcome this problem, some programmed temperature GCs use a carrier gas flow controller instead of a pressure regulator.
		An upward slope in the baseline may be due to column bleed from a packed column. The volatility of some non-silicone liquid phases gradually increases as column temperature increases and the detector responds to the increasing column bleed:
		• If intolerable, use a less volatile liquid phase.
		Valve or column leak. Inward diffusion of air causes oxidation of liquid phase and subsequent column bleed, increasing as temperature rises:
		• Check for leaks with an electronic leak detector or by pressurizing the entire column system and confirming no loss of pressure in 15 minutes.
		Don't use a liquid bubble solution for leak-checking capillary columns!
	No	In isothermal operation, drift may be from a cumulative contamination of detector or injector:
		• Clean or replace the detector or sample injection valve.
		Otherwise, it's probably an electronic problem.

Diagnosing wander

Wander is a distinct form of baseline instability in which the actual baseline is moving, whereas the other disturbances are deviations imposed on an otherwise flat baseline. A wandering baseline will randomly move up or down at random intervals resulting in no repetitive pattern, as in Figure 12.4. The baseline changes depicted in this figure truly are random, yet this curve seems to have adopted a cyclic overtone which doesn't exist in the actual data. It shows how easy it is to misread the chromatogram. Be careful in your interpretation.

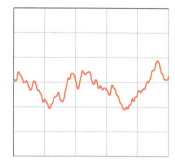

Figure 12.4 Baseline Wander.

True wander is random in both time and extent, so the upward movements of the baseline will tend to be balanced by equivalent downward movements. If a wandering baseline moves continually up or continually down, baseline drift is also present and should be diagnosed separately.

Common causes of wander include contamination of the sample injector and its outlet tube, inconstant pressure at the detector vent, or unstable detector electronics. Otherwise, the detector is probably at fault. The jet of a FID may be dirty and spontaneously releasing carbon into the flame. A TCD may also respond to the presence of solids or become damaged in abnormal operation. Table 12.2 may be helpful when troubleshooting a wandering baseline.

Table 12.2 Diagnosing a Wandering Baseline.

Observation		Indication
Baseline is wandering up and down. *Does wander disappear when you temporarily disconnect the detector vent line from the PGC?*	Yes	Minor pressure changes in the vent system are changing the pressure in the detector and shifting the baseline. This symptom is more likely to occur with concentration sensors like the thermal conductivity detector. *Is detector vent connected to a common vent header?* Flow changes of other gases in the common header are causing pressure variations: • Separate the detector vent from other vent flows. *Is detector vent connected to a flare header?* Flare pressure can change dramatically during flaring operations, having a much larger effect on the baseline: • Never connect a detector vent to the flare; it must go to atmosphere. *Is detector vent line connected to an outside vent?* Baseline wander is evident on some days and not on others. Wind pressure on the outside vent is causing pressure variations in the detector and shifting the baseline: • Shield the vent from the wind or install a vent diffuser.
	No	If using a flame detector, suspect a contaminated vent tube. Traces of chlorine compounds form hydrochloric acid that corrodes and gradually blocks the vent tube. • Inspect the detector and its vent tubing and clean as necessary. Wander might be caused by solid or oily deposits contaminating the sample injector or first column.

(continued)

Table 12.2 (continued)

Observation		Indication
		• Clean or replace the injector valve and tube connecting to the first column.
		• Backflush the first column for an extended time period, preferably at a higher temperature.
		• Replace the first column.
		Wander might be due to inconstant quality of carrier or utility gases.
		• Replace the gas cylinder.
		• If using an absorbent trap to remove impurities, recondition or replace the trap.
		Don't connect PGC gases with polymer tubes as they are permeable to atmospheric gases. Yes, permeation leaks can contaminate a higher-pressure gas stream!
		• Use cleaned stainless steel tubing (new tubing has an oily film inside).
		• Clean the tubing before installation by flushing with pure hexane followed by methanol, then dry with a pure nitrogen purge.
		• Clean the tubing again during annual maintenance.
		• Replace any gas pressure regulator that has a polymeric body or diaphragm with an all-stainless-steel regulator.
Wander is jerky and erratic.	Yes	A gas leak at one of the detector connections may cause baseline instability:
		• Check the gas connections for leaks.
		Liquid in the vent system may cause irregular backpressure on the detector:
		• Remove any liquid trap points in the detector vent line or vent header. Or heat the lines to avoid condensation.
	No	Continue.
Wander is more pronounced when column valve is on or off.	Yes	Carrier gas leak in valve or column.
		• Check for leaks with an electronic leak detector or by pressurizing the entire column system and confirming no loss of pressure in 15 minutes.
		Don't use a liquid bubble solution for leak checking!
	No	Continue.
Wander becomes more rapid when detector flow rates are temporarily increased.	Yes	Loose solid material is present in the detector:
		• Clean or replace the detector.
	No	The fault is likely to be electronic, due to a damaged detector, loose wire connection, dry solder joint, or component malfunction:
		• Check all electronic connections.
		• Repair or replace the detector.

Diagnosing cycles

A cyclic baseline looks like a low-frequency sine, square, or sawtooth waveform imposed upon a flat baseline, as in Figure 12.5. If the baseline is also rising or falling, treat it as a combination of drift and cycling errors. Cycling is always due to something oscillating at the same frequency as the observed baseline. The troubleshooting task is to find the synchronous source:

> If using a TCD, the most likely causes are cyclic variations in the detector temperature or its vent pressure.

> If using a FID, the most likely cause is a regular variation in column bleed due to inadequate control of column temperature. For this kind of bleed to occur, the column feeding the detector would most likely be a packed column containing a volatile liquid phase.

> Ionization detectors are high-impedance devices that can be susceptible to radio-frequency interference due to inadequate shielding.

> Photometric detectors are extremely sensitive to light and may respond to diurnal ambient light variations or changes in local illumination if inadequately shielded.

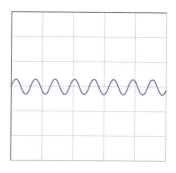

Figure 12.5 Baseline Cycles.

Otherwise, the root cause could be something outside the analyzer such as a regular pressure variation on the detector vent. A square wave could be due to electrical interference impressed upon the electricity supply by the intermittent operation of an air conditioner or a nearby process motor.

Table 12.3 provides guidance for troubleshooting a cycling baseline.

Table 12.3 Diagnosing a Cycling Baseline.

Observation		Indication
Baseline exhibits a sinusoidal waveform.	Yes	If using a thermal conductivity detector, its temperature is cycling: • Check the detector temperature control. • If in air-purged oven, ensure the detector is not in the direct air flow from the oven heater. If using an ionization detector, it's responding to variations in column bleed due to column temperature cycling: • Check the column oven temperature control. • If in air-blown oven, ensure the columns are not in the direct air flow from the oven heater. A faulty electronic pressure controller (EPC) is causing a ripple in the pressure of a detector gas. • Tune or replace the EPC.
	No	Continue.
Baseline exhibits a square waveform.	Yes	Most likely an electric interference. The up and down movements may be synchronous with an electrical device switching on and off, such as a local air

(continued)

Table 12.3 (continued)

Observation		Indication
		conditioner, motor, or heater:
		• Check PGC enclosure is properly grounded.
		• Check detector wiring shield is properly grounded.
		• Find the synchronous device by shutting down potential sources of interference.
		• Connect the PGC to an independent and clean power source.
Baseline exhibits a sawtooth waveform.	No	Continue.
	Yes	Pressure may gradually build up in the detector vent system, then suddenly release. Test for a vent problem by temporarily disconnecting the detector vent line from the vent header. If the waveform disappears you have a vent problem:
		• Check for liquid in the local vent header – more likely with a flame detector. Remove any low spots where liquid can collect.
		• Check for ice in the external vent pipework. Heat the vent line, as necessary.
		• Look for a restriction in the detector vent system that could cause pressure buildup. Remove a found device such as a backpressure regulator, pressure relief valve, check valve, bug protector, incinerator, or water trap – the detector must vent freely to atmosphere.
		• Never connect a detector vent to a flare header or sample recovery system.
Baseline exhibits regular spiking. *Are the spikes evenly spaced like the positive spikes in Figure 12.6?*	No	Continue.
	Yes	Liquid in vent header:
		• Check for liquid traps in the vent pipework.
		• Eliminate low pockets.
		With an FID, condensate may be dripping from a vent line:
		• Install a short length of wider-diameter tubing to prevent the formation of drips that block the flow of vent gases.
		Electronic interference from on–off switching of heater, air conditioner, fan, or pump:
		• Find synchronous device and suppress contact arcing or run it on a different power circuit.
		• Consider installing a mains power conditioner for the PGCs.
	No	Any other waveform is probably due to an electronic fault.

Diagnosing noise

While it's possible to define all modes of baseline irregularity as noise, with each having its own characteristic frequency (Annino and Villalobos 1992, 252–258), chromatogram noise usually refers to a continuous pattern of fast, random, and erratic deviations from a smooth baseline. The noise depicted in Figure 12.6 also includes synchronous and random spiking.

All measurements contain noise, defined as random variation in the measurement signal caused by myriads of uncontrollable variations in the **measurement environment**. Instrument designers strive to minimize noise by specifying high-quality components and by finely controlling the physical and electronic variables that affect the detector signal. Yet some noise always remains.

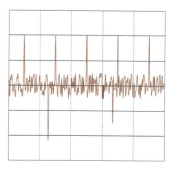

Figure 12.6 Baseline Noise.

Noise can be minimized but not eliminated. Focus on the noise that is adversely affecting the detection and measurement of analyte peaks.

We can measure noise simply as the peak-to-peak limits of its random variation or more precisely by assessing a statistical property such as standard deviation (σ). The peak-to-peak estimate is about 6σ.

Noise limits the ability of an instrument to detect the measurand. For reliable detection, the signal due to the measurand must be at least twice the noise level and some scientists double that. Standard GC methods for laboratory analysis often specify their **limit of detection** (LOD) on this basis. PGCs rarely specify their LOD because they usually don't attempt to measure trace quantities of analytes with concentrations close to the LOD. PGC users generally consider **measurement range** to be more important – together with an implicit expectation of accuracy.

All process measuring devices suffer some instability in their baseline signal. This instability is still present when the instrument is measuring an upscale value, so it adds to the sensor output signal and reduces the precision of measurement. The loss of precision is particularly noticeable when measuring low values of the measurand, due to the reduced signal-to-noise ratio at that time.

Like all direct sensors, early PGCs measured peak height above a baseline level thereby incorporating the baseline noise into their measurement signal. In addition, the detector response was itself imperfect, adding more random variation not present in the baseline noise.

PGCs now measure peak area and this has a different relation to noise. When the PGC integrates the area of a chromatogram peak, it accumulates dozens of data points. Each of these data points is a height above the baseline and includes a sample of the baseline noise level at that instant. Peak integration sums all the data points which tends to cancel the positive and negative deviations due to signal noise – for random deviations the noise cancels out. Yet, as mentioned previously, the baseline noise at the start and end of peak integration still affect the calculated zero line used to quantify the integrated area of the peak.

Signal noise reduces the precision of the chromatographic measurement, but a more troublesome effect is the unwanted and untimely triggering of commands to start or stop peak integration.

A more troubling effect of PGC baseline noise is its ability to trigger peak integration commands.

A modern PGC can reduce the signal noise level by applying a data smoothing technique. The PGC digitizes the analog detector signal by sampling its voltage at a preset rate, typically up to 40 times per second. It can then dampen the noise level by averaging data points. There are two ways to average the data. The PGC may average an odd number of successive data points, thereby reducing the noise and the data storage requirement, or calculate the moving average and retain the same number of data points (Waters 2020, 281–283). The technique employed depends on the model of PGC in use.

If carefully applied, data smoothing can reduce the noise level, but it also reduces the speed of response and might cause inaccurate integration of fast peaks. Most PGC applications don't need it and chromatographers often disable the function:

➤ Don't use data smoothing to hide a noisy signal; find the cause of the noise and fix it.

➤ Don't even use data smoothing on a smooth signal because it will hide a future noise problem that you need to be aware of.

Most PGC signal noise is caused by gas leaks in valves, columns, or detector; by contamination of the detector; or by electronic interference. Table 12.4 may help with troubleshooting a noisy baseline.

Table 12.4 Diagnosing a Noisy Baseline.

Observation		Indication
Disconnect the display device and short its signal input connections. *Does the baseline noise disappear?*	Yes	The display device is not causing the noise. • Reconnect the display and continue.
	No	The noise is from the display device.
Disconnect the detector vent tubing from the analyzer and allow the detector to vent directly to the atmosphere. *Does the baseline noise diminish when the detector vents directly to atmosphere?*	Yes	Pressure pulses on the vent: • If noise is variable day-by-day, check for wind buffeting on an atmospheric vent. Install vent shield or muffler. • Eliminate pressure variations due to the operation of a vent scrubber or incinerator. • Look for liquid accumulation in the vent pipe that forces the vent gas to bubble through. Install an effective liquid drain. • Check the entire vent system and rectify any vent restriction that might cause a variable backpressure on the detector.

(continued)

Table 12.4 (continued)

Observation		Indication
		• Replace the vent connection and confirm an improved baseline.
		Never connect a detector vent to a flare header – it must vent to atmosphere.
	No	Continue.
Reduce the carrier gas flow rate into the detector. *Does this cause a change in the level or frequency of the baseline noise?*	Yes	The noise is flow-sensitive:
		• Noise sources may include too much flow in detector, solid matter or involatile liquid in detector, or a detector leak.
		• A TCD sensor element may be damaged.
		A high air flow into an FID may cause noise by flame flutter.
		• Reduce the air flow.
	No	The noise is mechanical:
		• Check for excessive vibration at detector.
		The noise is electronic:
		• Sources may include a loose connection or dry joint, ungrounded cable shield, insulation breakdown of the high voltage source (for ionization detector), RF interference from process motors, etc.
		• Consider a power conditioner to smooth the PGC power supply.
Noise varies in duration or intensity (not uniform and continuous as in Figure 12.6d).	Yes	A gas leak in the connection between the column and detector or on a detector fuel gas line:
		• Check all detector gas connections for leaks.
		• Fix the leaks.
	No	Noise is probably electronic.
For a flame detector, shut off the fuel supply to extinguish the flame. *Has the baseline noise lessened?*	Yes	Solid material in detector is causing spurious signals:
		• Clean or replace detector and its inlet tubing.
		• Check internal surfaces and vent tube for corrosion.
		Liquid water is collecting in the detector body:
		• Increase the detector body temperature.
		Incorrect gas flows:
		• Check the flow rates of fuel and air in detector.
		• Reset as necessary.
		A capillary column inserted too far into the detector jet, causing overheating of the column end:
		• Remove the column from the detector, cut off and discard a few centimeters, then carefully reinstall in the correct position.

(continued)

Table 12.4 (continued)

Observation		Indication
For a thermal conductivity detector, switch off the detector and check for leaks with an electronic leak detector or by pressurizing the entire column system, isolating the gas supply, and confirming no loss of pressure in 15 minutes. *Leaks observed?*	No	Continue.
	Yes	A gas leak anywhere in the system; valves, columns, or detectors: • Find and fix the leaks. Don't use a liquid bubble solution for leak-checking capillary columns!
The noise includes rapid spikes of similar amplitude and duration. *Are the spikes equally spaced in time?*	No	Continue.
	Yes	Check the duty cycle of any local electrical device internal or external to the PGC that might be synchronous with the spikes such as a thermostat, relay, light switch, level controller, conveyer, elevator, process motor, or sample recovery system: • If possible, add arc suppression devices to identified sources. • Connect the PGC to an independent and clean power supply. • Check the electrical shielding of signal wires is effective.
	No	Random spikes may be due to detector vent gas bubbling through condensate in the detector vent piping: • Eliminate any low points in vent lines where liquids can collect. Vent pressure pulses by wind buffeting on detector vent outlet: • Install vent shield or muffler. Solids contamination in detector, possibly emitted from a solid-phase column such as a PLOT column. • Clean or replace detector and its inlet tubing.

Overall effect

Figure 12.7 illustrates the cumulative effect of baseline instability on the detector signal and the observed chromatogram. The four detector disturbances combine with a true chromatogram signal to yield a noisy, drifting chromatogram signal that is difficult to interpret. In an extreme case, the noise can hide a true peak or create a false one. The baseline noise will also cause errors in peak integration and a consequent loss of measurement accuracy.

The four kinds of baseline disturbance must be minimized before attempting to diagnose a chromatographic malfunction by the techniques outlined in the next chapter. Indeed, eliminating the baseline anomalies may suffice to solve the problem and obviate the need for further diagnosis.

Diagnosis 323

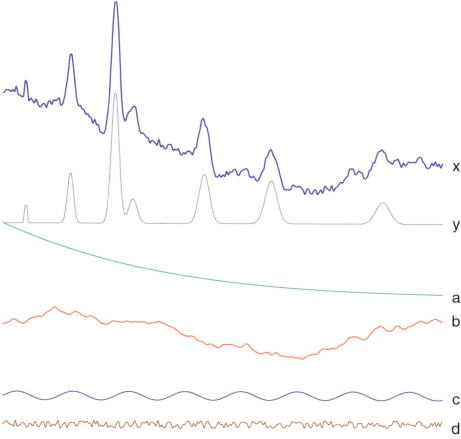

Four types of detector baseline instability can distort the true chromatogram signal to produce the observed chromatogram from a PGC:

x: Observed (blue curve) is the actual chromatogram output from the PGC.
y: Real (gray curve) is the true chromatogram produced by the columns.
a: Drift (green curve) is a gradual change in the baseline level, usually downward.
b: Wander (red curve) is random medium-term changes in the baseline level.
c: Cycling (purple curve) is a regular and smooth oscillation of the baseline level.
d: Noise (brown curve) is short-term random variation in an otherwise flat baseline.

Note how the baseline instability changes the shape of the peaks and makes their true area difficult to measure.

Figure 12.7 Chromatogram with Unstable Baseline.

Knowledge Gained

About validation

❖ *Don't rush into troubleshooting until you're certain that the problem resides in the PGC.*

❖ *Issues with the sampling system are the root cause of most PGC problems.*

❖ *A validation program may use several procedures to collect data and discover faults.*

❖ *Watching process data and logging calibration factors are low-cost validation procedures.*

- *Control charting establishes known performance data and can predict incipient failure.*
- *Analyzing a constant sample without statistical analysis is a simpler way to detect failures.*
- *Comparing PGC results with laboratory results often reveals sampling errors.*
- *Comparing redundant process analyses is a very effective way to gain confidence in the data.*

About troubleshooting

- *Use binary logic to find potential root causes of a problem without omitting any possibility.*
- *All chromatogram symptoms indicate either instrumentation or chromatography problems.*
- *Instrumentation faults prevent the PGC from outputting a flat and smooth baseline.*
- *Chromatographic faults prevent a PGC from accurately measuring the analytes.*
- *Random settings or hardware swaps are not a reliable approach to troubleshooting.*
- *An ideal detector would respond only to analyte molecules, so its baseline would be true zero.*
- *Real detectors respond to other molecules and to changes in their operating conditions.*
- *The baseline is the chromatogram signal that would exist if the analyte peak was not there.*
- *Any error in the estimated baseline level causes a larger error in the analyte measurement.*
- *Always test and confirm the baseline is flat and smooth before diagnosing other symptoms.*
- *If the baseline is improving, wait. It may take hours or days for it to recover.*
- *Since the baseline signal comes from the detector, it's natural to suspect the detector.*
- *The TCD responds to the concentration of analyte molecules in the carrier gas.*
- *The FID responds to the rate-of-arrival of analyte molecules in the detector.*
- *Another thought: there may be nothing wrong with the detector.*
- *Divide the possible errors into mutually exclusive groups; for example, the detector IS or ISN'T at fault.*
- *Divide each group into mutually exclusive subgroups, four categories are probably enough.*
- *Having understood the possibilities, evaluate the symptoms visible on the PGC baseline.*
- *The baseline may exhibit drift, wander, cycling, or excessive noise, or any combination of these.*
- *The four kinds of baseline instability combine to form a complex pattern that is difficult to interpret.*

About drift

- *Drift is a continuous upward or downward change in the chromatogram baseline.*
- *Drift is often caused by low-volatility molecules slowly eluting from a column.*
- *Drift can be caused by column bleed, particularly if columns are temperature programmed.*
- *Drift can be caused by a flow change, particularly if columns are pressure programmed.*

About wander

- *Wander is a slow and random upward and downward change in the baseline signal.*
- *Wander is random and forms no repetitive pattern of baseline disturbance.*
- *Wander is often due to pressure variations at the detector vent.*
- *Wander may be due to a leak in the carrier gas connection to the detector.*
- *Wander could indicate a dirty or contaminated detector.*
- *Wander may indicate a poor electrical connection or an electronic fault.*

About cycling

- *Cycling is a low-frequency waveform impressed upon an otherwise flat and smooth baseline.*
- *A cycling baseline is due to something oscillating at the same frequency as the baseline.*
- *The cycling baseline may exhibit a sinusoidal, square, sawtooth, or spikey waveform.*
- *The baseline cycle may be synchronous with the temperature variation of a column or detector.*
- *The baseline cycle may be due to cyclic pressure variation at the detector vent.*
- *The cycle may be from the duty cycle of a power load transmitted via the main power supply.*
- *The cycle may be due to RF interference picked up by a poorly shielded high-impedance detector.*
- *The cycle may be due to light leakage onto a poorly shielded photometric detector.*

About noise

- *Noise is erratic high-frequency variation in an otherwise flat and smooth baseline.*
- *The precision and accuracy of all measurements are limited by their baseline noise.*
- *Baseline noise may be electronic or from random flow or pressure variations in the detector.*
- *Peak integration tends to cancel noise, but the imputed baseline is still dependent on noise.*
- *A noise error in baseline level measurement is magnified by the peak integration procedure.*
- *A more troubling effect of noise is its ability to trigger peak integration commands.*
- *PGCs can smooth data to reduce noise but this also limits the detector response to sharp peaks.*
- *Noise may be due to random pressure changes on the detector vent from wind or trapped liquid.*
- *Noise may be due to spurious signals from a dirty or leaky detector.*
- *Noise may include random spikes from electrical switching of relays, lights, or power devices.*
- *Noise can hide a true peak or create a false one and limit measurement accuracy.*

Did you get it?

Self-assessment quiz: SAQ 12

Q1. What are the two types of problem that may be visible on a chromatogram?

Q2. Consider a symmetrical peak that's 40 % high on the display range with a forced integration window timed to start one base width before the peak apex and end one base width after the peak apex. What measurement error would be caused by a constant error in the baseline level equal to 1 % of the display height? Express your answer as a percentage of the true peak area.

Q3. True or false?

  A. The fuel gas added to the carrier gas entering an FID dilutes the peaks and reduces detector sensitivity.

  B. The area of a peak from an FID doesn't change when the carrier flow rate changes. For this purpose, assume the detector sensitivity is unchanged.

  C. The area of a peak from a TCD doesn't change when the carrier flow rate changes. For this purpose, assume the detector sensitivity is unchanged.

  D. The TCD baseline is more sensitive to environmental factors like temperature and pressure than an FID baseline is.

Q4. What are some typical causes of baseline drift?

Q5. What are some typical causes of baseline wander?

Q6. What are some typical causes of baseline cycling?

Q7. What are some typical causes of baseline noise?

References

Further reading

The principles of metrology including statistical concepts of precision, accuracy, confidence, and control charting are detailed in Waters (2013, 612–650).

The statistical validation procedure is described in Horst and van Burgh (2022) and in Waters (2020, 304–306). When getting into detail, refer to the latest editions of ASTM D3764 (2022a) and ASTM D6299 (2022b).

The foremost troubleshooting guide for gas chromatographers is the cited book by Dean Rood (2007). While aimed specifically at capillary columns in laboratory gas chromatographs, this useful book provides a concise review of GC theory and gives countless hints and tips for resolving problems with GC hardware. It may therefore be helpful for diagnosing PGC baseline problems.

An older but comprehensive guide is *Part II: Gas Chromatography* in the maintenance and troubleshooting book by Walker et al. (1977, 95–304).

Cited

Annino, R. and Villalobos, R. (1992). *Process Gas Chromatography: Fundamentals and Applications*. Research Triangle Park, NC: Instrument Society of America.

ASTM D3764 (2022a). *Standard Practice for Validation of the Performance of Process Stream Analyzer Systems*. West Conshohocken, PA: ASTM International.

ASTM D6299 (2022b). *Standard Practice for Applying Statistical Quality Assurance and Control Charting Techniques to Evaluate Analytical Measurement System Performance*. West Conshohocken, PA: ASTM International.

Horst, D. and van Burgh, M. (2022). *Practical Guide for Validation of Process Analyzer Systems According to the International Standards*. Proceedings of the Analyzer Technology Conference (ATC 2022) on 2022/05/16-18. Galveston, TX: AT Conference International Inc.

Rood, D. (2007). *The Troubleshooting and Maintenance Guide for Gas Chromatographers*, Fourth Edition. Weinheim, Germany: WILEY-VCH Verlag GmbH & Co., KGaA.

Walker, J.Q., Jackson Jr., M.T., and Maynard, J.B. (1977). *Chromatographic Systems: Maintenance and Troubleshooting*, Second Edition. New York, NY: Academic Press Inc.

Waters, T. (2013). *Industrial Sampling Systems: Reliable Design and Maintenance for Process Analyzers*. Solon, OH: Swagelok Company.

Waters, T. (2020). *Process Gas Chromatographs: Fundamentals, Design and Implementation*. Chichester, UK: John Wiley and Sons.

Tables

12.1   Diagnosing a Drifting Baseline.
12.2   Diagnosing a Wandering Baseline.
12.3   Diagnosing a Cycling Baseline.
12.4   Diagnosing a Noisy Baseline.

Figures

12.1   Diagnosing Process Data Trends.
12.2   Four Patterns of Baseline Disturbance.
12.3   Baseline Drift.
12.4   Baseline Wander.
12.5   Baseline Cycles.
12.6   Baseline Noise.
12.7   Chromatogram with Unstable Baseline.

New technical terms

If you read the whole chapter, you should now know the meaning of these new technical terms.

agreed reference value
baseline
baseline cycling
baseline drift
baseline noise
baseline wander
calibration sample
chromatographic symptoms
control chart

instrumentation symptoms
limit of detection
measurement environment
measurement range
root cause
validation
validation fluid
validation program

For more information, refer to the Glossary at the end of the book.

13

Troubleshooting chromatograms

"All chromatographic faults are visible in the chromatogram if you know where to look. Learn how to read a chromatogram and you'll be better prepared to diagnose problems – or even predict them in advance!"

Reading the chromatogram

Your ultimate skill

Successfully troubleshooting a PGC is the ultimate skill of the process analyzer specialist. It will take all the knowledge you gained from prior chapters and all your practical experience. You will have to think deeply about the symptoms you see and what they really mean.

> *Successfully troubleshooting a PGC is the ultimate skill of the process analyzer specialist.*

Troubleshooting is like detective work, and all the clues are in the chromatogram. Learn to read the chromatogram and you will be a better troubleshooter. Gather as many clues as you can: some may implicate the perpetrator and some will exonerate the innocent. Eventually, the evidence will reveal the guilty party. You'll enjoy the elation of solving some difficult problems. And you will never be without a job!

The chromatogram is the unprocessed signal from a detector and contains all the information generated by the chromatographic system. This information includes the desired analysis but also reveals signs of inadequate chromatography or faulty hardware. If anything is wrong with the measurement the error must be visible along the chromatogram. The challenge is to find it.

> *All chromatographic faults are visible in the chromatogram if you know where to look.*

Process Gas Chromatography: Advanced Design and Troubleshooting, First Edition. Tony Waters.
© 2025 John Wiley & Sons Ltd. Published 2025 by John Wiley & Sons Ltd.

This is the **first rule of PGC troubleshooting**:

> **Rule #1:** If there's nothing wrong with the chromatogram there's nothing wrong with the chromatograph.

This general principle doesn't apply to calibration errors nor electronic failures that shut down the chromatogram display. Some might also doubt it applies to peak integration and identification errors, but even these data processing faults leave clues on the chromatogram from a modern PGC.

A competent analyzer technician can perform a calibration or diagnose and repair an electronic problem – the manufacturer's training courses provide good instruction on those procedures. But it's not enough. To fully understand the PGC, you must know how to **read the chromatogram** and to use the information found there to diagnose an existing or incipient problem that others might not see. Therefore, this last chapter of the book focuses on the chromatogram and what it can teach us.

The previous chapter introduced two different kinds of symptoms that may be seen on a chromatogram: symptoms of *instrumentation* or *chromatographic* malfunction. Refer to that chapter for the diagnosis and repair of instrument problems – those that disrupt a flat and smooth baseline. Once the baseline is good, we can focus on the chromatogram and the information it contains.

This is the **second rule of PGC troubleshooting**:

> **Rule #2:** Get a flat and smooth baseline.

Diagnosing chromatographic faults

Noticing a problem

After obtaining a flat and smooth baseline, the focus moves to the chromatogram itself. A full diagnosis and repair will require four actions:

> Recognizing that a problem exists.
> Seeking the most likely cause.
> Confirming that the diagnosis is correct.
> Rectifying the problem.

The hardest of these is recognizing that a problem exists. Many PGCs suffer from chronic issues that would be easy to fix if only someone noticed. This is why in Chapter 12 we strongly recommended a validation program to monitor analyzer performance and react to insipient faults before they affect the quality of the PGC data.

You can't fix a problem you don't know about.

Recognizing a malfunction

Recognition depends on your ability to notice small deviations from normal. Of course, this assumes you know what normal looks like. Expert chromatographers instantly notice slight abnormalities in peak shape or position and intuitively know whether a baseline disturbance is worrisome or not. These skills come from experience, but don't despair because there's a logical procedure to reach the same conclusions. Start by comparing the current chromatogram with an older **reference chromatogram** – one that exemplifies perfect performance.

Modern PGCs can store one or more reference chromatograms and superimpose them over a current chromatogram on the display – a powerful diagnostic tool. It's a good idea to keep two reference chromatograms, one of the calibration sample and one of the process fluid. The two samples will have different amounts of the analytes and some different background components. These differences can be helpful when diagnosing a problem with your current chromatogram.

Look at the typical example in Figure 13.1. The peaks in a current chromatogram will not be the same height as those on the reference chromatogram, so focus instead on any change in their retention time or width. Look closely at their shape. Is any peak rising more rapidly than on the reference chromatogram? Or falling more slowly and tailing as it approaches the baseline? Or is any part of a peak missing? Look also at the baseline. Do both chromatograms exhibit the same baseline disturbances or has something new appeared?

Be skeptical about your reference chromatograms – watch out for imperfections that have been present from the start. The following **SCI-FILE:**

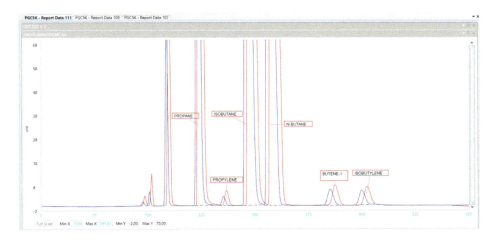

Overlaying the current chromatogram with a reference chromatogram stored when the PGC was performing well is a powerful method for detecting small changes. Here we see that the retention times of all the peaks have shifted. But look closely. Since the earliest peaks have moved by the same factor as the latest peaks, we diagnose a change of carrier gas flow rate. A change in column temperature or liquid phase performance would not affect the holdup time, so the early peaks would not have shifted as much.

Figure 13.1 Overlaid Reference Chromatogram. *Source:* With permission from ABB, Inc.

On Diagnosis gives an example of a chromatogram problem missed during factory inspection and still present during commissioning at the jobsite. If you solve a problem like that, don't forget to save the new reference chromatogram.

Seeking the cause

The need to troubleshoot a chromatogram may be due to an acute malfunction that is disabling the PGC and urgently needs attention. Or the PGC is working well but you notice something that doesn't seem right and wonder what's causing it. In both situations, it's helpful to know the **prime cause of everything**.

Yes, there is one (and only one) cause for everything on the chromatogram! Consider this logic:

> When the PGC is in standby condition with no programmed events occurring, the baseline is flat and smooth.

> When the Method is running, programmed events occur.

> Therefore, everything on the chromatogram that's not flat and smooth is caused by a programmed event.

Stop for a moment and think about that: it's powerful.

Events that affect the chromatogram include turning chromatographic valves ON or OFF, making baseline adjustments like autozero, adjusting display sensitivity to keep peaks onscale, and inserting marker spikes to indicate event timing. Of these, **valve actions** are the cause of all the peaks and baseline disturbances. We include pressure-balance column switching in the generic phrase, *valve actions*.

> *Valve actions are the key to PGC troubleshooting because they cause everything on the chromatogram, including all the peaks and all the baseline disturbances.*

This key role of valve actions is peculiar to process gas chromatographs, since laboratory gas chromatographs rarely use valves. It's captured by the **third rule of PGC troubleshooting**:

> **Rule #3:** Everything on the chromatogram is caused by valve actions.

Except the inserted marker spikes, yet even those small marks reveal important information about event timing.

At first, it seems there must be exceptions to this simple rule. It assumes that the columns are performing well, and the pressures, temperatures, and flow rates are properly set. True, but we shall see that applying the third rule also leads to deductions about those important variables, so it can help to diagnose them too. For instance, checking the time between a valve action and a peak can reveal a difference in carrier flow rate or a column failure and can even point to which column is at fault (Waters 2017).

Identifying artifacts

Now we know valve actions cause everything on the baseline, it's natural to ask what kinds of thing can we expect to see on a PGC chromatogram? And what do they look like? Let's call them **artifacts**.

Four categories of artifact might appear on a flat and smooth baseline:

> **Peaks:** hopefully the familiar detector response to real injected components – but they might be alien to the sample. We'll call them artifacts until they're confirmed as real peaks.

> **Spikes:** momentary baseline deflections due to detector malfunction or electronic interference.

> **Bumps:** the fleeting detector response to transient pressure or flow variations.

> **Steps:** the sustained detector response to a changed flow rate or a newly contaminated carrier gas.

Every deviation from flat and smooth must fit into one of these categories, so the initial diagnostic task is to correctly assign each artifact to one of the four. To do that, you'll need to discover and evaluate two key properties of each artifact: its *overall shape* and its *retention time*.

If a fast diagnosis is important, it's okay to focus on the anomaly that is interfering with a measurement and urgently needs resolution. The best practice, though, is to evaluate every artifact on the chromatogram. There may be a chronic problem or a developing fault that no one has yet noticed. See the **SCI-FILE:** *On Diagnosis* for a good example of how that can happen.

> *The approach to troubleshooting outlined here is rapid and conclusive. Instead of guessing and making arbitrary changes – hoping to find the problem by blind luck – this logical procedure always leads to a solution.*

The first task is to pinpoint the valve switching times, usually identified by marker spikes. In a simple backflush system using a 10-port valve there are only two valve actions, valve ON and valve OFF. It's perhaps surprising to realize that these two actions are responsible for every artifact on the chromatogram. There's no need to come up with fanciful notions; these two valve actions are *the prime cause of everything*. Look no further, one of them must be causing the anomaly you are trying to diagnose.

More complex column systems will have more valve actions and it's important to locate them all because one of them must be the cause.

Carefully estimate the **elapsed time** from each prior valve action to the suspect artifact. For a peak, note the time to its apex; for other artifacts note the time to their onset. There might have been several valve actions prior to the artifact and each one is a potential cause, so you may have several times to consider.

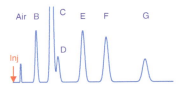

True peaks exhibit a gradual increase of width and may display the doubling rule as evident with Peaks B, C, E and G above.

Figure 13.2 Typical Peaks.

A spike that occurs at the same time on successive chromatograms is the result of a scheduled event. The spike occurs immediately after the event with no time delay.

Figure 13.3 Typical Spikes.

A bump due to a pressure wave reaches the detector immediately after the event that caused it. A bump due to a flow disturbance suffers delay due to the holdup time of a column. Both kinds of bump may look like a small peak. Make a temporary change in the event time to confirm it's a bump and not a peak.

Figure 13.4 Typical Bumps.

Taken together, the shape and elapsed time of each artifact determine whether it's a peak, spike, bump, or step. Each deviation from the baseline must be one of these, including the anomaly itself. Later in the chapter we discuss the key properties of these four artifacts, which will help to confirm their **root cause**. For a tentative initial diagnosis, though, here are the main properties to look for:

➢ A peak has a familiar near-Gaussian shape and a known retention time that is longer than the holdup time of each column. On liquid-phase columns, peak widths follow a distinct pattern increasing by the square root of their retention time, as in Figure 13.2. Any peak that doesn't fit that pattern must have entered the column system at a different time than the other peaks. However, peak widths can be irregular on solid-phase columns. For instance, a branched paraffin peak may be wider than the normal paraffin that follows it.

➢ A spike can be due to transient detector vent pressure disturbances, particles from a PLOT column, or electronic interference. Spikes synchronous with the PGC Method are associated with programmed events, per the example in Figure 13.3. These spikes are instantaneous with no delay from their causative event. A typical spike has zero rise time but might exhibit a slightly dampened decay, sometimes going a bit negative before returning to the baseline. Any spikes asynchronous with the Method will also occur during baseline testing and those tests should have diagnosed and eliminated them. Electronic spikes rarely occur in digital measurement systems but were common in the older analog systems.

➢ A bump is due to a pressure or flow disturbance in the carrier gas passing through the detector. Bumps are more prevalent with thermal conductivity and other concentration detectors. Figure 13.4 shows a typical example. A valve action can generate a pressure wave that travels at the speed of sound and reaches the detector immediately after that event. Typically, the bump will look like a small, distorted peak that rises more rapidly than it falls. Flow disturbances are temporary flow upsets typically caused by reversing the flow in a column. They often look like a dampened sine wave and travel at carrier speed, delayed only by the holdup time of the column.

➢ A step is a shift in the baseline due to a sustained change in carrier gas flow rate or purity. Figure 13.5 shows the typical short delay before a flow step appears. This delay is typically equal to the holdup time of the last column. An equal delay occurs before the baseline returns to its original level. A contamination step can take much longer to appear as the delay is equal to the retention time of the contaminant molecules in the last column. A good example is the retention of a remnant peak after a heart-cut valve action.

This is the **fourth rule of PGC troubleshooting**:

> **Rule #4:** Identify a chromatogram artifact by comparing its shape with the time elapsed since each valve action that might have caused it.

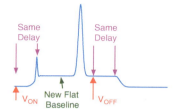

Confirming the diagnosis

In most cases, you can confirm a valve action is causing the artifact by making a small temporary change to its event time. The artifact should follow the change, staying the same time after the event – but remember the affected artifact may be in the next analysis cycle! If the artifact doesn't move, you are looking at the wrong valve action.

This is the **fifth rule of PGC troubleshooting**:

> **Rule #5:** Confirm the diagnosis by making a small change to the suspect event time and seeing the same change in the time of the artifact.

A step due to a flow change will occur a short time after the event that caused it – usually the holdup time of the last column. It takes the same time to return to the original baseline. Steps due to introducing a contaminated carrier gas may take much longer to reach the detector.

Figure 13.5 Typical Steps.

Another useful test is to rig the PGC to inject its own carrier gas. For a PGC with a thermal conductivity detector, simply divert the reference vent flow into the sample inlet. Then, run the Method as usual. All the sample peaks will disappear leaving the valve artifacts in full view.

The logical approach has identified the artifact and the valve action causing it. The following **SCI-FILE:** *On Diagnosis* provides a worked example of this powerful procedure.

SCI-FILE: *On Diagnosis*

Worked example

This case study outlines the logic used to diagnose the "Transient" artifact seen in Figure 13.6. We obtained this chromatogram at an onsite startup in Sweden.

Checking the history

The first step was to compare this chromatogram with the original one obtained six months ago during the factory acceptance test. We found the artifact was also present on that original chromatogram and was about the same shape and height and in the same position as on the current chromatogram. The spikey artifact was strangely reproducible.

When chromatographers call an artifact a *Transient*, it means they don't know what it is.

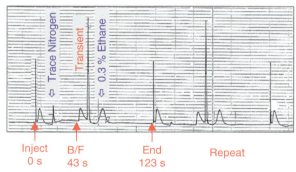

This chromatogram has a large spikey artifact that occurs repeatably just before the ethane peak on each chromatogram. The column designers didn't know what it was, so they labeled it "Transient." Here we demonstrate a definitive diagnosis by logical troubleshooting direct from the chromatogram record.

Figure 13.6 Troubleshooting Example.

This artifact was not affecting the measurement and was doing no harm, yet we were curious about what it was and why it was there. Our curiosity led to a good example of logical troubleshooting.

Finding the valve actions

Only two valve actions occur on the chromatogram, so it's probable that this column system uses a single valve to inject a sample and backflush the first column. The first programmed event turns the column valve ON at time zero to inject the sample. The second event turns the valve OFF to start backflush at 43 s into the 123 s analysis cycle.

This PGC uses hydrogen carrier gas and measures the ethane content of a propane stream. The propane peak is backflushed along with any later components.

Studying the shape

The artifact is very narrow compared with the adjacent ethane peak and we can't expand the timebase to examine its shape in more detail. At first glance it looks like a spike, but it might be a true peak.

Estimating elapsed times

The given cycle time is 123 s, so each division on the chromatogram chart is 20 s and the estimated elapsed times are:

From valve action:	#1	#2
• Nitrogen peak:	20 s	n/a
• Suspect artifact:	55 s	12 s
• Ethane peak:	70 s	27 s

Seeking the cause

The nitrogen peak comes out in 20 s, so if the artifact is from sample injection it has an additional retention of 35 s. Of the 4 artifacts, only a peak can have a longer retention than nitrogen, but the artifact is far too narrow to be a 55 s peak when compared with the adjacent 70 s ethane peak.

Deduction: the artifact may be a peak but it didn't come from the first valve action.

As an aside, the most likely extra peak is methane and by the doubling rule (discussed in the main text) this would elute at 45 s, about halfway between nitrogen and ethane. Even if we disregard the width disparity, the artifact can't be methane.

Perhaps the artifact is a spike caused by electronic interference? No, the artifact is synchronous with the analysis and no event occurs at that time. An electronic interference can't wait for 12 s before it spikes.

Deduction: the artifact is not a spike.

The only viable conclusion is that the artifact is a peak that is introduced into the column system by the second valve action. But that raises another issue: how can a peak reach the detector in only 12 s when the nitrogen peak takes 20 s? There's only one way; the artifact peak doesn't pass through both columns.

Deduction: the artifact is a true peak that enters the second column when the valve turns OFF.

Since it takes only 12 s to traverse the second column the peak is highly likely to be air or nitrogen. Nothing else could travel that fast. The peak is probably air, as there is no constant supply of nitrogen available. It can't be hydrogen because this PGC runs on hydrogen carrier gas.

Deduction: a reproducible volume of air enters the second column each time the valve turns OFF.

This PGC uses a slide valve. Inspection of the valve diagram in Figure 13.7 reveals that the valve slider traps and holds a small volume of carrier gas during the valve ON duration – a common feature of all linear valves. Our diagnosis suggests that the trapped hydrogen leaks out and air leaks in to replace it. When the valve turns OFF, this precise volume of air enters the second column – which explains why the artifact is so reproducible.

Final diagnosis: the valve has a damaged slider or gland allowing atmospheric air to replace the trapped volume of hydrogen carrier gas.

Diagnosing chromatographic faults 337

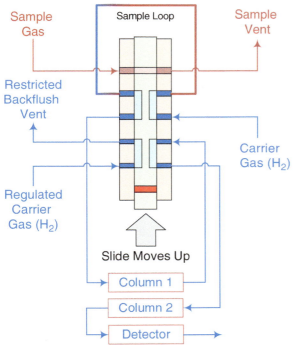

This is the 10-port gas sample injection and backflush valve used to produce the chromatogram in Figure 13.6. When in backflush mode the slider is up allowing the hydrogen carrier gas to pass through the hole in the slider. Then, when the slider comes down to inject the sample it traps carrier gas in the hole. The trapped hydrogen (colored red above) leaks out and air leaks in to replace it. When the slider moves up to backflush the first column the precise volume of air in the hole enters Column 2 and quickly reaches the detector, forming the sharp peak seen on the chromatogram in Figure 13.6.

Figure 13.7 Valve Diagram.

Confirming the diagnosis

Set the valve OFF time two seconds earlier and run a chromatogram. A change in artifact elution time from 55 to 53 s (staying 12 s after the OFF time) will confirm the diagnosis.

Another confirmation is to purge the sample gas from the valve with a stream of carrier gas, then run the PGC to inject only carrier gas. If the artifact remains even without sample gas it must be an air peak.

The solution is to repair or replace the column valve and reset the valve timing. The artifact will disappear.

Other observations

The chromatogram includes artifacts due to pressure waves formed by the reversal of the first column. Their typical shapes appear immediately after valve actions. Look closely at how they differ from the true shape of the ethane peak.

The first column is in forward flow mode for 43 s and in backflush mode for 80 s, which shows good design. Since the backflush runs for 65 % of the time, there is no need for a high backflush flow. If the backflush flow is high, reduce it to match the forward flow. The lower flow rate will minimize the pressure waves and reduce the artifacts they generate during column switching.

Fixing a chromatographic fault

We know that the *prime cause of everything* on a chromatogram is valve actions, but to discern the root cause of a particular artifact you need to know:

➤ Which kind of artifact is it?

➤ Which valve action caused it?

➤ How did it get to that location in the chromatogram?

Now you do. You know what the artifact is and how it got there. That knowledge may be enough to reveal the root cause of the problem. If so, fix the problem and exit this troubleshooting procedure.

But knowing what the artifact is – peak, spike, bump, or step – and discovering how it got into the system might not be enough. The final task is to look closely at the artifact:

- Why is it that shape?
- Why is it in that location?
- Where did it come from?

These can be difficult questions. The answers will come from a keen understanding of the properties of the different artifacts and why they may not behave as expected. It's impossible to catalogue all possible faults, so we list below the key properties of peaks, spikes, bumps, and steps. We hope this listing will help you discover the root cause of the problem you are trying to solve and thereby reveal the proper solution.

Key properties of peaks

What to know about peaks

A peak is the detector response to component molecules in the carrier gas. Only molecules experience more delay than the carrier gas as they pass through the columns. It follows that any baseline artifact that experiences a delay more than the expected carrier gas delay must be due to molecules whether it looks like a peak or not.

Peaks from liquid phase columns form a pattern. Their retention time is predictable, and their width increases with their time in the columns. Knowing this, we expect a peak to fit in with the peaks around it. A peak that doesn't fit this pattern is an anomaly. It might be a true peak that entered the column system at a different time than the other peaks – or it might not be a peak at all. Either way, it shouldn't be there.

Carefully examine each peak on the chromatogram to confirm it really is a peak and it fits the pattern, then check that its identification conforms with all the visible evidence. Keep in mind that the retention time of a peak can't tell you for certain what the peak is, but it can certainly tell you what it isn't.

Retention time pattern

When several substances have chemical structures that differ from each other only by their number of methylene groups ($-CH_2-$), they are known as **homologs**.

Homolog peaks form a distinctive pattern on the chromatogram. Each additional methylene group in a homolog doubles the adjusted retention time of its peak. Thus, an n-pentane peak has double the adjusted retention time of an n-butane peak. This is **the doubling rule** (Waters 2020, 61–64) and it works for every liquid phase. If you know the retention times of any two n-paraffins (or one n-paraffin and the air peak), this simple rule predicts the location of every n-paraffin on that chromatogram.

The doubling rule also works for the iso-paraffins, alcohols, ketones, or any other homologous series of peaks on an isothermal column. The multiplier may not be exactly 2.0 but it's usually in the range 1.9–2.0.

Temperature programming of the column eliminates the doubling. The locations of the homolog peaks are still predictable; they are equally spaced when they experience a uniform rate of temperature increase.

This is a very useful rule when troubleshooting peak identities. If a peak is at its predicted location, there's a good chance that the identification is correct. And if the peak is somewhere else it's almost certain the peak is incorrectly identified.

Retention time shift

Process chromatographs identify peaks by their retention times so it's important for the retention times to be reasonably constant. If the peaks on your current chromatogram don't line up exactly with the reference chromatogram there's something wrong. Look closely:

- Did all the peaks move by the same absolute amount? This would be unusual and could indicate that the injection time marked on one of the chromatograms is incorrect.
- Did all the peaks move by the same retention multiplier? This would indicate a change in the carrier gas flow rate. You can see this effect in Figure 13.1.
- Did all the peaks move by the same retention multiplier starting from the estimated holdup time? This would indicate a change in column temperature or column damage (such as contamination or loss of liquid phase), which always causes a reduction in retention times.
- Did one or more peaks move, and the others stayed put? This would indicate a flow change (or a temperature change) in only one of the columns. Or perhaps the time of a valve action changed?
- Was peak movement random? Suspect a carrier gas leak in a column valve or one of its column connections. A midway leak increases the flow through upstream columns and reduces the flow through downstream columns.

Peak width

A key property of a peak from a liquid-phase column is the relationship between retention time and width. From Equation 2.23, we know that the base width of a peak varies by the square root of its retention time. Any peak on the chromatogram that doesn't fit that relationship is suspect:

> Any peak narrower than an earlier onscale peak probably hasn't been in the columns as long as the other peaks – it didn't enter the column system at sample injection time. Look for another valve action that could have injected it.

> Any onscale peak wider than a later peak has been in the column system longer than the other peaks. Most likely it's a peak left over from the previous analysis. Or perhaps it's a composite of two or more overlapping peaks, although it's rare for a combined peak to have a perfect shape.

> A misshapen peak may be naturally wide, as often happens on solid-phase columns. Severely tailing or fronting peaks are often wider than those around them. This is also true of offscale peaks.

Peak height

If all peaks are higher than expected, suspect an abnormal detector response or, for a gas sample, a pressurized sample volume. All peaks lower can be due to a partly plugged sample volume, an inadequate detector response, or a leak:

> The leak might be in the sampling system; purge nitrogen leaking into and diluting the sample gas flow. Install block-and-bleed isolation valves on the nitrogen inlet.

> An internal injector valve leak allowing carrier gas or air to leak into the sample volume would dilute the injected sample, reducing all the peak heights.

> An outward leak anywhere in the path between injector and detector would reduce the height of all peaks traveling that path.

Leaks are a common problem with PGCs. Carrier gas leaks can occur in the injector valve, column valve, or detector, or in their column connections. A leak can reduce the peak heights and change the carrier flow rate, affecting retention times. In addition, a column valve leak might upset the flow balance of the columns, possibly spoiling the separation. A leak at the detector inlet would allow air ingress, probably causing baseline noise.

An enhanced detector response from a TCD might be due to a higher element current or lower block temperature. With an FID, it might indicate a change in the flow rate of carrier, makeup, or fuel gas going to the detector. With an FPD, it could also be due to a flow change or to an increased excitation voltage on the photomultiplier tube. A reduced detector signal would indicate the inverse of these effects.

If some peaks are affected and some are not, it's most likely a chromatography problem. A retention time shift may have moved some peaks out of sync with the fixed timing of a column valve. If the PGC uses parallel chromatography and all the affected peaks are in one column train, troubleshoot that train. If the fault affects only one type of analyte – olefins or acetylenes, for instance – suspect a chemical reaction, especially with hydrogen carrier gas.

Randomly variable peak heights (loss of precision) can be due to many of the causes for higher or lower peaks cited in Table 13.1a.

As noted in Chapter 12, it's a very good practice to keep records, electronically or manually, of the calibration factors for each PGC measurement. A gradual or sudden change in calibration factor is likely an indication of something wrong and must be investigated.

Unusual peak shape

True peaks have a characteristic shape that makes them easy to recognize. Most are approximately Gaussian in shape, slightly skewed toward their direction of travel.

There are a few exceptions. A severely tailing or triangular peak may occur when the polarity of its molecules is grossly different from the polarity of the stationary phase. Such tailing is frequently seen with active solid columns. A fronting peak is more likely due to injecting a sample volume that exceeds the maximum feed volume or sample capacity of the column – especially with capillary columns. For more information on peak shape, review Chapters 7 and 8.

Newly appearing peak shapes that demand investigation are **overlapping peaks**, **sliced peaks**, and **flat-topped peaks**. Table 13.1b provides some advice for troubleshooting these malfunctions.

A valve action event should never occur during or just before the integration window of a measured peak as the associated baseline upset could distort the peak shape and cause errors in peak detection and quantification.

Unexpected peaks

When a new peak appears on the chromatogram, the task is to discover what it is and why it is there. If the new peak fits the existing pattern, it's probably just an extra component in the process sample or a normally present one that should not reach the detector – perhaps it should have been backflushed? It might be easy to identify a peak like this.

Are you sure it's a peak? Some baseline disturbances look like peaks. The main difference is in their timing. Most baseline upsets occur at or soon after the event that caused them. A peak is different, it has a significant retention time from its causative event.

If the rogue peak doesn't fit the existing pattern, carefully examine its shape and measure its retention time from any prior valve actions. If one of those retention times is consistent with the width of the peak, examine that valve closely. It's likely that an internal leak in the valve is allowing

Table 13.1a Procedure for Diagnosing Peak Size Problems.

Observation		Indication
No peaks on chromatogram. *Is the chromatogram signal attenuated to zero?*	Yes	No signal is reaching the display: • Adjust the signal attenuation or display sensitivity to suit. • Repeat the observation.
	No	No sample was injected: • Check sample flow and adjust as necessary. • Check sample injector valve operation – repair or replace as necessary. • Check for a severe leak of carrier gas into the sample volume. Detector failure: • Check detector is powered. • For a flame detector, check flame is alight. • For a thermal conductivity detector, check for broken element. All the peaks were backflushed: • Check the backflush valve timing and reset as necessary. Refer to Chapter 10. The heartcut valve failed to operate so no peaks entered the second column: • Check operation of the column valve – repair or replace as necessary. Refer to Chapter 11. No sample components reached the detector due to: • Broken fused-silica capillary column. • Blocked valve, column, or connecting tube. • Incorrect valve connections (after service). • Severe carrier gas leak.
The size of all chromatogram peaks from the same detector has <u>increased</u> by the same factor. *Was the detector sensitivity increased by a change to the carrier or fuel gas flow rates or by electronic means?*	Yes	If there was a good reason for changing the flow rates or electronic sensitivity, the PGC must be recalibrated: • Recalibrate the PGC and repeat the observation.
	No	Gas sample pressure in the injector valve has increased. This has no effect on liquid sample injection: • Check the PGC has an atmospheric referencing valve (ARV) installed. If not, install one. • Check the ARV is operating correctly. • Ensure the ARV vent goes to atmosphere and not to flare.
One or more (but not all) peaks has increased in size. *Check the calibration. Is it still correct?*	Yes	The extra peak height is a correct response to a process change: • Check process data to confirm a change.

(continued)

Table 13.1a (continued)

Observation		Indication
		Not an analyzer problem. The process sample contains a new component that is coeluting with the analyte: • Check process data to confirm a change. • If a permanent change, consider a redesigned column system. If a multistream PGC, a stream valve is leaking and contaminating the analyzed stream: • Repair or replace the leaky valve. • Ensure the stream selector uses double-block-and-bleed switching. Catalytic reactions in the analyzer or sample system are making more of the analyte: • Reduce temperature of sample lines and injector, while avoiding condensation. • Clean entire sample pathway to remove catalytic deposits. • Replace with surface-deactivated items. • Change from hydrogen to helium carrier gas.
The size of all chromatogram peaks from the same detector has <u>decreased</u> by the same factor. *Was the detector sensitivity reduced by a change to the carrier or fuel gas flow rates or by electronic means?*	No	Calibrate the analyzer and repeat observation.
	Yes	If there was a good reason for changing the flow rates or electronic sensitivity, the PGC must be recalibrated: • Recalibrate the PGC and repeat the observation.
	No	Carrier gas is leaking into the sample volume and displacing some of the sample (more likely with gas sample injection): • Repair or replace the injector valve, as necessary. If process sample is under vacuum, sample may be diluted by an air leak: • Check full sampling system for leaks. When PGC uses an ARV to compress sample to atmospheric, ARV failure would result in a subambient sample injection: • Check ARV function. A partial loss of sample via a carrier gas leak to atmosphere at the exit of the sample injector valve or at its connection to the first column. Column pressures and flows are unaffected: • Check and secure any leaks from the valve or its connections. A leak at the valve joining two columns might cause a reduction in size of all peaks, but it would also affect the flow rates in both columns and change the timing of the valve action: • Check and secure any leaks from the valve or its connections.

(continued)

Table 13.1a (continued)

Observation		Indication
		The sample injector is partially blocked – more likely with liquid sample injection, due to the small sample volume: • Clean or replace the injector valve. • Ensure a liquid sample injector is at the correct temperature to vaporize the injected sample. There is a carrier gas leak at the detector inlet connection, so the same fraction of each peak is leaking to atmosphere. The leak may also be causing baseline noise: • Check for leaks and seal as necessary. Detector sensitivity has declined. The detector may still have about the same flow rate due to the makeup gas – particularly with capillary columns: • Check flows of carrier, makeup, and fuel gases into the detector and adjust as necessary. • Check the detector power supply (voltage or current) has not changed and reset as necessary. • Clean, repair, or replace the detector.
One or more (but not all) chromatogram peaks are reduced or completely missing. *Are you sure the missing peak(s) are present in the sample?*	Yes	Incorrect column valve timing or a flow or temperature change in the first column: • Check the carrier flows and valve timing – reset as necessary. • For an air-bath oven, ensure the columns are not in the direct air flow from the heater. If using backflush, the missing peak may be backflushed: • Check for a leak of blockage in the column valve or its connections that's causing reduced flow in the first column. • Find the leak or blockage and fix it. • Reset the flow rates and backflush times. • Review the backflush procedures in Chapter 10. If using heartcut, you may not be capturing the missing peak. If the heartcut remnant peak has also disappeared from the chromatogram: • Check the heartcut valve is operating correctly on command. If the remnant peak is still present: • Check the heartcut valve timing and adjust if necessary. • Review the heartcut procedures in Chapter 11. Certain analytes may be reacting with hydrogen carrier or with other components of the sample: • Switch to helium carrier.

(*continued*)

Table 13.1a (continued)

Observation		Indication
		Small polar peaks might be absorbed by liquid or solid deposits in the sample or carrier flow paths:
		• Clean the entire sampling system.
		• Clean the sample injector valve and its connecting tube to the first column.
		• Replace the first column.
		• Clean or replace the connecting tubes between valves and columns.
		• Clean the detector and its connecting tube.
		If the problem has persisted since initial commissioning, you may need to replace all contact materials with surface-deactivated items.
		If using a flame detector, a peak not detected by that detector may be coeluting with an analyte peak. The presence of the undetected peak may quench the detector's response to the analyte:
		• Check whether the column valve functions are supposed to eliminate the rogue peak and adjust the timing if necessary.
		• Consider a change in column design to prevent the coelution.
		Using a nitrogen-based calibration, sample may not yield an accurate calibration:
		• Use a calibration sample similar in composition to the process fluid.
	No	Spike the sample with the missing component and repeat the observation.
One or more additional peaks are present on the chromatogram. *Does the width of the additional peak match the normal progression of peak widths along the chromatogram?*	Yes	An extra component is present in the sample:
		• Check for a process change that might have created the extra component.
		Valve timing error. One or more extra peaks were allowed into the last column.
		• Adjust backflush, heartcut, or dual-column timing to remove the unwanted peaks.
	No	The alien peak was not injected at the same time as the other peaks.
		• Evaluate the width of misfit peaks by comparing their retention from other valve actions to determine when and where they entered the columns.
		A single peak was split and the fragments appear in two places. For instance, part of a peak could escape backflush and appear with the analytes, while the other part is backflushed to a detector.
		• Check the column flow rates and valve event times.
		Ghost peaks from a previous analysis due to sample being absorbed by polymer material (or contaminants) and then released into a subsequent injection:
		• Clean or replace the injector – particularly polymer slides or rotors.

Table 13.1b Procedure for Diagnosing Peak Shape Problems.

Observation		Indication
Some peaks are negative, going down from the baseline instead of up. *Was a new carrier gas cylinder recently installed?*	Yes	The sample is purer than the carrier gas. The new carrier gas contains significant amounts of the analytes and cannot produce accurate measurements: • Replace the carrier gas cylinder with an uncontaminated one. • Replace the carrier gas cylinder regulator with a non-venting two-stage regulator containing no rubber or polymer parts. • Replace any polymer tubes with cleaned stainless-steel tubing. • Replace any cleanup device used to purify the carrier gas. The negative peaks are known as vacancy peaks and appear at the correct position on the chromatogram for each substance. They go downwards because their concentration in the sample is less than their concentration in the carrier gas.
	No	The sample has recently become purer than the carrier gas. This reveals that the carrier gas has always contained traces of the analytes causing the PGC to operate with a false zero and reduced measurement accuracy: • Fix the carrier gas purity problem as noted above. • Recalibrate the PGC. The detector response to those analytes is naturally negative. Usually, the data processor reverses the polarity of a naturally negative peak, so it appears positive on the chromatogram: • Review the reference chromatogram to see if the peaks were positive before. • Check if there is a Method setting for a negative peak – maybe it was changed? A large upset in carrier gas flow rate – often when a column valve switch and EPC pressure change not synchronized. • Eliminate the flow change.
All peaks are now tailing. *Was work recently done on columns or detector?*	Yes	Excessive or unswept volume inside GC devices, connection tubing, or fittings anywhere between injector and detector inlet: • Reduce connection volumes. • Eliminate unswept volumes. Contamination of contact surfaces, for example, by oil, fingerprints, or graphite ferrule fragments: • Clean or replace GC devices and their connection tubing. Inadequate flow of detector makeup gas when using capillary columns:

(continued)

Table 13.1b (continued)

Observation		Indication
		• Check and reset flow rate.
		Polymeric items in the injected sample pathway are adsorbing and releasing the sample components.
		• As possible, remove or replace polymer items.
	No	Sample is leaking from injector into carrier gas stream:
		• Service or replace injector valve.
		Contamination by solid or condensed liquid deposits from process samples:
		• Clean or replace GC devices and their connection tubing.
		• Replace the first column or if a long capillary cut 50 cm off the front end.
		If using a sample splitter, inadequate split vent flow rate.
Some (but not all) peaks have started to tail as they return to the baseline. *Are all the newly tailing peaks polar (active) components like olefins and alcohols?*	Yes	If the PGC uses nonpolar columns, this tailing effect might have been present from the start:
		• Compare with the reference chromatogram to see if a real change has occurred. If not, skip this observation.
		Contamination in the valves, columns, or detector is adsorbing the polar molecules and causing their peaks to grow tails. If the tailing becomes intolerable:
		• Eliminate oil or condensate from injected gas sample.
		• Clean or replace the sample injector valve and replace the tubing (if any) connecting the injector valve to the first column, then repeat the observation.
		• Clean or replace the detector and replace the tubing (if any) connecting the column to the detector, then repeat the observation.
		• Replace the columns and repeat the observation.
	No	Refer to Chapter 8 for more information on tailing.
Some peaks have split, forming a single **bimodal peak** or double peaks. *Has the process changed?*	Yes	Another peak has appeared in the process sample:
		• Process change? Consult with Operations.
		• Consider a new column design.
	No	Inadequate sample vaporization, liquid sample is entering column:
		• Increase injector temperature.
Some peaks have flat tops – generally observed only with capillary columns. *Has the sample volume recently increased?*	Yes	The sample volume is too large and is exceeding the maximum feed volume of the column (usually occurs only with capillary columns):
		• See Chapter 8 for a discussion on feed volume.

(continued)

Table 13.1b (continued)

Observation		Indication
		• Reduce sample volume or change the capillary column for one that has a wider bore, thicker film thickness, or both.
	No	The sample composition has changed and some peaks now exceed their sample capacity
		• See Chapter 8 for a discussion on sample capacity.
		• Reduce the sample volume and recalibrate the PGC.
		Extra volume might be due to a damaged injector valve:
		• Repair or replace the injector valve.
		If a sample splitter is used, the extra volume might be due to a decreased split flow:
		• Reset the split flow.
		The peak height may have exceeded the range of the detector or its amplifier:
		• Reduce sample size or detector sensitivity.
One side of a peak that should be normal in shape is distinctly sharper than the other side. The peak has an unusually fast rise or fall time.	Yes	Part of the peak is missing – it has been cut by column switching.
		• Check the column valve timing and reset as necessary.
Problem solved?	Yes	Exit.
	No	Refer to Chapter 7 and Chapter 8 (Table 8.1) for more help to diagnose a misshapen peak.

a small amount of air or process sample to enter the carrier gas stream as the valve actuates.

To identify a rogue peak, compare its retention time with the known times of the other peaks. Bear in mind that the extra peak may not have passed through the same columns, so it might be narrower and faster than those other peaks.

Disappearing peaks

When an expected peak partly or wholly disappears from the chromatogram it's nearly always due to a timing problem. The valve timing hasn't changed, but the peak has moved. The main suspects are a changed carrier flow rate (likely due to a valve leak) or the deterioration of a column. This has changed the elution time of the peak and it's now getting backflushed or eluding heartcut. Or the peak hasn't moved, and the valve timing is wrong.

Certain analytes might react with hydrogen carrier gas or with each other. This is more likely when a catalyst is present. For example:

The original flow tube in the Live Tee valveless column switch was a platinum capillary tube that catalyzed reactions, so it was replaced with a stainless-steel capillary.

A high-temperature injector may also cause trouble, especially when contaminated with catalyst dust from the process. The best defense is to use helium carrier instead of hydrogen.

Negative peaks

The signal polarity of a concentration detector like the TCD is arbitrary, so if all chromatogram peaks are negative, simply reverse signal polarity in the software or swap the detector output connections. Infrequently, a PGC will use nitrogen or argon carrier gas and a few peaks might be negative. If so, their polarity reversal is accomplished in the Method.

The strange behavior of a hydrogen peak in helium carrier is well documented. As expected, the hydrogen peak is negative at low concentrations relative to all other peaks, but it turns inside out as the concentration increases, soon becoming positive. Such extreme nonlinearity is impossible to correct by calibration so the only solution is to use a different carrier gas. In the laboratory, a mixed carrier gas of about 7.5 % hydrogen in helium gives a positive peak for all concentrations of hydrogen. But it doesn't work on a PGC because the mixed gas separates when the carrier flow rate changes during column switching, particularly when backflushing. When a column experiences a sudden inrush of carrier gas, the hydrogen gets ahead of the helium due to its low viscosity, leading to severe baseline upsets. One solution would be to use helium carrier gas for the columns and add a makeup flow of hydrogen at the detector inlet. Another method sometimes used in PGCs is to use the reference element of a thermal conductivity detector (TCD) to measure hydrogen in nitrogen carrier. These days, it's much easier and more reliable to analyze the hydrogen by a separate applet using nitrogen or argon carrier gas.

Unexpected negative peaks are a prime symptom of impure carrier gas. When the carrier gas contains a higher concentration of a component present in the injected sample, the peak for that component will appear on the chromatogram at the expected retention time, but will be negative. This kind of peak is known as a **vacancy peak** and is a sure indication of contaminated carrier gas. To test for contamination, inject a sample of ultra-pure carrier gas; all the vacancy peaks will then be visible and will provide a measure of the identity and amount of each contaminant. To confirm, rig the PGC to inject its own carrier gas: all the vacancy peaks should disappear.

Diagnosing peaks: An example

Figure 13.8 shows a chromatogram with two valve actions and seven artifacts which have symmetrical Gaussian shapes and look like peaks, albeit idealized. The ones identified as propane, n-butane, and n-hexane have adjusted retention times that follow the doubling rule. They also have gradually increasing widths. This visible evidence supports their assigned identities. But there are serious doubts about the identities of the peaks labeled isobutane, n-pentane, and unknown.

The "isobutane" peak is too narrow to be part of the injected sample. The width of this artifact tells us it has spent less time in the columns than the

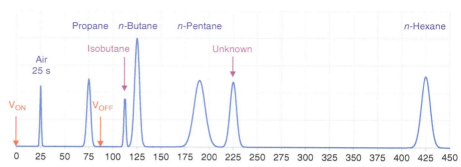

Figure 13.8 Diagnosing Chromatogram Peaks.

adjacent *n*-butane peak; thus, it did not enter the column system during sample injection. Look for a valve action that could have caused it. The clue here is the VALVE OFF function that occurs at 85 s. The rogue peak is as narrow as the air peak and elutes at 110 s, exactly 25 s after V_{OFF}. Since 25 s is the holdup time, there is a strong indication that the injection valve is allowing air to enter the columns as it switches off.

The "unknown" peak has an adjusted retention time of 200 s, exactly the position we would expect for *n*-pentane. This peak probably is n-pentane, but caution is appropriate here; although the peak is in the exact position that we expect to find n-pentane, it doesn't confirm that the peak really is n-pentane. Another peak might have the same retention time. On the other hand, the peak eluting at 175 s is *certainly not* n-pentane. If the identity of the propane, n-butane, and n-hexane peaks are correct, n-pentane cannot come out at 175 s on any column.

Did you notice another problem with this peak? Yes, it's too wide. It's not normal for a wider peak to turn up on a chromatogram before a narrower peak. There are two possible explanations:

> The observed peak is two or more overlapping peaks – unlikely, since there is no visual evidence of shape distortion.

> The peak doesn't belong to this sample injection at all. Being wider, we know it has spent more time in the columns than the other peaks, so it's probably from a previous sample injection and may have escaped backflush.

A test for a wide peak is to run a few chromatograms on the process sample, then inhibit sample injection while continuing to run the Method. This test should reveal the true retention time of the wide peak and confirm it's from a previous sample injection. A similar test is to extend an analysis cycle by 10 s and see if the suspect peak appears 10 s earlier on the next chromatogram.

A test for a narrow peak is to rig the PGC to inject samples of its own carrier gas. This test will eliminate all the sample peaks, but if the narrow peak at 110 s is due to air ingress it will appear on every chromatogram – even without sample injection!

Tables 13.1a, 13.2b and 13.1c provide procedures for diagnosing problems with peaks.

Table 13.1c Procedure for Diagnosing Peak Separation Problems.

Observation		Indication
The resolution between peaks has decreased and some analyte peaks are starting to merge making measurement difficult. *Are the carrier gas flow rates correctly set?*	Yes	Column temperature is too high: • Check the application data and reset oven temperature as necessary. An active solid column has been deactivated by absorbing polar substances. • Regenerate column in laboratory oven per manufacturer's instructions. Columns are damaged or depleted, at the end of their useful life: • Replace the column set.
	No	• Reset the carrier flow rates to the original settings and repeat the observation.
Another peak has appeared and is overlapping an analyte peak. *Check with Operations – Is the new component expected in the process fluid?*	Yes	Choose to tolerate the new interference or redesign the column system: • Tolerate the interference, or • Redesign the column system to separate the new peak.
	No	The extra peak is normally diverted by column switching like backflush or heartcut: • Reset the valve timing to eliminate the peak.
The remnant peak in a heartcut system has become very large and is merging with the analyte peak. *Are the carrier gas flow rates correctly set?*	Yes	The heartcut timing is too wide: • Retune the heartcut valve timing. • Refer to Chapter 11 for more details. If not effective, the first column of the heartcut system has lost its separating power (a common fault): • Install a new first column and retune the heartcut timing.
	No	• Reset the flow rates and repeat the observation.
An analyte peak in a heartcut system is not fully resolved from its remnant peak. *Are the carrier gas flow rates correctly set?*	Yes	The heartcut timing is too wide: • Retune the heartcut valve timing. • Refer to Chapter 11 for more details. If not effective, the second column of the heartcut system has lost its separating power (an unusual fault): • Replace the second column, reset the flow rates, and retune the heartcut timing.
	No	• Reset the flow rates and repeat the observation.

(continued)

Table 13.1c (continued)

Observation		Indication
The backflush event time is now much earlier than it used to be. *Are the carrier flow rates correctly set?*	Yes	The backflush column has lost its separating power (a common fault): • Replace the first column, reset the flow rates, retune the backflush timing, and repeat the observation.
	No	• Reset the flow rates, retune the backflush timing, and repeat the observation.
One or more of the poorly resolved peaks is much larger than it used to be. *When the large peak was smaller, was it unmeasured and partly removed by backflush or heartcut?*	Yes	Too much of the large peak is escaping backflush or heartcut: • Adjust the column valve timing to remove most of the unwanted peak.
	No	The process has changed, increasing the concentration of the large peak and reducing the resolution: • Replace the column set with a new set designed to resolve the new sample concentrations.

Key properties of spikes

What we know about spikes

Spikes have no appreciable width, but they can look like narrow peaks due to the limited time response of the display device. The giveaway is that an electronic pulse travels at the speed of light and is virtually instantaneous with the event that caused it. A peak always has a retention time delay after the event that caused it.

There are two possible reasons for spikes in the chromatogram display:

- A spike that is synchronous with the Method is due to an event during the analysis and occurs at the same place on each chromatogram – immediately after the event. Such spikes rarely occur in modern equipment. Older PGCs may use relays to switch solenoid valves and these may need arc suppressors to eliminate induced spikes.

- Asynchronous spikes are caused by induction from switching electrical loads that are external to the PGC or by pressure fluctuation in the detector vent. They are baseline problems and should be cured by the methods of Chapter 12 prior to chromatogram troubleshooting.

Table 13.2 gives a procedure for diagnosing spikes on a chromatogram.

Key properties of bumps

What we know about bumps

Bumps on the baseline are usually due to pressure or flow upsets and occur most often with thermal conductivity detectors. A pressure wave forms when

Table 13.2 Procedure for Diagnosing Chromatogram Spikes.

Observation		Indication
Spikes in the chromatogram occur at random times.	Yes No	Refer to baseline troubleshooting in Chapter 12. • Continue.
A spike appears at the same time on each chromatogram. *Does the spike occur at the same time as a chromatogram event?*	Yes	The chromatogram event is causing the spike, possibly by switching an inductive load: • Install an arc suppressor on each inductive device that operates from the chromatogram event. The column valve is operating slowly and interrupting the carrier flow, causing a pressure wave in the detector: • Check the valve actuation pressure is adequate and not leaking. • Service or replace the column valve. The valve action is causing an instant pressure pulse on the detector: • A backflush system using Live Tee switching might connect a pressurized column directly to a detector. • Review the column system design for the possibility of a pressure pulse, perhaps due to valve failure. • If necessary, install a capillary restrictor to moderate the pulse and reprogram the analysis. The valve action is briefly depriving a flame detector of combustion air causing the spike: • Separate the valve air and detector air supplies.
	No	The spike is electronic and generated by the data processing system. If the spikes are asynchronous with the Method, refer to the diagnosis of baseline problems in Chapter 12.
The event is a valve action and there's a significant delay (several seconds) between the event and the spike. *Does the artifact exhibit the same delay from the valve action when you move the valve event by a few seconds?* *Try each prior valve event in reverse order.*	Yes	The artifact is a peak that entered a column during the valve action and has spent less time in the columns than the other peaks. • Continue with the peak diagnostics above:
	No	The artifact is not a peak: • Continue.
To confirm the diagnosis, rig the PGC to inject its own carrier gas. Run the Method as usual. *Is the artifact still present?*	Yes No	The artifact is an air peak, probably from a leaky valve. The spike is a baseline problem: • Rectify by the methods of Chapter 12.

a valve action connects a column to a different pressure source. Pressure waves travel at the speed of sound and hit the detector immediately after the valve action. A pressure wave consists of compressed carrier gas molecules and will affect only a concentration detector – in the absence of column bleed, ionization detectors are immune.

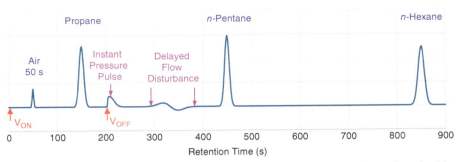

Transient pressure or flow changes can cause bumps on the baseline from a thermal conductivity detector. A pressure wave causes a baseline bump immediately after the responsible valve action (V_{OFF} in this example). V_{OFF} might also cause a flow transient that arrives at the detector after the holdup time of an intervening column. A flow bump can take the form of a dampened sine wave and might not go negative as shown above; it sometimes looks like a peak. Pressure and flow bumps are transitory and don't affect the overall chromatogram. They are problematic only when they occur close to an analyte peak and interfere with baseline detection.

Figure 13.9 Diagnosing Chromatogram Bumps.

On the chromatogram, a pressure wave looks like a small, distorted peak – as shown in Figure 13.9. Note the exact shape of the disturbance: it has a rapid onset and a slower recovery. Be careful with the diagnosis; even a five-second delay is impossible at the speed of sound. If you have a delay followed by a bump, it can't be a pressure wave and might be a real peak. On the other hand, if the bump is immediately after sample injection and before the air peak, it can't be a peak and is certainly a pressure wave.

Flow disturbances are due to reversing a column during backflush or inserting a column in series with other columns. The transient flow change reestablishes the pressure gradient in the column and is unavoidable. For instance, when a column returns from backflush its high-pressure end connects to the lower pressure at the head of the second column, causing a flow surge into the second column. Meanwhile, the pressure at its low-pressure end suddenly increases, causing a flow surge into that column. The flow surges take a little time to dissipate as they reestablish the pressure gradients in both columns. A flow-sensitive detector will respond with a baseline disturbance that's not instantaneous and may end with a small negative excursion, reminiscent of a dampened sine wave. Figure 13.9 illustrates a typical flow disturbance; in most column systems it lasts about 30 s. Sometimes the positive deviation is not followed by a noticeable negative one and it may look very much like a peak.

A simple test is to temporarily move the valve action time and see if the bump follows it. The ultimate test is to set up the chromatograph to inject a sample of pure carrier gas; another option is to inhibit sample injection. In both cases, all the real peaks will disappear, but the pressure bumps will remain.

Since baseline bumps are due to valve actions they can't be moved at will. If a bump is interfering with the baseline under a peak, the only solution is to change the column lengths to move the valve actions and peaks farther apart.

Table 13.3 is a procedure for diagnosing chromatogram bumps.

Table 13.3 Procedure for Diagnosing Chromatogram Bumps.

Observation		Indication
A bump on the chromatogram that was not present during baseline tests. *Does the PGC use a concentration-sensitive detector, such as a thermal conductivity detector (TCD)?*	Yes	The artifact is likely to be a pressure bump caused by column switching: • Continue.
	No	The artifact is unlikely to be a pressure bump and is probably a true peak: • Continue.
The bump immediately follows a valve action. *Does the bump stay close to the valve action when you move the event by a few seconds?*	Yes	The artifact is a pressure bump caused by the valve action. Most likely, the valve reverses the flow direction in a column: • If backflushing, consider reducing the backflush flow rate to minimize the pressure wave upon switching. • Before changing anything, refer to Chapter 10 for more details.
	No	Continue.
The bump is located many seconds after the previous valve action. *Does the bump stay the same time after the valve action when you move the event by a few seconds?*	Yes	The artifact is caused by the valve action. It's a flow bump or a true peak injected by the valve action.
	No	The artifact is not caused by this valve action. It's caused by a previous valve action and may be a flow bump or a true peak which enters the column system when that valve actuates: • Move the previous valve action by a few seconds and observe whether the artifact follows the move. • If so, continue to determine whether it's a peak or a flow bump.
Consider the width of the bump. *Is the bump wider than a true peak would be if injected at the time of the valve action now known to be its cause?*	Yes	A relatively wide bump is a transient flow disturbance due to column switching: • It's probably unavoidable. If the flow bump interferes with a peak measurement: • Increase the length of the last column to move the peak away from the bump.
	No	A narrow artifact is a peak that enters the columns when the valve actuates. It's probably an air peak: • Continue with the above procedure for diagnosing peaks.
To confirm the diagnosis, rig the PGC to inject its own carrier gas. Run the Method as usual.	Yes	The "bump" is a true peak due to a component in the sample: • Study the peak width and retention time compared with other peaks on the chromatogram:

(continued)

Table 13.3 (continued)

Observation		Indication
Has the artifact disappeared from the chromatogram?		• If the peak is too narrow, it may be a slice of a peak cut by column switching.
	No	The artifact is not a peak from the sample. It's a flow bump or an air peak caused by column switching, as determined above.

Key properties of steps

What to know about steps

There are two main reasons for baseline steps:

> A valve action has caused a change of carrier gas flow rate through a flow-sensitive detector.

> A valve action has introduced a continuous flow of contaminated carrier gas into a detector.

The left side of Figure 13.10 illustrates the effect of a carrier flow change. A short time after the $V1_{ON}$ valve action, the baseline moves to a new stable position. When the valve is deactivated, a similar delay occurs before the baseline returns to its original position. This often happens with a TCD when a flow balancing restrictor is incorrectly set, causing different detector flow rates in the two column valve states. The elevated or depressed baseline

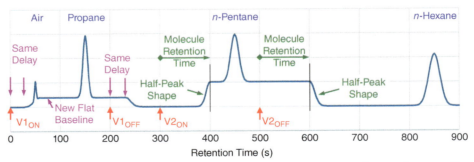

This figure illustrates the features of two different kinds of baseline step. The step between 50 s and 250 s is caused by a flow change. The short delays before its start and end are equal and due to the holdup time in the column feeding the detector. The step between 400 s and 600 s is caused by contaminated carrier gas. The longer delays before its start and end are also equal, but due to the retention time of the contaminating substance in the column feeding the detector. Each step starts with vertical sides that soften as they pass through the column, becoming half-peaks whose width reflects their time spent in the last column

Figure 13.10 Diagnosing Chromatogram Steps.

is not detrimental to measurement and would not affect the propane measurement.

At the instant of valve action the flow change is near instantaneous, but it takes time for that change to reach the detector (the holdup time of the intervening column), delaying the appearance of the step on the chromatogram. The same delay occurs when the valve action is reversed returning the columns to their original conditions. As the step passes through the column, gas diffusion effects soften the vertical rise and fall of the baseline and each adopts a half-peak shape, as depicted on the left side of Figure 13.10.

The other cause of steps is contamination. The detector is responding to a new and constant level of foreign molecules in the carrier gas. Since the carrier gas supply hasn't changed, the additional molecules must be from a contaminated column, a column with higher bleed rate, or a leaking sample injector. Then, a valve action such as $V2_{ON}$ in Figure 13.10 applies a constant flow of the contaminated carrier gas to a column connected to the detector. After the retention time delay, the detector responds to the arrival of the foreign molecules, elevating the baseline. Since the concentration of foreign molecules is constant, the baseline stays at the elevated level until the valve turns off.

A similar effect always occurs in a heartcut system used to measure parts-per-million of an analyte. The contamination is due to a major component saturating the first column. After this contaminated carrier gas is switched into the second column, there's a short delay before a step change occurs. The delay is equal to the retention time of the contaminating molecules on the second column. The same delay occurs before the baseline returns to zero after pure carrier replaces the contaminated supply. We call this step a remnant peak. Its top is not flat but slopes downward, reflecting the slope of the major component tail. Consult Chapter 10 for more information.

In diagnosis, remember this: only molecules experience significant delay in the column system. The other baseline artifacts are over within 30–60 s. Molecules can take a long time to get through a column, but that time is predicable; for the baseline step on the right side of Figure 13.10, the contaminant molecules take 100 s to reach the detector. They take the same time to clear the column after $V2_{OFF}$. The shift in baseline has no effect on the measurement of n-pentane.

The stepping response to a sustained sample input is a well-known phenomenon called **frontal chromatography**. When the carrier gas contains several significant contaminants, the intervening column will delay each contaminant by its own unique retention time, forming multiple and cumulative steps on the baseline. The chromatogram then shows a whole staircase of steps, the height of each one reflecting its concentration in the carrier gas.

The diagnosis of component steps can be more difficult than flow steps. The best clue is when the retention time for each step to fall is the same as its time to rise and exceeds the holdup time of the column. That delay can help to identify the contaminant.

Note the different rise and fall profiles of the step. Each is a true half-peak shape, but the fall profile is wider than the rise profile reflecting its extended time in the column.

Table 13.4 is a procedure for diagnosing chromatogram steps.

Table 13.4 Procedure for Diagnosing Chromatogram Steps.

Observation		Indication
Baseline suddenly steps to near-zero at the same time in each analysis cycle.	Yes	Method autozero command. Not related to column performance.
	No	• Continue.
The baseline changes to a new constant level partway through the analysis and returns to the original level before the next analysis starts. *Does the baseline go to a constant lower level?*	Yes	Step is due to a flow change: • Carefully balance the carrier flow rates into the detector so no change occurs with valve switching. • Repeat the observation.
	No	A raised step in the baseline is due to a flow change or to contaminated carrier gas: • Continue.
The step rises to a higher level. *Does the step rise a short time (about 30–60 s) after the previous valve action.*	Yes	Step is probably due to a flow change: • Carefully balance the carrier flow rates into the detector so no change occurs with valve switching. • Repeat the observation.
	No	A delay of more than 60 s is most likely due to contaminated carrier gas.
The time duration of the step is equal to the elapsed time between a previous valve action and the reversal of that valve action.	Yes	Confirms that the step width is due to the actuation and deactuation of the same valve.
	No	The step rise and fall are due to different valve actions.
Observe the elapsed time between the previous valve action and the top of the rise. *Is the elapsed time equal to the expected retention time in the last column for a major component in the sample?*	Yes	The step is due to contaminated carrier gas entering the column when the valve action occurs: • The retention time identifies the contaminant. • This always occurs in heartcut systems.
	No	The contaminant is unknown: The carrier gas is coming from the same source as before, so contamination must occur within the column system: • If using a backflush system, ensure the sample injector is backflushed along with the first column. • Clean or replace the injector valve and its connection to the first column. • Clean or replace the column valve and its tubing. • Replace the columns with a new column set.

Epilog

Last words

When giving advice on how to select the right column for a gas chromatograph, three distinguished professors came up with a sage recommendation which also applies to PGC troubleshooting:

Ask someone who knows. (McNair et al. 2019, 69)

How true that is. A few words with an experienced person can save many hours of struggle. Don't hesitate to ask for help; it doesn't make you weak, and you'll likely learn a lot more than you expect.

Chromatogram troubleshooting

Enjoy your work with PGCs. Chromatogram troubleshooting can be frustrating, but it's always fun. We'll leave you with a summary of the rules: you can add your own insight and inspiration:

- **Rule #1:** If there's nothing wrong with the chromatogram there's nothing wrong with the chromatograph (*discounting calibration error or electronic malfunction*).
- **Rule #2:** Start with a flat and smooth baseline.
- **Rule #3:** Everything on the chromatogram is caused by valve actions.
- **Rule #4:** Identify a chromatogram artifact by comparing its shape with the time elapsed since each valve action that might have caused it.
- **Rule #5:** Confirm the diagnosis by making a small change to the suspect event time and seeing the same change in the time of the artifact.

Knowledge Gained

On chromatograms

- *All chromatographic faults are visible on the chromatogram if you know where to look.*
- *If there's nothing wrong with the chromatogram, there's nothing wrong with the chromatograph.*
- *The difficult part is noticing that a problem exists – you can't fix a problem you don't know about.*
- *To spot minor changes, overlay the current chromatogram on a reference chromatogram.*
- *If you improve PGC performance, don't forget to update your reference chromatogram.*
- *Valve actions are the cause of everything visible on the chromatogram.*

On artifacts

- *Artifacts are deviations from the flat baseline that occur only when the PGC Method is running.*
- *Artifacts may be true peaks, spikes, bumps, or steps – all caused by valve actions.*
- *Identify an artifact by noting its shape relative to the elapsed time from the event that caused it.*
- *Confirm by moving the event a few seconds and seeing the same time change in the artifact.*
- *To see the artifacts without the peaks, rig the PGC to inject its own carrier gas.*
- *This logical procedure identifies the root cause of unfamiliar artifacts on the PGC chromatogram.*
- *Fixing a problem may require full knowledge of the properties of peaks, spikes, bumps, and steps.*

On peaks

- *A peak has a characteristic shape and retention time longer than the column holdup time.*
- *Peaks have a predicable retention time and homolog peaks follow the doubling rule.*
- *The retention time of a peak can't certainly tell you what it is but can certainly tell you what it isn't.*
- *If the peaks on your chromatogram have moved, evaluate the percent movement of each peak.*
- *If all peaks moved by the same percentage, suspect a flow change.*
- *If early peaks move by a smaller percentage than later peaks, suspect a temperature change.*
- *If not a flow or temperature change, suspect a valve timing error.*
- *If not a valve timing error, the fault is most likely due to a column failure.*
- *Peak width increases with the square root of retention time.*
- *A peak that doesn't fit the pattern was not injected at the same time as the peaks around it.*
- *A very narrow peak might indicate air ingress in a faulty column valve.*
- *An excessively wide peak is probably from the previous sample injection.*
- *Some peaks from certain columns are naturally wide and misshapen.*
- *New tailing of polar peaks may indicate a contaminated injector valve, column, or detector.*
- *Abnormal peak heights may be due to carrier gas leaks or a change in gas sample-loop pressure.*
- *A negative peak may be a vacancy peak that indicates contaminated carrier gas.*
- *The vacancy shows a lower concentration of the component in the sample than in the carrier gas.*
- *The TCD gives an anomalous response to hydrogen with helium carrier gas.*

On spikes

- *Spikes are voltage pulses induced by electromagnetic waves that travel at light speed.*
- *There is no delay between the source and the appearance of the spike on the chromatogram.*
- *Most random or regular spikes should be eliminated during baseline troubleshooting.*
- *New spikes are due to Method events that switch inductive devices such as relay coils or solenoids.*

On bumps

- *A bump in the baseline is due to a pressure wave or flow upset reaching the detector.*
- *Pressure waves form when a valve connects a column to a different pressure source.*
- *Pressure waves hit the detector immediately after the valve action that causes them.*
- *Concentration detectors are more sensitive to pressure waves than mass-flow detectors.*
- *A flow bump is usually due to a valve action reversing the flow in a column.*
- *Flow bumps have an elapsed time from the valve action due to the holdup time of the last column.*
- *A flow bump may look like a dampened sine wave but the negative curve may not be visible.*

On steps

- *A baseline step is due to a valve causing a flow change or introducing contaminated carrier gas.*
- *A flow step rises or falls to a new baseline after the holdup time of the last column.*
- *A component step rises after the retention time of that component in the last column.*
- *Steps return to their original baseline when the valve action is reversed.*
- *The delay in a step returning to baseline is the same as the delay in its original formation.*
- *The width of the half-peak shapes at step begin and step end reflects their time in the column.*
- *Component steps experience more delay after the causal valve action than flow steps do.*
- *If the new carrier gas contains several contaminants, they will form a staircase of steps.*

Did you get it?

Self-assessment quiz: SAQ 13

Q1. Classify the chromatogram features listed below as symptoms of instrumentation or chromatographic problems:
   A. Signal noise.
   B. Bumps on the baseline.
   C. Baseline steps.
   D. Random spikes on the chromatogram.
   E. Drifting baseline.

Q2. What is a reference chromatogram? Choose the one best definition:
- **A.** It's a chromatogram constructed from an online database showing typical fault conditions.
- **B.** It's a chromatogram supplied by corporate management to illustrate the required performance of the PGC.
- **C.** It's a chromatogram stored at a time in the past when the PGC was performing to expectations.
- **D.** It's a chromatogram from the internet that shows the predicted separation of the analytes.

Q3. What is the prime cause of everything on a PGC chromatogram?

Q4. What are the four kinds of artifact that may occur on a PGC chromatogram?

Q5. What two visible patterns may be useful for diagnosing issues with peaks on a chromatogram?

Q6. On the chromatogram in Figure 13.6, why is it impossible for the transient to be a spike?

Q7. State the five rules of chromatogram troubleshooting.

References

Further reading

The prime troubleshooting guide for gas chromatographers is the cited book by Dean Rood (2007). While aimed specifically at capillary columns in laboratory gas chromatographs, this useful book provides a concise review of GC theory and gives countless hints and tips for resolving problems with GC hardware. As such, it may also be helpful for diagnosing PGC chromatogram problems, except those related to packed columns or column switching systems.

Vendor troubleshooting guides

The user guides published by instrument manufacturers and column suppliers also relate mostly to laboratory GC applications using capillary columns. The many issues with manual and automatic syringe injection dominate this field, so the advice given is of limited use for troubleshooting PGC chromatograms. However, you may find some useful hints in the booklets by the manufacturers listed below, particulary in the packed column troubleshooting guide by Supeloc (1999a):

Agilent (2018). *Agilent 8860 Series Gas Chromatograph: Troubleshooting*. Booklet, part number G2790-90016. Wilmington, DE: Agilent Technologies, Inc.

Elmer, P. (2022). *Quick Reference Guide to Troubleshooting Your Chromatography*. Waltham, MA: PerkinElmer Inc.

Kofel, M.S.P. (2005). *Gas Chromatography Troubleshooting and Reference Guide*. Zollikofen, Switzerland: MSP Kofel.

Phenomenex (2014). *GC Troubleshooting Guide*. Torrance, CA: Phenomenex Company.

Shimadzu (undated). *Gas Chromatography Troubleshooting Guide*. Kyoto, Japan: Shimadzu Corporation.

Sinnott, M. (2017). *Practical Steps in Troubleshooting: Techniques, Tips, and Tricks.* Santa Clara, CA: Agilent Technologies.

Stenerson, K. (1999). *Capillary GC Troubleshooting: A Practical Approach.* Bellefonte, PA: Supelco Division of Sigma-Aldrich Company.

Supelco (1999a). *Packed Column GC Troubleshooting Guide: How to Locate Problems and Solve Them.* Bulletin 792C. Bellefonte, PA: Supelco Division of Sigma-Aldrich Company.

Supelco (1999b). *Capillary GC Troubleshooting Guide: How to Locate Problems and Solve Them.* Bulletin 853B. Bellefonte, PA: Supelco Division of Sigma-Aldrich Company.

Cited

McNair, H.M., Miller, J.M., and Snow, N.H. (2019). *Basic gas chromatography*, Third Edition. Hoboken, NJ: John Wiley & Sons, Inc.

Rood, D. (2007). *The Troubleshooting and Maintenance Guide for Gas Chromatographers*, Fourth Edition. Weinheim, Germany: WILEY-VCH Verlag GmbH & Co., KGaA.

Waters, T. (2017). The fine art of chromatogram reading. In: *Proceedings of the ISA Analyzer Division Symposium, Pasadena, California.* Research Triangle Park, NC: International Society of Automation.

Waters, T. (2020). *Process Gas Chromatographs: Fundamentals, Design and Implementation.* Chichester, UK: John Wiley & Sons Ltd.

Tables

13.1a Procedure for Diagnosing Peak Size Problems.
13.1b Procedure for Diagnosing Peak Shape Problems.
13.1c Procedure for Diagnosing Peak Separation Problems.
13.2   Procedure for Diagnosing Chromatogram Spikes.
13.3   Procedure for Diagnosing Chromatogram Bumps.
13.4   Procedure for Diagnosing Chromatogram Steps.

Figures

13.1   Overlaid Reference Chromatogram.
13.2   Typical Peaks.
13.3   Typical Spikes.
13.4   Typical Bumps.
13.5   Typical Steps.
13.6   Troubleshooting Example.
13.7   Valve Diagram.
13.8   Diagnosing Chromatogram Peaks.
13.9   Diagnosing Chromatogram Bumps.
13.10 Diagnosing Chromatogram Steps.

New technical terms

You should now know the meaning of these technical terms:

artifact	**homolog**
chromatogram bumps	**prime cause of everything**
chromatogram peaks	**read the chromatogram**
chromatogram spikes	**reference chromatogram**
chromatogram steps	**sliced peak**
doubling rule	**vacancy peak**
elapsed time	
flat-topped peak	
frontal chromatography	
ghost peaks	

For more information, refer to the Glossary at the end of the book.

Answers to self-assessment questions

SAQ-01 Answers

1Q1. Correct answer: B. In gas–liquid chromatography, the liquid phase is held stationary while the carrier gas moves down the column.

1Q2. Correct answers: B, C, and D. Oxygen would not be used as carrier gas because it might react with and damage the liquid phase inside the column.

1Q3. Correct answer: D. For a gas–liquid chromatographic column, only a nonvolatile liquid is suitable as a stationary phase.

1Q4. Correct answer: C. While all parts of the chromatographic system are necessary, it's the contact with the stationary phase that causes the separation.

1Q5. Correct answer: B. It's only necessary to separate the analytes designated to be measured, not the unmeasured components.

1Q6. Correct answer: B. The detector generates a continuous signal proportional to the instantaneous number of component molecules leaving the column. This chromatogram signal is processed to measure the concentrations of the analytes.

1Q7. Correct answer: D. The chromatogram shows the baseline, peak shape, and peak separation.

SAQ-02 Answers

In the following, mm* designates a value in time units.

2Q1. By Equation 2.26, the plate number for peak A is:

$$N = 16\left(\frac{t_R}{w_b}\right)^2 = 16\left(\frac{370}{19}\right)^2 = 6068$$

2Q2. By Equation 2.26, the plate number for peak B is:

$$N = 16\left(\frac{t_R}{w_b}\right)^2 = 16\left(\frac{410}{20}\right)^2 = 6724$$

Then, by Equation 2.14, the plate height for peak B is:

$$H = \frac{L}{N} = \frac{4000 \text{ mm}}{6724} = 0.59 \text{ mm}$$

The answer is in distance units (mm).

Process Gas Chromatography: Advanced Design and Troubleshooting, First Edition. Tony Waters.
© 2025 John Wiley & Sons Ltd. Published 2025 by John Wiley & Sons Ltd.

2Q3. By Equation 2.5, the peak separation (S) is:

$$S = 410 - 370 = 40 \text{ mm}^*$$

By Equation 2.6, the average base width is:

$$\overline{w}_b = \frac{19 + 20}{2} = 19.5 \text{ mm}^*$$

Then, by Equation 2.8, the resolution is:

$$R_s = \frac{40 \text{ mm}^*}{19.5 \text{ mm}^*} = 2.05$$

2Q4. From Equations 2.11 and 2.12:

$$\frac{w_h}{w_b} = \frac{2.354\,\tau}{4\,\tau} = 0.5885$$

Therefore:

$$w_h(A) = 0.5885 \times w_b(A) = 11.8 \text{ mm}^*$$

2Q5. Equation 2.11 gives the standard deviation (τ) for peak B:

$$\tau = \frac{w_b(B)}{4} = 5.0 \text{ mm}^*$$

Therefore, the variance (τ^2) of peak B is:
$\tau^2 = 25 \text{ mm}^{*2}$

2Q6. From Equation 2.21:

$$H_t = H \cdot \frac{t_R}{L}$$

Using the values for peak B calculated above:

$$H_t = 0.59 \text{ mm} \times \frac{410 \text{ mm}^*}{4000 \text{ mm}} = 0.060 \text{ mm}^*$$

The distance units cancel out leaving the answer in time units, though expressed in mm.

2Q7. From Answer 2Q3 above, the resolution on a 4 m column is 2.05. Resolution is proportional to the square root of the column length. Therefore, for a new resolution of 1.5, the minimum new column length (L), rounded up, needs to be:

$$L = 4 \text{ m} \times \left(\frac{1.5}{2.05}\right)^2$$

$$L = 2.15 \text{ m}$$

SAQ-03 Answers

3Q1. True or false?
- **A.** False – capillary columns can be much longer than packed columns.
- **B.** False – packed columns have larger internal diameters.
- **C.** False – any column with a lower plate height would be more efficient.
- **D.** False – capillary columns generally have higher plate numbers.
- **E.** True – a packed column can accept a larger sample size.
- **F.** False – packed columns rarely employ a temperature program.

3Q2. True or false?
- **A.** True – capillary columns can be much longer than packed columns.
- **B.** True – capillary columns have smaller internal diameters.
- **C.** False – any column with a lower plate height would be more efficient.
- **D.** True – capillary columns generally have higher plate numbers.
- **E.** False – a packed column can accept a larger sample size.
- **F.** True – capillary columns are more likely to employ a temperature program.

3Q3. True or false?
- **A.** False – micropacked columns are limited to shorter lengths.
- **B.** True – micropacked columns have smaller internal diameters.
- **C.** False – any column with a lower plate height would be more efficient.
- **D.** True – micropacked columns generally have lower plate heights.
- **E.** False – a regular packed column can accept a larger sample size.
- **F.** False – micropacked columns rarely employ a temperature program.

3Q4. Order of elution:
- **A.** On a nonpolar column we would expect the components to elute in boiling-point order: first isobutane, then 1,3-butadiene, and then n-butane.
- **B.** On a polar column, the polarizable butadiene peak will be retained relative to its boiling point location, probably changing the elution order to: first isobutane, then n-butane, and then 1,3-butadiene.

3Q5. Phase ratio:
No. It would be better to use a column having a low phase ratio because it contains more liquid phase and would increase the retention ratio of light components and hence their separation.

3Q6. Fast operation:
Two reasons for running a column much faster than its optimum carrier velocity are (A) to achieve a shorter analysis time and (B) to reduce plate number and thus allow a larger sample injection volume.

3Q7. Plate number:
The peak retention factor: $k = t_R/t_M = 5$.
From Equation 3.5, the theoretical minimum plate height (H_{min}) is:

$$H_{min} = d_c \cdot \sqrt{\frac{1 + 6k + 11k^2}{12(1 + k)^2}}$$

$$H_{min} = 0.53\sqrt{\frac{1 + 30 + 275}{12 \times 36}} = 0.45 \text{ mm}$$

Therefore, the theoretical maximum plate number (N_{max}) is:

$$N_{max} = \frac{L}{H_{Th}} = \frac{30{,}000 \text{ mm}}{0.45 \text{ mm}} = 66{,}700$$

SAQ-04 Answers

4Q1. Correct Answer: B
   B. The intermolecular forces are electronic, due to asymmetric arrangements of electrons.

4Q2. Correct answers: B, C, D, and E
   A. No. The polarity of the analyte molecule would not affect its retention on a nonpolar column.
   B. Yes. Dispersion forces are the only cause of retention on a nonpolar column and the affinity is proportional to the size of the analyte molecules. This is why an iso-paraffin always elutes before the normal paraffin.
   C. Yes. The analytes with lower vapor pressure spend more time in the liquid phase.
   D. Yes. The analytes with higher boiling point have lower vapor pressures and spend more time in the liquid phase.
   E. Yes, of course. See SAQ 4Q6.
   F. No. The McReynolds constants are not properties of the analyte!

4Q3. Correct Answer: A
   A. The paraffin isomers would have different molecular sizes and different boiling points, so a nonpolar column such as OV-1 is likely to separate them. The n-paraffin would come out last.
   B. The paraffin isomers are all nonpolar. Although a polar column like Carbowax 20M has an underlying boiling point retention, it has only a weak affinity for nonpolar solutes and is unlikely to separate them well.

4Q4. Correct Answers:
   A. The polymer has one methyl group and one phenyl group attached to each silicon atom.
   B. The copolymer has two methyl groups randomly attached to half the silicon atoms and two phenyl groups attached to the other silicon atoms.
   The two formulations both have 50 % phenyl content and the different structures have a negligible effect on performance.

4Q5. Correct Answer: 10
The average of the beeswax McReynolds constants is:

$$\text{MRC}_{Avg} = \frac{43 + 110 + 61 + 88 + 122}{5} = 84.8$$

The 0-100 polarity scale is based upon squalane = 0 and TCEP = 100. From Table 4.8, the average MRC for TCEP is 832. Therefore, the integer polarity number ($P_\#$) for beeswax is:

$$P_\# = 84.8 \times \frac{100}{832} = 10$$

With only 10 % polarity, beeswax is a slightly polar liquid phase.

4Q6. Correct Answer: 475 s
The adjusted retention time t_R' for n-pentane (C5) is 300 s:

$$t_R' = t_R - t_M = 350 - 50 = 300 \text{ s}$$

The adjusted retention time for n-hexane (C6) should be about 600 s (i.e., double that of n-pentane).

Answers to self-assessment questions 369

To find the adjusted retention time of a component i having $I = 550$ use Equation 4.1:

$$I_i = 100 \left[n + \frac{\log_{10}(t_R')_i - \log_{10}(t_R')_n}{\log_{10}(t_R')_{n+1} - \log_{10}(t_R')_n} \right]$$

$$550 = 100 \left[5 + \frac{\log_{10}(t_R')_i - \log_{10}(300)}{\log_{10}(600) - \log_{10}(300)} \right]$$

$$0.50 = \frac{\log_{10}(t_R')_i - 2.477}{2.778 - 2.477}$$

$$\log_{10}(t_R')_i = 2.628$$

$$(t_R')_i = 425$$

$$(t_R)_i = 425 + 50 = 475 \text{ s}$$

4Q7. Answers:
  B. For a linear isotherm, the peaks should be symmetrical, the Gaussian peak shape. In practice, some other effects would tend to cause slight tailing – see Chapter 8.
  C. For an isotherm exhibiting a downward curve, the higher analyte concentrations at the center of the peak would migrate faster than the lower concentrations at the peak base, causing a tailing peak.
  D. For an isotherm exhibiting an upward curve, the higher analyte concentrations at the center of the peak would migrate slower than the lower concentrations at the peak base, causing a fronting peak.

SAQ-05 Answers

5Q1. Retention factor (k) for Peak $X = 2/1 = $ **2.0**
Retention factor (k) for Peak $Y = 3/1 = $ **3.0**

5Q2. Separation factor (α) for Peaks X & Y $= 3/2 = $ **1.5**

5Q3. For Peak Y: $t_R = 240$ s, $N = 3600$

$$N = 16 \left(\frac{t_R}{w_b} \right)^2$$

$$w_b = 4 \cdot \frac{t_R}{\sqrt{N}}$$

$$w_{bY} = 4 \times \frac{240}{60} = \mathbf{16 \text{ s}}$$

For Peak X: $t_R = 180$ s

$$w_b = 4\sqrt{H \cdot t_R} \quad (Eqn. \ 2.23)$$

Under constant conditions, plate height (H) is constant:

$$w_{bX} = w_{bY} \cdot \sqrt{\frac{H \cdot t_{RX}}{H \cdot t_{RY}}} = 16 \times \sqrt{\frac{180 \text{ s}}{240 \text{ s}}} = \mathbf{13.86 \text{ s}}$$

Note that peak width is proportional to square root of retention time.

5Q4. Resolution is separation (S) divided by average base width ($\overline{w}_b$).

$$R_s = \frac{2(240 - 180)}{16 + 13.86} = \mathbf{4.02}$$

5Q5. Let the new column length be a fraction (x) of old column length.
Under constant conditions, the separation of two peaks is proportional to column length. The old separation was 60 s, therefore, the new separation in seconds is:

$$S_{new} = x \cdot S_{old} = 60\,x$$

From 5Q3, we see that under constant conditions, average base width is proportional to the square root of column length. The old average base width was:

$$\overline{w}_{b(old)} = \frac{29.86}{2} = 14.93\ \text{s}$$

Also from 5Q3, peak width is proportional to the square root of retention time. Therefore, the new average width in seconds is:

$$\overline{w}_{b(new)} = \sqrt{x} \cdot \overline{w}_{b(old)} = 14.93\,\sqrt{x}$$

The new resolution is 2.0 so:

$$R_{s(new)} = \frac{S_{new}}{\overline{w}_{b(new)}} = 2.0$$

$$\frac{60\,x}{14.93\,\sqrt{x}} = 2.0$$

$$x = \left(\frac{14.93}{30}\right)^2$$

$$x = 0.248$$

The conditions are the same so the plate height is unchanged and the new plate number is directly proportional to the column length:

$$N_{new} = 3600 \times 0.248 = \mathbf{893}$$

5Q6. Under constant conditions (of temperature and average carrier gas velocity), retention time is proportional to column length.

$$t_{R(new)} = x \cdot t_{R(old)}$$

For Peak X:

$$t_{R(new)} = 0.248 \times 180 = \mathbf{44.64\ s}$$

For Peak Y:

$$t_{R(new)} = 0.248 \times 240 = \mathbf{59.52\ s}$$

As a check, this new separation is 14.88 s which is equal to 60 x.

Answers to self-assessment questions 371

5Q7. From 5Q5 above:

$$w_{b(new)} = \sqrt{x} \cdot w_{b(old)}$$

For Peak X:

$$w_{b(new)} = \sqrt{0.248} \times 13.86 = \mathbf{6.90\ s}$$

For Peak Y:

$$w_{b(new)} = \sqrt{0.248} \times 16 = \mathbf{7.97\ s}$$

As a check, the new resolution is:

$$R_{s(new)} = \frac{2 \times 14.88}{6.90 + 7.97} = 2.00$$

SAQ-06 Answers

6Q1. $N_B = 20{,}000$
6Q2. $t_M = 40\ s$
6Q3. Yes.
6Q4. $(w_b)_{Peak\ B} = 5.66\ s$
6Q5. $(N)_{Peak\ B} = 13{,}300$
6Q6. In Figure 6.2, the B/u term has about the same contribution to the minimum plate height as the Cu term does.
6Q7. These changes might be effective to reduce the slope of the column efficiency curve at high carrier gas velocities.
Correct answers:
  C. Use a lower-viscosity liquid phase.
  D. Increase column temperature (doubtful).
  E. Change to hydrogen carrier gas.

SAQ-07 Answers

7Q1. If the PGC is well-constructed the extracolumn contributions to dispersion due to connecting tubes and electronics should be minor. The most likely dispersions remaining are those caused by the sample injection procedure and those caused by the internal volume of the detector.
7Q2. The first reason would be because the injected sample volume exceeds the sample capacity of the column. The amount of a sample component is higher than the linear range of its partition isotherm.
The second reason is because the injected sample volume exceeds the maximum feed volume of the column. The total volume of the sample is occupying too many plates.
7Q3. **A.** Yes, a fronting peak may be the outcome of having a solute volume that exceeds the linear portion of its partition isotherm. The higher concentrations at the peak top are less soluble and move faster than the lower base concentrations, thereby distorting the peak.
  B. No, low concentrations are unlikely to cause fronting.
  C. No, later peaks tolerate larger sample volumes and are less likely to suffer fronting.
7Q4. Equation 7.10 exemplifies the difference between plate height and variance:

$$\sigma^2 = L \cdot H$$

The variance (σ^2) of a peak is a measure of the total longitudinal dispersion of its molecules during its passage through the column. From Equation 7:10, plate height (H) is the amount of variance created per unit length of column. Variance and plate height are both measured in length units within the column and in time units on the chromatogram.

7Q5. Existing peak has a base width (w_b) of 4 s. Per Equation 2.16:

$$w_b = 4\tau$$

So existing peak standard deviation (τ) is 1.0 s and existing peak variance is 1.0 s^2.

A. Equation 7.14 gives the additional time variance from a connecting tube:

$$\tau_{ot}^2 = \frac{1}{12} \cdot \left(\frac{V_{ot}}{\dot{V}_c}\right)^2$$

$$\tau_{ot}^2 = \frac{1}{12} \cdot \left(\frac{0.2 \text{ mL}}{2 \text{ mL/min}} \cdot \frac{60 \text{ s}}{\text{min}}\right)^2 = 3 \text{ s}^2$$

Extra variance is 3 s^2.
New total variance is 4 s^2.
New standard deviation is 2 s.
New peak base width is 8 s. Twice as wide!

7Q6. Variance increases linearly with empty tube length but by the fourth power of empty tube diameter! Peak width increases with the square root of tube length and the square of diameter.

7Q7. The resolution achieved by the capillary column is:

$$R_s = \frac{3 \text{ s}}{2 \text{ s}} = 1.5$$

The standard deviation (τ) of each peak is:

$$\tau = \frac{w_b}{4} = 0.5 \text{ s}$$

The variance (τ^2) of each peak is:

$$\tau^2 = 0.25 \text{ s}^2$$

The peak variance after a twofold change in diameter is given by Equation 7.23:

$$\tau^2 = 0.25 \times 2^4 = 4 \text{ s}^2$$

Their standard deviation is now:

$$\tau = 2 \text{ s}$$

And their base widths are:

$$w_b = 4\tau = 8 \text{ s}$$

So their resolution is now only one-fourth of what it was when leaving the column:

$$R_s = \frac{3\,s}{8\,s} = 0.375$$

Lesson learned: Never use connecting tubing having a larger internal diameter than your column!

SAQ-08 Answers

8Q1. A
8Q2. D
8Q3. F
8Q4. E
8Q5. B
8Q6. A, C, D, and E
8Q7. F

SAQ-09 Answers

9Q1. The best choice is (B). Although parallel column systems will most likely include columns in series, that arrangement of columns is not what gives them an advantage. Their advantage comes from using simple column arrangements (reliable, easy to design and maintain) for concurrently analyzing selected sets of analytes (fast analysis). Parallel chromatography would have those two advantages whatever columns they used.

9Q2. The housekeeping rule requires a column system to surely remove all injected sample molecules from the columns before injecting another sample. The backflush column system meets this requirement by reversing the flow direction in the first column and flushing strongly retained components to a vent or detector.

9Q3. The general elution problem expresses a range limitation of gas chromatography. Any column that separates the early peaks will take an excessively long time to elute the strongly retained components. And vice versa: a column that separates the later peaks in reasonable time will not be able to separate the early peaks. The two options for ameliorating the problem are column switching (mainly used by process GCs) and temperature programming (mainly used by laboratory GCs).

9Q4. The basic answer is that PGCs don't need them. Capillary columns are more expensive to buy and more challenging for the maintenance technician. Most PGCs measure only a few known components, a task often accomplished faster and more reliably by a column-switching system using packed columns. Some would disagree, and it's probable that technical improvements will make capillary columns more attractive in the future.

9Q5. When columns in series use the same packing or coating, the system does not exploit the additional separating power of two different stationary phases. The retention factor is the same for each column. For some easy separations, a single stationary phase is adequate, but more difficult separations benefit from the combination of two phases. A second stationary phase might move peaks to unoccupied segments of the chromatogram, improving resolution and reducing analysis time.
A second disadvantage of identical columns is their equal lengths. Because of their different average pressures, the carrier gas velocity is much higher in the second column and components typically will get through it in about half the time they spent in the first column. Often, this doesn't give enough time for the next process in the analytical method – such as further separation or backflush. To avoid this issue, the second column is often about twice the length of the first one.

9Q6. A. The inlet/outlet pressure ratio is: $\dfrac{4\text{ bara}}{1\text{ bara}} = 4.0$

The junction distance down the column: $Z = 0.333\ L$

To find the junction pressure (P_z), apply Equation 9.3:

$$P_{rz} = \sqrt{P_r^2 - \frac{Z}{L}(P_r^2 - 1)}$$

$$P_{rz} = \sqrt{16 - \frac{1}{3}(16 - 1)} = \sqrt{11}$$

$$P_{rz} = 3.32$$

From Equation 9.2:

$$p_z = 3.32 \cdot p_o = 3.32 \text{ bara}$$

B. The pressure ratios (P_r) come from Equation 9.1:

Column 1: $P_r = 4.00 \text{ bara}/3.32 \text{ bara} = 1.20$
Column 2: $P_r = 3.32 \text{ bara}/1.00 \text{ bara} = 3.32$
Both Columns: $P_r = 4.00 \text{ bara}/1.00 \text{ bara} = 4.00$

The compressibility factors (j) come from Equation 9.5:

$$j = \frac{3}{2} \cdot \frac{(P_r^2 - 1)}{(P_r^3 - 1)}$$

Column 1: $j = 0.907$
Column 2: $j = 0.422$
Both columns: $j = 0.357$

The average pressures ($\bar{p}$) come from Equation 9.4:

$$\bar{p} = \frac{p_o}{j}$$

Column 1: $\bar{p} = 3.66$ bara (2.66 barg) = answer to **9Q6. B**
Column 2: $\bar{p} = 2.37$ bara (1.37 barg) = answer to **9Q6. C**
Both Columns: $\bar{p} = 2.80$ bara (1.80 barg) = answer to **9Q6. D**

9Q7. The retention time in Column 1 is 100 s. Get the retention time in Column 2 by rearranging Equation 9.14:

$$t_{R2} = t_{R1} \cdot \frac{L_2}{L_1} \cdot \frac{\bar{p}_2}{\bar{p}_1}$$

$$t_{R2} = 100 \cdot \frac{2}{1} \cdot \frac{2.37}{3.66}$$

$$t_{R2} = 130 \text{ s}$$

SAQ 10 Answers

10Q1. A, C, and D are correct:
- **A.** Yes, backflush does the housekeeping.
- **B.** No, this would need a heartcut system.
- **C.** Yes, by backflushing into a detector.
- **D.** Yes, by removing unmeasured components, backflush enables faster analyses.
- **E.** No, this would be an example of parallel chromatography.

10Q2. A weakly retained peak might travel almost the length of the first column before being backflushed. It then experiences about the same average column pressure as it returns to the front end of the column, so its travel time is about the same in both directions. However, a strongly retained peak travels slowly into the column at high pressure and faster when exiting at low pressure, so it elutes from the front end of the first column before the weakly retained peak gets there.

10Q3. An intercolumn detector is a low-volume thermal conductivity detector installed between two columns to track the peaks eluting from the upstream column. Seeing the peaks emerge from the column can assist in setting column-switching times, particularly in backflush systems.

10Q4. A regrouped peak may contain several unique substances that have different detector sensitivities. Unless they are always present in the same proportions, the detector response cannot be accurately calibrated.

10Q5. To ensure the net flow in the first column is backward, the column must be backflushed at a higher flow rate or for a longer time than in forward flow.

10Q6. The two columns may contain the same packing or coating, a polar column may follow a nonpolar column, or a nonpolar column may follow a polar column.

10Q7. Since the first column is at higher pressure, its average carrier velocity is lower. A peak just escaping backflush will get through the second column in about half the time it spent in the first column. But the backflush will not be complete for twice as long. To avoid waiting for the backflush, the backflush flow must be higher than the forward flow, wasting carrier gas and subjecting the column to larger pressure pulses.

SAQ 11 Answers

11Q1. A. Is the best description.

11Q2.
- **A.** Two
- **B.** The second column
- **C.** One
- **D.** One
- **E.** Two – one detector vent and one column vent
- **F.** Only one - all flow rates are the same

11Q3.
- **A.** Two
- **B.** Three
- **C.** Four

11Q4.
- **A.** Yes, if the peak gets larger, part of it was missing. If the peak stays constant, don't forget to reset the timing back to what it was.
- **B.** No, this might cut the peak and doesn't reveal whether a part of it was missing.
- **C.** No, this might cut the peak and doesn't reveal whether a part of it was missing.
- **D.** Yes, if the peak gets larger, part of it was missing. If the peak stays constant, don't forget to reset the timing back to what it was.

11Q5.
- **A.** This is the least difficult method.
- **B.** This is a more difficult method.
- **C.** This is the most difficult method.

11Q6. A. The first column must separate the two analyte peaks from the major component and from each other – enough to create a sufficient gap for the first analyte peak to fit between the two remnant peaks.

- **B.** The second column cannot increase the separation between the two remnants because they are the same component.
- **C.** The second column must be exactly long enough to place the first analyte peak between the two remnants, or it must be so powerful that it retains both analytes more than both remnants.
- **D.** The first cut may interfere with the timing of the second cut.
- **E.** The second analyte peak is likely to be strongly retained, extending the analysis time.

11Q7.
- **A.** Ethene exits the first column while the valve is ON (between 33 and 52 s), so 42.5 s is an acceptable answer but allowing for normal peak asymmetry 41 s is a better estimate.
- **B.** Estimate the total retention time of ethene on the chromatogram (about 106 s), then subtract its time on the first column (41 s) giving its retention on the second column as 65 s.
- **C.** The methane remnant enters the second column while the valve is on (centered at 42.5 s) and its center on the chromatogram is about 67.5 s, giving a retention of 25 s on the second column.
- **D.** The ratio of times for ethene on the two columns is 41/65 and the columns have the same packing, so the retention time of methane on the first column is 25*41/65 s = about 16 s.

NOTE: your math is not important – but your logic is!

SAQ 12 Answers

12Q1. Instrumentation problems that affect the baseline and chromatographic problems that affect the measurement of analyte peaks.

12Q2. The given baseline error is 2.5 % of the peak height. According to the chapter text, a baseline error of 1 % of peak height causes 3.2 % error in the area measurement when integrating for a duration of two peak base widths. Therefore, a baseline error of 2.5 % of peak height is 8.0 % of the peak area.

12Q3. True or false:
- **A.** False: the FID is a rate detector. Provided analyte molecules arrive at the same rate per microsecond the detector response doesn't change, it ignores the presence or absence of diluent molecules.
- **B.** True: the FID gives an elemental signal for every molecule in the peak. If the flow rate into the detector is halved, the detector outputs half the signal for twice the time resulting in the same peak area.
- **C.** False: the TCD gives a signal proportional to concentration and the flow rate doesn't affect the concentration. If the flow rate into the detector is halved, the detector outputs the same signal for twice the time resulting in the twice the peak area.
- **D.** True.

12Q4. Some causes of baseline drift include molecules being flushed from the columns, inadequate detector temperature control, and contamination of carrier gas, injector, or detector.

12Q5. Some causes of baseline wander include a varying pressure at the detector vent, a detector leak, and contamination of the detector.

12Q6. Some causes of baseline cycling include inadequate temperature control of detector or column, wind buffeting on the vent outlet, and interference from a synchronous electrical load switching.

12Q7. Some causes of baseline noise include pressure pulses in the detector vent line, solid or liquid material in the detector, leaking detector gas connections, random switching of electrical loads, electronic component failure, and dirty electrical connections.

SAQ 13 Answers

13Q1. (A), (D), and (E) are symptoms of an instrumentation problem.
(B) and (C) are symptoms of a chromatographic problem.

13Q2. (C) is the best definition.

13Q3. Valve actions.

13Q4. Peaks, spikes, bumps, and steps.

13Q5. Visible patterns:
   A. Peak width increases with the square root of retention time.
   B. Peaks from homologous compounds follow the doubling rule.
13Q6. The 12 s delay between the prior valve action and the artifact rules out a spike. Spikes occur instantaneously with the event that causes them.
13Q7. Here are the rules:
Rule #1: If there's nothing wrong with the chromatogram there's nothing wrong with the chromatograph (except calibration or electronic failures).
Rule #2: Start with a flat and smooth baseline.
Rule #3: Given a flat and smooth baseline, valve actions cause everything on the chromatogram.
Rule #4: Identify a chromatogram artifact by comparing its shape with the time elapsed since each valve action that might have caused it.
Rule #5: Confirm the diagnosis by making a small change to the suspect event time and seeing the same change in the time of the artifact.

Subject index of SCI-FILEs

The more-scientific information is captured in **SCI-FILES** throughout both of our Wiley books on Process Gas Chromatographs. For quick reference, the below tabulation of **SCI-FILE**s includes those from this book and our previous book: *Process Gas Chromatographs: Fundamentals, Design and Implementation* (Waters 2020).

SCI-FILE	Book[a]	Chapter	Pages
On Analytic Units	1	7	123
On Asymmetric Peaks	2	8	212–214
On Chemical Names	1	4	60–61
On Column Types	1	1	12–13
On Detectors	1	10	187–189
On Diagnosis	2	13	335–337
On Distribution	2	5	125–128
On Extracolumn Variance	2	7	190-196
On Mutual Affinity	2	3	43–45
On Plate Theory	2	2	24–27
On Pressure Drop	2	9	238–240
On Rate Theory	2	6	153–159
On Resolution	2	5	121–122
On Response Factors	1	15	312–313
On Retention Data	2	4	98–102
On Retention Factor	2	5	129–131
On Solubility	1	2	36–37

[a] Book 1 refers to Waters (2020) and Book 2 is this volume.

Glossary of terms

The glossary defines the new technical terms listed at the end of each chapter and some general terms used in gas chromatography. In addition, the list includes the names and formulae of components that often appear on chromatograms, including all the C_1–C_4 hydrocarbons, and the chemical elements involved. The cryptic notation shown for some of these is a shorthand version of their name often used by chromatographers. The included atomic mass or molar mass (M) of each substance may be useful when converting measurement units.

Cross references within the glossary are in **bold** font. To find more information within the book on any technical term of interest, please use the index.

A

Absolute pressure	The real fluid pressure on a scale starting at absolute zero (a complete vacuum) and measured in **pascals** (Pa), kilopascals (kPa), megapascals (MPa), or bar absolute (**bara**). Must be used in calculations.
Acetylene	See **ethyne**.
Acetone	See **2-propanone**.
Acetylenes C_nH_{2n-2}	The **alkyne** family of hydrocarbons.
Active solid	A granular solid that has a large surface area capable of adsorbing gas molecules. Used as a stationary phase in early PGCs but now mostly replaced by synthetic column packings. Examples include molecular sieves, silica gel, charcoal, and alumina.
Adipate	An ester of adipic acid: $HO_2C \cdot (CH_2)_4 \cdot CO_2H$
Adjusted retention time (t'_R)	Peak retention time as measured from the air peak, equal to the time component has spent in the stationary phase.
Adsorption	The entrapment of gas molecules onto the surface of an active solid due to physical or electronic attraction.
Adsorbent solid	A granular solid used as a stationary phase in gas–solid chromatography, having a large surface area that adsorbs gas molecules.
Adsorption isotherm	A graphical plot of molecules adsorbed on a surface versus their partial pressure in the gas phase when at equilibrium at a specified constant temperature.
Affinity	A general term to indicate the degree of attraction between a solute and the stationary phase which is a function of three forces: dispersion, polar, and ionic.

Agreed reference value (ARV)	The concentration of an analyte in a **calibration sample** or **validation fluid** deemed correct by all interested parties in Operations, Maintenance, Laboratory, and Process Automation.
AID	See **argon ionization detector**.
Air peak	An unretained peak. Any component that is insoluble in the liquid phase and therefore travels through the column at the same speed as the carrier gas.
Alcohol [—OH]	A member of the alcohol family of organic chemical compounds, containing the [OH] functional group.
Aldehyde [—CHO]	A member of the aldehyde family of organic chemical compounds, containing the [CHO] functional group.
Alkane C_nH_{2n+2}	The generic name for a member of the **paraffin** family; that is, noncyclic hydrocarbons without double or triple bonds.
Alkene C_nH_{2n}	The generic name for a member of the **olefin** family; that is, noncyclic hydrocarbons with at least one double bond.
Alky group C_nH_{2n+1}	The generic name for an alkane **radical**, usually an attachment to an organic molecule. Not an independent chemical substance.
Alkyne C_nH_{2n-2}	The generic name for a member of the **acetylene** family; that is, noncyclic hydrocarbons with at least one triple bond.
Allene	See **propadiene**.
Allotrope	A different physical form of the same chemical element; for example, diamond and graphite are allotropes of carbon.
Alumina Al_2O_3 $M = 101.960$ g/mol	An active solid capable of adsorbing gas molecules, previously used as a solid stationary phase.
Amide [=NCO]	An organic chemical compound containing the [N—C=O] functional group.
Amine [—NH$_2$]	A member of the amine family of organic chemical compounds, containing the [NH$_2$] functional group.
Amount of substance	The number of molecules present. A more fundamental measure than the weight or volume of the molecules. The actual number is enormous, so chemists often count them in multiples of a **mole**.
Ampere (A)	The SI base unit of electric current (I).
Analysis time	The time from sample injection to final presentation of results. May be less than **cycle time**.
Analyte	A substance or a group of substances whose concentration in the sample fluid is the target of an analytical measurement. A more general term applicable to any measured quantity is the **measurand**.
Analytical measurement	A measurement of composition. May be **qualitative** (to discover what is present) or **quantitative** (to assess how much is present).
Analytical units	The fraction of molecules in a sample that conform to a desired identity or quality, usually expressed as a percentage by volume, mole, or weight.
Anion	An atom or a fragment of a molecule that has gained an electron, becoming a negatively charged ion. Attracted to an anode.

Anode	A positively charged electrode that attracts anions.
Applet	In PGC, a simple valve and column arrangement that injects, separates, and measures some of the desired analytes in a sample, but not all. To perform the complete analysis, two or more independent applets are necessary.
Application engineering	The art of configuring a PGC to perform a desired analysis. Central to the art is the design of a **column system** and the specification of the individual columns.
Argon Ar $M = 39.95$ g/mol	The chemical element with atomic number (Z) = 18. A monatomic gas, the third of six noble gases. An extremely inert gas sometimes used as a carrier gas. As an analyte, argon is difficult to separate from oxygen.
Argon ionization detector (AID)	A GC detector that is highly sensitive to most organic compounds; uses a radionuclide to ionize argon gas, which in turn ionizes the analyte molecules. Now obsolete in PGCs, mostly superseded by the flame ionization detector.
Aromatic compound	Any organic compound that contains a structure comprising six carbon atoms in a ring, apparently with alternating single and double bonds. **Benzene** [C_6H_6] is the founding member of the category.
Aromatics	A group of aromatic compounds.
Artifact	Any disturbance to the **baseline** other than a genuine peak originating from the current sample injection. Artifacts often have unknown cause, but it's worth the effort to find the cause as it may be a valve or column malfunction that will eventually spoil the measurement.
Asymmetry ratio (A_s)	A measure of the asymmetry (or skew) of a peak, the ratio of its trailing to leading part-widths measured at an elevation of 10 % of the peak height.
Atmospheric pressure	The pressure of the atmosphere at the location of the measurement, which varies by elevation and local weather. For reporting data, the international standard reference pressure is 1.0 bara (about 14.5 psig). In the USA, the traditional reference pressure is 14.7 psig (101.325 kPa).
Atmospheric referencing	Allowing a gas sample to equilibrate to atmospheric pressure immediately before sample injection.
Atmospheric referencing valve (ARV)	A device to adjust the pressure of a gas sample volume to atmospheric pressure before sample injection. One or two valves stop the sample flow and connect the sample loop to atmosphere for a few seconds. Must vent to outside air, not to flare.
Atomic mass	The mass of one atom of a chemical element, measured in daltons or amu.
Atomic mass unit (amu)	One-twelfth of the mass of the carbon-12 atom. The **dalton** (Da) is the name now given to the amu.
Autosampler	A laboratory device that sequentially draws a sample from the next vial in a moving line of vials and injects that sample into a laboratory chromatograph.
Autozero	A timed event that resets the detector **baseline** signal to zero. The chromatogram baseline is often set slightly higher than zero to make negative excursions visible.
Availability	A measure of reliability. Availability is the fraction of time that an analyzer is properly calibrated and available to deliver the measurements, expressed as a percentage of the time the plant is running. Most PGCs achieve 95 %–98 % availability.
Average pressure ($\bar{p}$)	The pressure drop in a column is nonlinear, so the average pressure cannot be found by averaging the inlet and outlet pressures. Instead, divide the absolute outlet pressure by the **gas compression factor** (j).

382 Glossary of terms

Average velocity ($\bar{u}$)	The carrier gas velocity is inversely proportional to its absolute pressure, so it accelerates down the column. The average velocity occurs at the **average pressure**. It may be calculated by dividing the column length by the **holdup time** of an unretained component.
Axial diffusion	The spreading of molecules along the axis of the column tube due to the difference of molar concentration within the band of molecules and the pure carrier gas on either side of the band.

B

Backflush	A method of removing **heavies** from a column by reversing the direction of carrier flow. Used in most PGCs to remove all undesired components from the first column before injecting another sample.
Backflush column	The initial column in a column system designed to separate **heavies** from analytes so the heavies can be backflushed. The heavies emerge as a **composite peak** that is flushed to vent or sent to a detector for a measurement.
Backflush column system	A system of valves and columns to accomplish the **backflush** function.
Backflushed peak	A peak that has been backflushed from a column into a detector and may comprise several components fully or partially regrouped together by the reversed backflush flow.
Backflush-to-detector	A backflush column system that includes a detector to measure the backflushed peak.
Backflush-to-vent	A backflush column system that does not include a detector to measure the backflushed peak.
Backflush vent	An unrestricted outlet to atmosphere, not connected to a flare. Emits a small flow of carrier gas plus a recurrent but miniscule amount of **heavies**.
Band	A group of component molecules traveling inside a column: an embryo peak.
Bar	An international unit of pressure defined as 100 kPa and adopted as the standard reference pressure for measurements that depend on pressure. 1 bar is approximately equal to 14.5 psig.
Bara	Absolute pressure in bar, equal to the gauge pressure plus the local barometric pressure.
Barg	Gauge pressure in bar. Most pressure gauges measure differential pressure relative to the local atmospheric pressure.
Base width (w_b)	The width of a chromatogram peak in distance or time units, measured between the intersections of the triangulated peak sides with the extended baseline.
Bargraph	An obsolete method of displaying concentration data by the length of vertical bars on a strip-chart recorder.
Baseline	The detector signal when no peak is passing through the detector. The PGC computes peak heights or peak areas relative to the baseline. When displayed on a screen or printed on a chart, the baseline should be flat and smooth.
Baseline cycling	Any repetitive up and down change in the baseline signal. Generally applied to baselines that exhibit a rectangular, triangular, or sinusoidal waveform oscillating around a constant level with a fixed wavelength. Usually caused by imprecise temperature control of the detector block or synchronous interference from the switching of nearby equipment.

Baseline drift	Gradual change in the baseline signal in one direction, up or down. A downward drift is often observed after startup due to absorbed impurities purging from the columns, and also following a large offscale peak.
Baseline noise	Jagged, irregular, and high-frequency variations impressed upon a flat baseline. Often due to a carrier gas leak, contaminated detector, or electronic malfunction. May include spikes from RF interference or from pressure pulses in the detector vent due to wind buffeting or vent gas bubbling through condensate in vent line.
Baseline wander	Slow, random changes in the baseline around a constant level. Causes include pressure variations at the detector vent due to wind buffeting or variations in vent manifold pressure. Never connect a detector vent to a flare header. Also caused by injector valve or detector contamination, or by air drafts impinging on a detector.
Benzene C_6H_6 $M = 78.114$ g/mol	A colorless, flammable, liquid hydrocarbon. The benzene molecule has six carbon atoms arranged in a ring structure. Simply stated, the bonding comprises alternate double and single bonds between carbon atoms. Benzene is the founding member of all aromatic compounds and an important intermediate in the synthesis of many useful chemicals. Therefore, it often occurs on chromatograms. Selected as the first **McReynolds probe**.
Benzene ring	The six-carbon ring structure of benzene which is present in all aromatic compounds.
Bimodal peak	A peak that has split and exhibits two apexes. Often due to condensation of the vaporized sample on cooler surfaces in the injector or column. When the vapor peak passes, the condensate revaporizes creating a second peak. Another possibility is a chemical reaction that creates a new chemical species with a different retention factor. This might cause peak tailing or a bridge between two apexes.
Bleed	See **column bleed**.
Bonded phase	A stationary phase polymerized onto the inner surface of a capillary column thus forming a polymer film with low vapor pressure and low **bleed**.
Bridge component	A component measured by two or more parallel applets to link their measurement sensitivities and thus enable full sample computations such as **normalization**.
Bump	A brief disturbance to the **baseline** due to a pressure wave or flow upset caused by valve action. Pressure waves form a bump immediately after the valve action that caused them but flow upsets are delayed by the holdup time of an intervening column.
1,2-Butadiene C_4H_6 1,2-$C_4''\,''$ $M = 54.092$ g/mol	A hydrocarbon having four carbon atoms and two double bonds, with the carbon skeleton C=C=C—C. Often measured in the analysis of 1,3-butadiene streams and in other light olefin samples.
1,3-Butadiene C_4H_6 1,3-$C_4''\,''$ $M = 54.092$ g/mol	A hydrocarbon having four carbon atoms and two double bonds, with the carbon skeleton: C=C—C=C. 1,3-butadiene is an important monomer used in the production of synthetic rubber, notably for vehicle tires. It tends to polymerize in sample systems to form the notorious **green oil**, particularly at temperatures above 60 °C.
1,3-Butadiyne C_4H_2 $C_4'''\,'''$ $M = 50.060$ g/mol	Also called diacetylene. A hydrocarbon having four carbon atoms and two triple bonds, with the carbon skeleton: C≡C—C≡C. Often an analyte in light hydrocarbon samples.

Butane C_4H_{10} nC_4' $M = 58.124$ g/mol	Also called n-butane. A straight-chain **paraffin** having four carbon atoms that commonly occurs in light hydrocarbon gas streams.
1-Butanol C_4H_9OH $M = 74.123$ g/mol	Also called n-butyl alcohol. Occurs as an industrial intermediate. Selected as the second **McReynolds probe**.
1-Butene C_4H_8 $1\text{-}C_4''$ $M = 56.108$ g/mol	Also called 1-butylene. One of the four butene isomers. A straight-chain **olefin** having four carbon atoms with the carbon skeleton: C=C—C—C. Commonly occurs in light olefin gases such as ethylene, propylene, or butadiene plant streams.
cis-2-Butene C_4H_8 cC_4'' $M = 56.106$ g/mol	One of the four butene isomers. A straight-chain **olefin** having four carbon atoms with the carbon skeleton: C—C=C—C. The "cis" tag indicates that the terminal methyl groups are adjacent, on the same side of the double bond. This makes the molecule more compact than the "trans" version, so this peak elutes first on a chromatogram. Commonly occurs in light olefin gases such as ethylene, propylene, or butadiene plant streams.
trans-2-Butene C_4H_8 tC_4'' $M = 56.106$ g/mol	One of the four butene isomers. A straight-chain **olefin** having four carbon atoms with the carbon skeleton: C—C=C—C. The "trans" tag indicates that the terminal methyl groups are on opposite sides of the double bond. This makes the molecule less compact than the "cis" version, so this peak elutes after that component on a chromatogram. Commonly occurs in light olefin gases such as ethylene, propylene, or butadiene plant streams.
Butenyne C_4H_4 $C_4'''{}''$ $M = 52.075$ g/mol	Also called vinyl acetylene. A straight-chain hydrocarbon having one double bond and one triple bond with the carbon skeleton: C=C—C≡C. Commonly occurs in light olefin gases such as ethylene, propylene, or butadiene plant streams.
1-Butyne C_4H_6 $1\text{-}C_4'''$ $M = 54.091$ g/mol	Also called ethyl acetylene. A straight-chain hydrocarbon having an initial triple bond with the carbon skeleton: C≡C—C—C. Commonly occurs in light olefin gases such as ethylene, propylene, or butadiene plant streams.
2-Butyne C_4H_6 $2\text{-}C_4'''$ $M = 54.091$ g/mol	Also called dimethyl acetylene. A straight-chain hydrocarbon having a central triple bond with the carbon skeleton: C—C≡C—C. Commonly occurs in light olefin gases such as ethylene, propylene, or butadiene plant streams.

C

Cal gas	A gaseous calibration sample. Sometimes erroneously applied to a liquid calibration sample!
Calibration	The procedure for ensuring that a PGC will accurately measure the concentration of the selected analytes in a process sample.
Calibration factor	The mathematical relationship between measured peak area (or height) and the known concentration of an analyte in the injected calibration sample. Typically, the calibration factor stored by a PGC is the reciprocal of the **response factor**.
Calibration fluid	A gas or liquid calibration sample.
Calibration sample	A contained gas or liquid mixture having specified analyte concentrations accepted as correct. Often this acceptance will be contingent upon a certified chemical analysis by a trusted laboratory.
Candela (cd)	The SI base unit of luminous intensity (I_v).

Capillary column	General term for any narrow-bore open-tubular column, including **WCOT**, **PLOT**, or **SCOT** versions.
Carbon C $M = 12.011$ g/mol	The chemical element with atomic number $(Z) = 6$. The sixth element, having six electrons and four chemical bonds. The foundation of all life on Earth and present in every organic molecule. Exists in several allotropic forms: charcoal, graphite, and diamond.
Carbon Dioxide CO_2 $M = 44.01$ g/mol	A colorless, asphyxiant gas. Usually separated by a porous polymer column and measured by a thermal conductivity detector. For low ppm measurement, a PGC may catalytically convert it to methane for measurement by a flame ionization detector.
Carbon Disulfide CS_2 $M = 76.139$ g/mol	A highly volatile and flammable liquid that usually has a pungent odor due to sulfurous impurities. A flame photometric detector can measure low-ppm concentrations of carbon disulfide.
Carbon Monoxide CO $M = 28.01$ g/mol	A toxic and flammable gas. Usually separated by a molecular sieve column and measured by a thermal conductivity detector. For low ppm measurement, a PGC may catalytically convert it to methane for measurement by a flame ionization detector.
Carbonyl Sulfide COS $M = 60.07$ g/mol	A toxic and flammable gas with an unpleasant odor. Has a linear molecule with the structure: O=C=S. A flame photometric detector can measure low-ppm concentrations of carbonyl sulfide.
Carrier gas	The mobile phase of a gas chromatograph that carries the component molecules through the column. A PGC maintains constant carrier gas pressure and flow, except in the rare applications of pressure and flow programming. Common carrier gases include hydrogen, helium, nitrogen, and argon.
Carrier gas velocity	See **velocity**.
Cathode	A negatively charged electrode that attracts positive ions.
Cation	An atom or a fragment of a molecule that has lost an electron, becoming a positively charged ion and attracted to a cathode.
Cavity ring-down spectroscopy	A powerful photometric technique capable of measuring parts-per-trillion of certain analytes in gas samples.
Celite	A granular solid with high surface area often used as the inert support in packed columns. Comprises the fossilized remains of diatoms – a common type of algae found in natural waters. See **diatomaceous earth**.
Charcoal	An amorphous granular form of carbon that is capable of adsorbing gas molecules; previously used as a solid stationary phase.
Chemical reaction	Many measurements in analytical chemistry involve a chemical reaction with the analyte. For process analysis, though, we prefer instrumental methods like PGC where no chemical reactions occur.
Chemometrics	Chemometrics is a mathematical procedure that discovers a set of characteristics (such as photon absorption wavelengths) that can predict a physical property of a fluid. It does that by comparing the spectra of dozens of known training samples. If the training samples are representative of the likely process variation, the instrument can predict the desired physical property or analyte concentration of a process stream with adequate accuracy – and can do so very much faster than conventional methods.

Chlorine Cl $M = 35.45$ g/mol	The chemical element with atomic number (Z) = 17. Usually encountered in PGC as the chlorinated derivatives of hydrocarbons, the lightest of which is chloromethane.
Chlorine gas Cl_2 $M = 70.90$ g/mol	Chlorine is a yellow-green, highly poisonous gas at room temperature and extremely reactive.
Chloromethane CH_3Cl $M = 50.49$ g/mol	Also called methyl chloride. A colorless, odorless, and flammable gas widely used in industrial chemistry. Has the structure of methane in which a chlorine atom replaces one hydrogen atom.
Chromatogram	A chart or display showing the variation in a GC detector output signal over time. Generally, the chromatogram shows a predicable number of peaks on a mostly flat baseline. The chromatogram is an essential aid for setting up or troubleshooting a PGC.
Chromatogram signal	The detector output signal. It may be a raw analog signal direct from the detector, a live digitized detector signal, or a representation of an analog chromatogram reconstructed from digital memory.
Chromatograph	An analytical instrument using the principles of chromatography.
Chromatographic symptoms	Symptoms of malfunction that appear on a flat and smooth baseline only when the PGC Method is running. Mainly due to valve operations.
Chromatography	The science and art of separating substances by a mobile phase moving across or through a stationary phase. For analysis, quantitative detection follows the chromatographic separation.
Colorimetry	In this very sensitive wet-chemistry method of laboratory and process analysis, the analyte reacts with a chemical to produce an intense color. The instrument then measures the color density using a photometer tuned to respond to that color.
Column	The heart of any gas chromatograph. A long tube that allows the carrier gas to transport the injected sample across a large surface area of solid or liquid stationary phase.
Column bleed	Detectable molecules coming from the stationary phase. With liquid phases, a combination of liquid-phase vapor pressure and products of reaction. The latter may be due to thermal cracking of the liquid phase or its reaction with impurities in the carrier gas, notably oxygen.
Column, capillary	See **capillary column**.
Column efficiency	The ability of the stationary phase to resolve solutes, indicated by **plate height**. For equal column lengths, a lower plate height produces improved resolution.
Column length (L)	The length of a column in meters.
Column, micropacked	See **micropacked column**.
Column, nonpolar	See **nonpolar column**.
Column, polar	See **polar column**.
Column oven	The column oven maintains a very constant column temperature or precisely varies the temperature during analysis to follow a predefined pattern.
Column overload	An excessive sample volume that results in a **fronting peak** on the chromatogram. See also **sample capacity** and **feed volume**.
Column, packed	See **packed column**.

Column switching	The technique of allowing selected components to pass through different columns, thus improving analyte resolution and analysis time.
Column system	An arrangement of two or more columns and column-switching devices capable of performing a designated separation of components.
Column vent	A carrier gas outlet that must flow to atmosphere, not to a recovery or flare line.
Come read	An instruction transmitted to a computer to inform that new data is available. The computer then reads the new value.
Component	Any unique chemical substance present in the sample fluid. A measured component is an **analyte**.
Composite peak	A peak or group of partially separated peaks containing two or more components. Either the column doesn't separate those components or the column system has intentionally regrouped them together. Since perfect synchronicity of peaks is unlikely, the composite peak may have an irregular shape.
Comprehensive method	Laboratory term for a GC analysis that identifies and measures all the components in a sample, as opposed to a **heartcut method**. Not used in PGC.
Concentration	See **mole fraction**.
Concentration detector	A gas chromatograph detector that responds to the concentration of the analyte in the column eluent.
Concurrent separations	The concurrent analysis of different components of the same process stream by two or more independent applets housed in a single process gas chromatograph, each applet having its own sample injector, columns, and detector.
Concurrent streams	The concurrent analysis of two or more process streams using independent applets housed in a single process gas chromatograph, each applet having its own sample injector, columns, and detector.
Confidence level	The estimated probability that the true value of the measurand falls within the **uncertainty interval**.
Continuous variable	A variable having an infinite number of possible values. It would require an infinite number of digits to exactly represent the value of a continuous variable, so a measurement can only estimate its approximate value.
Control chart	A plot of sequential measurements of a constant sample, represented as deviations from the mean value, that also exhibits warning and action limit lines. These limits are typically at ± 2 and ± 3 standard deviations from the mean, respectively. Applying the standard rules of Statistical Quality Control can warn of impending failure or dictate the need for calibration.
Control unit	An electronic device capable of controlling the time-dependent events in a process gas chromatograph and processing a chromatogram signal to produce useful information. All modern instruments use an integral microcomputer for these functions.
Corrected retention volume (V_R°)	The **retention volume** of a component peak corrected for the average pressure in the column by applying the **gas compressibility factor** (j).
Correlogram	Detector output signal from an advanced laboratory GC technique that splits a composite peak into short segments which are partially separated in a second column. The resulting correlogram is mathematically deconvoluted to measure the individual components.

Count	An exact value having zero uncertainty, but not necessarily an integer. Any defined quantity is a count, not a measurement. For instance, pi (π) is a count.
Cut peak	Can refer to a whole peak successfully cut from the tail of a major component or to an incomplete **sliced peak**, one with a piece removed by column switching.
Cutter column	See **heartcut**.
Cyano group [—CN]	Also called **nitrile**. An attachment to an organic molecule having a carbon atom bonded to a nitrogen atom. Not a separate chemical compound.
Cycle time	Usually the time elapsed between sample injections. May also refer to the time needed to complete the whole cycle of a multistream sequence.
Cyclobutane C_4H_8 cyclo-C_4 $M = 56.107$ g/mol	An isomer of butene. A hydrocarbon gas with a molecular structure comprising four carbon atoms connected in a circular formation by single bonds. Sometimes appears as a peak on chromatograms of light hydrocarbon samples.
Cyclohexane C_6H_{12} cyclo-C_6 $M = 84.16$ g/mol	A hydrocarbon liquid with a molecular structure comprising six carbon atoms connected in a circular formation by single bonds. Cyclohexane is a powerful industrial solvent. It often occurs on chromatograms separating aromatic compounds.
Cycloparaffins C_nH_{2n}	The family of cyclic hydrocarbons having only single bonds between carbon atoms. No double or triple bonds. Cyclopropane is the smallest member.
Cyclopentane C_5H_{10} cyclo-C_5 $M = 70.1$ g/mol	An isomer of pentene. A hydrocarbon gas with a molecular structure comprising five carbon atoms connected in a circular formation by single bonds. Rarely appears as a peak on chromatograms of light hydrocarbon samples.
Cyclopropane C_3H_6 cyclo-C_3 $M = 42.08$ g/mol	An isomer of propene. A hydrocarbon gas with a molecular structure comprising three carbon atoms connected in a circular formation by single bonds. Often appears as a peak on chromatograms of light hydrocarbon samples.

D

Dalton (Da)	The SI unit of atomic mass, defined as one-twelfth of the mass in grams of a carbon-12 atom. Thus, one **mole** of daltons is a mass equal to one gram.
Debye force	The molecular interaction between a permanent dipole and an **induced dipole**.
Dead leg	An unpurged cavity, recess, or tube that can retain prior concentrations of sample molecules and allow them to diffuse into the current analyzed sample.
Dead volume	See **dead leg**.
Decane $C_{10}H_{22}$ nC_{10}' $M = 142.286$ g/mol	Also written as n-decane. A normal paraffin with ten carbon atoms.
Detector	A device that generates a signal, usually electronic, proportional to the instantaneous number of component molecules in the effluent from a column. All PGC detectors are differential, responding to the difference between the baseline condition and the presence of a peak.
Detector vent	The outlet for gases discharged from a detector. It's important neither to restrict the detector vent nor to subject it to pressure pulses.
Diaphragm valve	A valve that applies pressure to a diaphragm in a manner that blocks one or more flow paths while opening others.

Diatomaceous earth	The fossilized skeletal remains of ancient marine algae. Retains the fine structure of the original microscopic organisms, usually as a pair of species-dependent half-shells finely perforated with micron-sized holes. Used as a solid support, its complex structure provides a large surface area and multiple pathways for gas flow.
Dielectric constant	Traditionally, the ratio of the capacitance of a capacitor filled with the measured material to the capacitance of an identical capacitor in a vacuum. Now called relative permittivity.
Diffusion	A movement of molecules that gradually ensures uniform composition in an enclosed space. For a gas, the driving force of diffusion is a difference in its partial pressure at separate locations in the enclosed space. The rate of diffusion is proportional to the square root of the gas density (Graham's law).
1,2-Dimethylbenzene C_8H_{10} o-Xy $M = 106.16$ g/mol	Also called ortho-xylene. Frequently measured by PGCs in an aromatics plant. The three xylene isomers are difficult to separate with liquid stationary phases.
1,3-Dimethylbenzene C_8H_{10} m-Xy $M = 106.16$ g/mol	Also called meta-xylene. Frequently measured by PGCs in an aromatics plant. The three xylene isomers are difficult to separate with liquid stationary phases.
1,4-Dimethylbenzene C_8H_{10} p-Xy $M = 106.16$ g/mol	Also called para-xylene. Frequently measured by PGCs in an aromatics plant. The three xylene isomers are difficult to separate with liquid stationary phases.
Dimethylether C_2H_6O DME $M = 46.07$ g/mol	Also called ether. The smallest member of the ether family having the structure CH_3—O—CH_3. A volatile and highly flammable liquid.
2,2-Dimethylpropane C_5H_{12} neo-C_5 $M = 72.15$ g/mol	Also called neopentane. An isomer of pentane with a central carbon atom connected to four other carbon atoms. Its compact shape gives neopentane a low boiling point and low retention time on liquid columns, being the first C_5 peak to appear on the chromatogram.
Diolefin C_nH_{2n-2}	A hydrocarbon having two double bonds between carbon atoms. The smallest diolefin is **propadiene**.
Dipole	An electronic charge within molecules due to an uneven distribution of electrons.
Dipole, induced	A temporary dipole created within a nonpolar molecule by the influence of an adjacent polar molecule.
Dipole, permanent	A sustained **dipole** due to electron displacement within a molecule containing two or more atoms of different **electronegativity**.
Dispersion force	A sustained attractive force between molecules due to myriad fleeting polarizations caused by the random motion of their electrons.
Distribution	In gas chromatography, distribution is the process by which solute molecules distribute themselves between a gas phase and a liquid phase to form a steady-state equilibrium.
Distribution column system	One that partially separates the analytes into two or more groups, then routes each group to a different secondary column optimized for the final separation of that group.
Distribution constant (K_D)	In effect, the **solubility** of a solute in the liquid phase. The distribution constant of a component is the ratio between its liquid-phase concentration and its gas-phase concentration, at equilibrium. The ratio is constant at low concentrations but may deviate as the concentrations increase. Such deviations are one of the causes of asymmetric peaks.

Disulfide	A chemical compound whose molecule contains two sulfur atoms. For organic disulfides, the two sulfur atoms bond together and have carbon chains attached at each end. The simplest example is dimethyl disulfide: $H_3C-S-S-CH_3$
Dodecane $C_{12}H_{26}$ nC_{12} $M = 170.340$	Also called n-dodecane. A straight-chain paraffin having 12 carbon atoms in each molecule.
Doubling rule	The adjusted retention time approximately doubles for each ($-CH_2-$) unit added to a member of a homologous series.
Dual column	A term that is not well defined but may indicate a **distribution column system**.
Dynamic equilibrium	A stable state where two opposing processes are occurring at the same rate, so the outcome is unchanging.

E

ECD	See **electron capture detector**.
EMG	See **exponentially modified Gaussian**.
Elapsed time	The duration between a valve action and its effect on the chromatogram baseline.
Electrochemical	A chemical reaction that produces or consumes electrons. Electrochemical devices often measure contaminants in water – such as dissolved oxygen, for instance.
Electrode potential	A voltage that forms on a wire immersed in an aqueous solution due to metal ions entering the solution. It may also occur between two solutions due to a difference in their concentrations. Electrode potentials can measure oxygen, pH value, and selective ion concentrations, among others.
Electropolishing	A technique for smoothing the inner wall of metal tubing. The tube is filled with electrolyte and an insulated wire drawn through. An electric current from the wire tip preferentially hits the high spots, wearing them down.
Electron capture detector (ECD)	A detector used in chromatography to measure low concentrations of gases with high electron affinity, like chlorine. It uses a radionuclide to ionize nitrogen carrier gas, releasing low-energy electrons. Two polarized electrodes then generate a constant baseline current that decreases when an analyte captures some electrons.
Electronegativity	A measure of the tendency for the nucleus of an atom to attract electrons, a function of the number of protons it has.
Electronic pressure controller (EPC)	An electronic device that actively controls gas pressure and may have a set point under the control of the PGC microprocessor.
Electronic pressure regulator (EPR)	See **electronic pressure controller**.
Electronic signal (4–20 mA)	The standard analog data-transmission signal for process control.
Elute	To carry a component peak out of a column. All PGCs use elution chromatography, where a carrier gas elutes a band of component molecules out of the column and into the detector, thereby forming a peak on the chromatogram.
Energy (E)	Defined as force multiplied by distance moved, the SI unit of energy is the joule (J) where: $J = kg\ m^2/s^2$ Electrically, a joule is equal to a watt-second (W s).

Ester	An organic compound made by replacing the hydrogen of an organic acid by an organic group. For example, ethyl acetate.
Ethane C_2H_6 C_2' $M = 30.07$ g/mol	A hydrocarbon gas, the second member of the paraffin homologous series. Occurs on chromatograms of natural gas and refinery gases.
Ethene C_2H_4 C_2'' $M = 28.05$ g/mol	Also called ethylene. A hydrocarbon gas, the first member of the olefin homologous series. Commonly occurs on chromatograms of samples from petroleum and petrochemical plants.
Ether [C—O—C]	A series of organic chemicals containing an oxygen atom between two carbon atoms. Colloquially, *ether* means the simplest member of the series: **dimethylether**.
Ethoxy group [—OC$_2$H$_5$]	An attachment to a long-chain molecule connecting an ethyl group via an oxygen atom. Found in many stationary phases, the ethoxy group instills additional polarity to the phase. Not a separate chemical compound.
Ethylene	See **ethene**.
Ethyl group [—C$_2$H$_5$]	An attachment to an organic molecule having two carbon atoms and five hydrogen atoms. Not a separate chemical compound.
Ethyne C_2H_2 C_2''' $M = 26.04$ g/mol	Also called acetylene. A hydrocarbon gas, the first member of the acetylene homologous series. Occurs on chromatograms of light hydrocarbon gases, particularly in olefin process streams.
Event time	Any timed function occurring during an analysis and controlled by the PGC clock. For most events, the user enters a time into the analysis **Method**. Ideally, the data entry is the time after injection the event must occur, but some PGCs inject the sample a few seconds into the analysis cycle, so the time entered is nominal.
Explosion-proof enclosure	See **flameproof enclosure**.
Exponentially modified Gaussian (EMG)	The combination of a symmetrical Gaussian function and an exponential function to model the effect of a minor degree of surface adsorption on the shape of a chromatogram peak. The model closely resembles a **normal peak**.
External standard	A calibration gas or liquid in a container, external to the process analyzer.

F

Feed volume	The total injected volume of sample and intermixed carrier gas, a measure of the width of an injection profile. Theory says feed volume should not exceed the volume of 50 plates.
FID	See **flame ionization detector**.
Film thickness	The thickness of a liquid phase film on the walls of a capillary column.
Flame arrestor	A safety device to prevent a flame from propagating into a flammable process environment. Comprises a set of narrow metal pathways that cool the flame below the gas ignition temperature.
Flame ionization detector (FID)	A popular PGC detector that uses a hydrogen flame to ionize hydrocarbon molecules. High-voltage electrodes then collect the ions and released electrons. The resulting electron current provides an extremely sensitive and linear measure of carbon content that follows the profile of each eluting peak. The detector responds only to organic compounds.

Flame photometric detector (FPD)	A PGC detector that uses a hydrogen flame to excite the fluorescence of infrared radiation by sulfur or phosphorus atoms. An optical filter selects the fluorescence due to sulfur, which then impinges on a photomultiplier tube. This device is very sensitive to light and outputs a signal that follows the profile of each eluting peak that contains sulfur.
Flame photometry	A method for the chemical analysis of certain metal elements that examines the wavelengths of light emitted by an analyte sprayed into a flame.
Flameproof enclosure	Also called an explosion-proof enclosure. An enclosure that complies with applicable codes for installation in a specified hazardous area. Quenches a flame by cooling it below the ignition temperature of the gas in question.
Flash point	The lowest temperature at which a flammable fluid will burn in air when exposed to an ignition source.
Flat-topped peak	A peak with its top flattened parallel to the baseline. May be due to an electronic range limit or a sample volume that exceeds the **sample capacity** or maximum **feed volume** of the column – more likely with capillary columns.
Fluoresce	To emit electromagnetic radiation.
Fluorescence	The emission of photons due to electrons returning to their ground state after being excited by heat, radiation, or chemical reaction. The emitted fluorescence may be in the infrared, visible, ultraviolet, or X-ray regions of the spectrum.
Fluorine F $M = 18.998$ g/mol	The chemical element with atomic number (Z) = 9. The lightest of the halogen group of elements.
Fluorine gas F_2 $M = 37.996$ g/mol	A yellow-green gas that is extremely reactive.
Force (F)	The physical influence that accelerates the motion of an object, defined as mass multiplied by acceleration. The SI unit of force is the newton (N) where: $N = \text{kg} \cdot \dfrac{\text{m}}{\text{s}^2}$
Forced integration	A method of chromatogram peak area integration. The integration proceeds continuously from the start command until the stop command. The accumulated integral includes all deviations of the baseline during that period.
Formaldehyde	See **methanal**.
Fourier transform	A mathematical procedure that converts a signal occurring in time to an equivalent set of sinusoidal waveforms that are easier to analyze.
FPD	See **flame photometric detector**.
Freon	One member of a family of fluorocarbon refrigerant gases.
Fronting peak	Sometimes called a *leading peak*. An asymmetric peak that falls more rapidly than it rises, typically due to a nonlinear adsorption isotherm or a **column overload**. The peak shape is due to an increase in solute affinity for the liquid phase at higher concentrations. In extreme cases, these peaks can look like right triangles, ramping up slowly and then falling rapidly to the baseline.
Fused peaks	Two or more partially separated chromatogram peaks, particularly those that exhibit no valley between them.

G

Galvanic current
An electric current produced solely by chemical reaction as in a dry cell battery. In process analyzers, the most common application for a galvanic cell is to measure oxygen at sub-ppm concentrations in a gas stream.

Gas chromatograph
The analytical instrument.

Gas chromatography (GC)
The technique of separating analytes prior to measuring them by injecting a vapor sample into a gas mobile phase that moves in contact with a liquid or solid stationary phase.

Gas compression factor (j)
A factor used to calculate the **average pressure** in a column, which allows for the expansion and increasing velocity of the carrier gas as it travels down the column.

Gas-liquid chromatography (GLC)
Gas chromatography using a liquid stationary phase.

Gas-solid chromatography (GSC)
Gas chromatography using a solid stationary phase.

Gate
See **peak gate**.

Gaussian peak
An ideal peak having the shape of a **normal distribution**, also known as the standard curve of error. When inside the column, peaks closely approximate to this shape. On the chromatogram, the peaks appear slightly skewed because the back of the peak stays in the column longer than the front, thus becoming a little wider.

General Elution Problem
The realization that any column to separate the light components would excessively retain the heavy ones. A consequence of the **doubling rule**, the General Elution Problem is solved by temperature programming.

Ghost peak
A small peak or set of peaks appearing in the next injection due to sample molecules adsorbing on surfaces in the sample injector. More common in laboratory chromatographs.

Grab sample
Also called a spot sample. A discrete sample of a process fluid contained in a balloon, cylinder, or bottle. Most grab sampling is manual and is subject to uncontrolled variations in procedure.

Gram mole
The mass in grams of one mole of molecules, numerically equal to their molar mass in atomic mass units.

Green oil
A polymer oil containing the dimer and tetramer of butadiene or other unsaturated C_4 hydrocarbons. Commonly occurs in gas samples containing C_4 olefins and acetylenes, particularly if they flow in heated lines. Tends to form more rapidly at temperatures above 60 °C.

Guard column
A column whose main function is to protect any following columns or devices from damage by reactive components in the sample. The guard column is backflushed to remove retained components so it also acts as a **housekeeping column**.
A common example is the use of a Porapak T column to protect a molecular sieve column.

H

Halogenated solvent
Excellent solvents capable of dissolving many compounds. They have a basic hydrocarbon structure but contain one or more halogen atoms in the place of hydrogen atoms.

Halogens	The group of similar elements that includes fluorine, bromine, chlorine, and iodine.
Hastelloy™	A nickel–molybdenum–chromium–tungsten alloy with excellent corrosion resistance in a wide range of severe environments.
Hazardous area	An area, or more correctly a volume, of space in a processing plant where a flammable atmosphere might exist.
Heartcut	A technique that enables the measurement of ppm-level analyte peaks following a large major component peak of high percentage concentration. Essentially, the technique concentrates the analytes by separating and removing most of the major component.
Heartcut column system	Uses an initial column to achieve a partial separation of the analytes while venting the bulk of the major component. A column switch then allows the analyte peak(s) and a small amount of the major component to enter a second column for final separation.
Heartcut method	Laboratory term for any kind of GC analysis that measures only selected analytes in a sample as opposed to a **comprehensive method**. Uses different techniques than those used in PGCs, including a **modulator** to transfer selected peaks to a second column. Not used in PGCs.
Heavies	Component peaks that elute from a column later than the analyte peaks. These are typically larger molecules with higher molar mass than the analytes, but not always so. Heavies are usually backflushed to vent or regrouped into a single peak for measurement.
Helium He $M = 4.0026$ g/mol	The chemical element with atomic number $(Z) = 2$. Also, a monatomic gas; the lightest of the six noble gases and extremely inert. Because of its safety and high thermal conductivity, it was the favorite carrier gas for PGCs in North America but is becoming scarce and expensive. PGCs are now more likely to use hydrogen carrier gas.
Helium ionization detector (HID)	A highly sensitive detector that can measure almost any analyte. A radionuclide ionizes the helium carrier gas, which then ionizes the analytes. Now obsolete in PGCs due to its instability and the odious regulations for operating a radioactive source.
Henry's law	At equilibrium, the amount of dissolved gas in a liquid is directly proportional to its partial pressure above the liquid.
Heptane C_7H_{16} nC_7' $M = 100.205$ g/mol	Also called n-heptane. A normal paraffin often measured in liquid process streams and in natural gas. The column system may group the heptane with heavies to give a composite C_6+ measurement.
HETP	Height equivalent to a theoretical plate. See **plate height**.
Hexane C_6H_{14} nC_6' $M = 86.178$ g/mol	Also called n-hexane. A normal paraffin often measured in liquid process streams and in natural gas. The column system may group the hexane with its many isomers and heavies to give a composite C_6+ measurement.
HID	See **helium ionization detector**.
Holdup time (t_M)	The time taken by the carrier gas to pass through a column, conveniently measured by the retention time of an **air peak**.
Homolog	A member of a **homologous series**.

Homologous series	A set of carbon compounds having identical molecular structure except for the length of the main carbon chain. A classic example is the paraffin series, whose members differ only by the incremental addition of a methylene group: $-CH_2-$.
Host computer	A computer, typically a process-control system, which receives analytical measurements from a PGC and may exercise some control over its operation.
Hot-wire detector	A thermal conductivity detector that uses heated wire elements, rather than thermistors.
Housekeeping column	A column that ensures the removal of all heavies before the PGC injects another sample. Typically, the first column in a backflush column system.
Housekeeping rule	A PGC column system must remove all components of an injected sample from the columns before it injects another sample.
Hydrocarbon C_xH_y	A chemical compound containing only carbon and hydrogen atoms.
Hydrogen H $M = 1.00794$ g/mol	The chemical element with atomic number $(Z) = 1$. The first element.
Hydrogen bond	The strong attractive force between two polar molecules, often but not always due to the weak **electronegativity** of a hydrogen atom.
Hydrogen gas H_2 $M = 2.01588$ g/mol	The lightest gas; transparent, odorless, and flammable. As a carrier gas, hydrogen is less expensive than helium. It generates a somewhat lower **plate number** than helium or nitrogen carrier gas but provides much faster separations.
Hydrogen chloride HCl $M = 36.46$ g/mol	A toxic gas that readily dissolves in water to form hydrochloric acid. When wet, it is highly corrosive. The molecule is covalent but highly polarized, so it tends to adsorb on surfaces.
Hydrogen cyanide HCN $M = 27.0253$ g/mol	A colorless and highly toxic liquid or gas with an almond odor. The liquid boils at 26 °C. The molecule is covalent and has the structure: $N\equiv C-H$.
Hydrogen fluoride HF $M = 20.006$ g/mol	A colorless and highly toxic liquid or gas. The liquid boils at 19.5 °C and is highly acidic. Dissolves in water to form the corrosive hydrofluoric acid.
Hydrogen sulfide H_2S $M = 1.00794$ g/mol	A colorless, flammable, and highly toxic gas with an odor of rotten eggs. Commonly occurs in samples of industrial gases and when at low ppm levels is often a designated analyte for PGC measurement.

I

Induced dipole	See **dipole, induced**.
Inert support	An inert solid like **Celite** that can hold a liquid phase immobile in a packed column. The liquid coats the large surface area of the support.
Infrared (IR)	The portion of the electromagnetic spectrum with wavelengths between 700 nm and 1 mm.
Injection profile	An imaginary plot of the concentration of the sample in the carrier gas as it leaves the injector. The ideal shape is often thought to be a square wave, rising instantly to 100 % concentration for a minimum duration, then rapidly returning to zero. Actually, a Gaussian profile would return the highest column efficiency.

Injector	A mechanism for injecting a precise volume of gas or liquid sample into the carrier gas and hence into the column, usually a mechanical valve.
Inorganic compound	Generally, a chemical compound that contains no carbon atoms, although we often classify the oxides of carbon, carbonates, and bicarbonates as inorganic.
Instrumentation symptoms	Symptoms of malfunction that upset a flat and smooth baseline when the PGC Method is not running and no valve operations occur.
Integration	The measurement of peak area.
Intercolumn detector (ITC)	A low-volume thermal conductivity detector placed between columns to allow visual indication of peaks transiting from one column into another and thereby to facilitate the setting of column switching times.
Internal standard	A selected component that is always present in the analysis. The calibration procedure uses an external standard to calibrate only that peak. The PGC then calculates the concentration of other peaks based on their response factors relative to the internal standard.
Intrinsically safe	An electrical device or circuit that is incapable of releasing enough energy to ignite an explosive atmosphere, under normal or abnormal conditions.
Ion	An atom or group of atoms that has gained or lost one or more electrons, becoming a negatively or positively charged entity.
Ionic liquid phase	An ionic and extremely polar phase that provides high stability and good peak shape when separating highly polar solutes.
Ionization	The process of becoming an ion.
Isobutane	See **2-methylpropane**.
Isobutene	Also called isobutylene. See **2-methylpropene**.
Isomers	Two or more compounds with the same chemical formula and the same molar mass, but having a different structure, leading to different physical properties and different chemical reactions. Hydrocarbon isomers having branched carbon chains tend to have a more compact molecule than their straight-chain homologs and therefore elute first on a gas–liquid column.
Isoparaffins	Branched chain isomers of the normal paraffins, of which there are many. When used to indicate a homologous series, typically refers to the 2-methyl isomers of the n-paraffins. For example, see **2-methylbutane**.
Isopentane	See **2-methylbutane**.
Isothermal analysis	A chromatographic separation performed at constant column temperature, followed by detection and measurement of the analyte concentrations.
ITC	See **intercolumn detector**.

J

Joule (J)	The SI derived unit of energy (E). An electrical definition: one joule is the energy dissipated as heat when an electric current of one ampere passes through a resistance of one ohm for one second.

K

Katharometer	Prior name for a thermal conductivity detector.
Keesom forces	The interactions between two permanent molecular **dipoles**.

Kelvin (K)	The SI base unit of temperature (T).
Ketone [—CO—]	A member of the ketone family of organic compounds, containing the [CO] functional group.
Kilogram (kg)	The SI base unit of mass (m).
Kinetic theory	The Bernoulli (1738) notion that gases consist of great numbers of molecules rapidly moving in random directions. This led to the realization that gas pressure is due to molecules colliding with the walls of the container and temperature is due to their average kinetic energy.
KISS	An acronym that emphasizes the relation between reliability and simplicity of design: *Keep It Simple, Stupid!*
Krypton Kr $M = 83.798$ g/mol	The chemical element with atomic number (Z) = 36. A monatomic gas, the fourth of six noble gases. Not used in PGC.

L

Laminar flow	A smooth kind of fluid flow in tubes that occurs at low flow velocity, in which all of the fluid is traveling parallel to the axis of the tube and no radial motion occurs. The velocity profile across the tube diameter is parabolic with the fluid at the center of the tube traveling at twice the average velocity and the fluid in contact with the tube walls hardly moving.
Langmuir isotherm	A simple equation representing the **adsorption isotherm** of an adsorbate on an adsorbent surface. The equation relates the partial pressure of the absorbate in the gas phase to its volume adsorbed at the surface at a specified constant temperature.
Leading peak	See **fronting peak**.
Limit of detection	The value in measurement units of the smallest peak that one can reliably distinguish from the background noise, often computed as twice the peak-to-peak noise level.
Liquid chromatography (LC)	The technique of separating analytes prior to measuring them by injecting a sample into a liquid mobile phase that flows over or through a solid stationary phase. A process liquid chromatograph was once available for sale but was not commercially successful. The technique is common in the laboratory for the analysis of food, drugs, and other complex mixtures, but rarely used online in the industrial process industries.
Liquid loading	In a packed column, the percentage by weight of liquid phase deposited on a solid support. In effect, a measure of liquid film thickness.
Liter	Also called a litre. A unit of volume equal to one-thousandth of a cubic meter.
Live Tee	A valveless method of column switching that controls the direction of carrier gas flow by manipulating the column pressures. A short capillary tube connects two columns. Electronic pressure regulators control the upstream or downstream pressure on the capillary, thereby dictating the direction of flow through the capillary. This flow reversal can be set up to perform backflush or heartcut functions.
	The main advantage of the Live Tee is its extremely low volume, making it suitable for use with capillary columns.
London force	See **dispersion force**.

M

Major component	The main constituent of an analyzed sample. Usually refers to a very large peak that tends to interfere with the analysis of ppm concentrations of other components in the sample.
Marker spikes	Small vertical upward or downward spikes impressed upon the chromatogram baseline by the processor to indicate event times such as valve actions, peak detection, and integration windows.
Mass-flow detector	A rate-sensitive detector that responds to the instantaneous population of analyte molecules present, thus effectively measuring their mass flow.
Mass percent	See **weight percent**.
Mass-sensing detector	See **mass-flow detector**.
Mass spectrometry (MS)	An analytical technique that ionizes the sample gas under vacuum and then accelerates the ions through a magnetic or electronic field, which defects their motion in proportion to their mass and electronic charge. Alternatively, the analyzer separates the ions according to their time-of-flight. The analyzer may detect specific ions by their mass-to-charge ratio or output a complete mass spectrum for further analysis.
McReynolds constants	A measure of liquid phase polarity based on the retention of five or more **probes**, standard solutes. One measures the **retention index** of each probe at 120 °C on the subject liquid phase and subtracts its retention index on **squalane** under the same conditions. Thus, each probe reveals different aspects of the liquid phase polarity.
Mean	The arithmetic average of a set of results.
Measurand	A general term used in metrology to identify the "quantity intended to be measured."
Measurement environment	The totality of variables, internal and external of the measuring device, that may affect and degrade the measurement in any way.
Measurement range	The span in measurement units from zero to the highest measurement limit. Usually denotes the current instrument setting rather than the maximum or minimum limit.
Measurement span	The span in measurement units from the lowest to the highest measurement limit. Usually denotes the current instrument setting rather than the maximum or minimum limit.
Megabore column	A capillary column with an internal diameter of 0.53 mm.
MEMS	Micro-ElectroMechanical Systems: refers to tiny integrated devices or systems that combine mechanical and electrical components. They are fabricated using integrated circuit batch processing techniques and can range in size from a few micrometers to millimeters.
Mercaptan [—SH]	A hydrocarbon derivative containing the —SH functional group. Used to odorize natural gas.
Mesh size	A measure of particle size based on the sieves used for grading the particles. For instance, 80–100 means the particles pass through an 80-mesh sieve and are blocked by a 100-mesh sieve. About 90 % of the granules in the product are within the mesh sizes of the two sieves. The US definition of mesh size is the number of openings in one square inch of a screen.

Meter (m)	Also called a metre. The SI base unit of length (L).
Methanal CH_2O $M = 30.026$ g/mol	Also called formaldehyde. The smallest aldehyde; a colorless, toxic gas with a pungent odor. Not detected by an FID, unless converted to methane by a methanator.
Methanator	A device that converts ppm amounts of carbon monoxide, carbon dioxide, or formaldehyde to methane, so a flame ionization detector can measure them. The column effluent mixed with hydrogen passes through a heated catalyst tube to convert the peaks to methane.
Methane CH_4 C_1 $M = 16.04$ g/mol	A hydrocarbon gas; initial member of the paraffin homologous series. Occurs on chromatograms of petroleum gases and comprises about 85 % of natural gas.
Methanol CH_3OH MeOH $M = 32.04$ g/mol	Also called methyl alcohol. The smallest member of the alcohol family of organic chemical compounds.
Method	In chemistry, a protocol for doing a chemical analysis. In PGC, a software tabulation of instructions specifying the timed events and calculations necessary to perform an analysis of one or more process streams. Method settings can be modified by the user so it's best practice to save a copy of the original file.
Methoxy group [—OCH_3]	An attachment to a long-chain molecule connecting an ethyl group via an oxygen atom. Found in many stationary phases, the methoxy group instills additional polarity to the phase. Not a separate chemical compound.
Methyl acetylene	See **propyne**.
Methylbenzene C_7H_8 $M = 92.141$ g/mol	Also called toluene. An aromatic compound having a single methyl group attached to a benzene ring. Often designated as an analyte for PGCs in aromatics plants.
2-Methylbutane C_5H_{12} iC_5' $M = 58.12$ g/mol	Also called isopentane. A paraffin having five carbon atoms with the second carbon atom connected to three other carbon atoms. An isomer of n-pentane.
Methyl chloride	See **chloromethane**.
Methylcyclopropane C_4H_8 $M = 56.108$ g/mol	An isomer of cyclobutane that often occurs in light hydrocarbon streams.
Methylene group [—CH_2—]	The repeating chain component in **paraffin** molecules and many polymers.
Methyl group [—CH_3]	A common feature of branched-chain hydrocarbon molecules. Not a separate chemical compound.
2-Methylpropane C_4H_{10} iC_4' $M = 58.12$ g/mol	Also called isobutane. The simplest branched chain hydrocarbon. Due to its more compact molecule, isobutane elutes before n-butane on all liquid columns. Frequently occurs on chromatograms of light hydrocarbon gases.
2-Methylpropene C_4H_8 iC_4'' $M = 56.106$ g/mol	Also called isobutene or isobutylene. The simplest branched chain **olefin**, and one of the four isomers of butene. Frequently occurs on chromatograms together with other light olefins.
Microliter (μL)	One millionth of a liter (=1 mm^3).
Micropacked column	A packed column <⅛-inch o.d., typically about 1 mm i.d.

Migration rate	The net rate of progress of a whole peak through the column. On a given column, the migration rate of a peak is inversely proportional to its retention factor.
Milliliter (mL)	One-thousandth of a liter (=1 cm^3). Often colloquially called a "cc".
Mixing chamber	A larger diameter space in the flow path where mixing can occur. When the carrier enters a wider tube, a peak experiences an increase in width in proportion to the square of the diameter change.
Mobile phase	For a PGC, this is the carrier gas. Common choices are hydrogen, helium, or nitrogen, but others are possible. The gas must be pure and dry.
Modulator	Generic laboratory term for any device that transfers component peaks from one capillary column to another. The terminology covers a host of resampling techniques including valves, pressure balance switching, peak slicers, and cold traps. Not used in PGC.
Molar weight	The force of gravity acting on one molar mass.
Mole (mol)	The SI base unit of quantity (n). A way to count molecules or other small entities. One mole is about 6.022×10^{23} of them.
Molecular sieves	A series of solid stationary phases capable of separating oxygen and nitrogen as well as other light gases such as hydrogen, methane, and carbon monoxide. Is prepared from a mineral known as a zeolite, which can selectively sort molecules based on their size, due to its very regular pore structure. The maximum size of molecule that can enter the pores of a zeolite depends upon the dimensions of the pores.
Molecular weight	See **molar mass**.
Molecule	The smallest amount of a chemical substance that can independently exist and still retain the properties of that substance.
Mole fraction	The fraction of the molecules in a sample that conforms to a specified identity or type.
Mole percent	The mole fraction expressed as a percentage. For a gas sample at low pressure, the mole percent and the volume percent are approximately equal.
Monel™	A group of nickel alloys, primarily composed of nickel (from 52 to 67 %) and copper, with small amounts of iron, manganese, carbon, and silicon. Monel alloys are resistant to corrosion by many agents, including rapidly flowing seawater.

N

Naphthalene $C_{10}H_8$ $M = 128.17$ g/mol	A dual-ring aromatic hydrocarbon with a strong mothball odor.
Naphthenes	An obsolete and confusing term for cyclic hydrocarbons, not related to the chemical substance naphthalene.
Near infrared (NIR)	The portion of the electromagnetic spectrum having wavelengths between 780 and 2500 nm.
Newton (N): $N = kg \cdot \dfrac{m}{s^2}$	The SI unit of **force**: the force required to give a mass of one kilogram an acceleration of one meter-per-second-per-second.

Glossary of terms

Nitrogen
N
$M = 14.007$ g/mol
: The chemical element with atomic number $(Z) = 7$, and a valency of 3.

Nitrogen gas
N_2
$M = 28.014$ g/mol
: The familiar, colorless, and odorless gas. Mostly inert. Sometimes used as a PGC carrier gas.

Nitro group
[—]
: An attachment to an organic molecule having a nitrogen atom bonded to two oxygen atoms. Not a separate chemical compound.

Nitropropane
$C_3H_7NO_2$
$M = 89.094$ g/mol
: Selected as the fourth **McReynolds probe**.

Neopentane
: See **2,2-dimethylpropane**.

Noble gas
: Generic name given to the elemental gases helium, neon, argon, krypton, xenon, and radon. All are chemically inert, occur in trace amounts in the atmosphere, and are often measured by gas chromatography.

Nonane
C_9H_{20} $nC_9{'}$
$M = 128.259$ g/mol
: A normal paraffin often measured in liquid process and natural gas streams. The column may group the nonane with its many isomers and heavies to give a composite C_9+ measurement.

Nonpolar
: Having an even distribution of electrons and no permanent dipole. However, a nonpolar substance might be polarizable by the influence of an external electromagnetic field.

Nonpolar column
: A column with a stationary phase that separates solutes solely by dispersion forces so the solutes elute in order of their boiling points. However, some nonpolar liquid phases can be polarized by and retain a polar solute.

Normal distribution
: A function that represents the distribution of many random variables, having a symmetrical bell-shaped graph.

Normalization
: A mathematical procedure to proportionally adjust all measured concentrations so they total 100 %. The technique is valid for percent-level measurements but may not work for ppm-level measurements. The PGC must measure all components of the sample that have a significant effect on the total – typically all those greater than 0.1 % by volume.

Normal peak
: Most chromatogram peaks are slightly asymmetric due to multiple processes that tend to cause a slightly slower return to baseline than the rise to apex. This asymmetry is considered normal and is more pronounced in wider peaks from packed columns than in narrow peaks from capillary columns.

O

Octane
C_8H_{18} nC_8
$M = 114.232$
: Also called n-octane. A straight-chain paraffin having eight carbon atoms in each molecule.

Octane number
: A measure of the quality of gasoline related to its tendency to pre-ignite under compression and "knock."
On the arbitrary octane number scale, 2,2,4-trimethylpentane has an octane number of zero and n-octane has an octane number of 100. Other values of octane number derive from simple binary mixtures of these standard liquids.

Olefin [>C=C<]
C_nH_{2n}
: A member of the **alkene** family of hydrocarbons having one or more double bonds between carbon atoms.

Open tubular column	A column that has no packing so there's an open path for gas flow. Such columns have little flow resistance, so they can be long and narrow, and thereby very powerful. The stationary phase is a thin film of liquid phase or a thin layer of solid particles attached to the wall.
Optical filter	A device that allows a selected band of photon frequencies to pass through but blocks any higher or lower frequencies.
Order of magnitude	To differ by a factor of ten; for example, a difference of three orders of magnitude is one thousandfold.
Organic compound	Any chemical compound containing carbon, except the carbon oxides, carbonates, bicarbonates, and sulfides.
Oxygen O $M = 15.999$ g/mol	The chemical element with atomic number (Z) = 8. Present in the molecular structure of many chemical compounds analyzed by PGCs.
Oxygen gas O_2 $M = 31.998$ g/mol	The familiar gas in the atmosphere and present in many gases analyzed by PGCs. Difficult to separate from argon by gas chromatography.

P

Packed column	A tube packed with small particles that may or may not have a thin film of liquid phase deposited on them. The original PGCs used packed columns and they are still popular today. Tube sizes have gradually changed from ¼-inch o.d. to $1/16$-inch o.d. The most common size found in a PGC is ⅛-inch o.d.
Paraffin [—C—] C_nH_{2n+2}	A member of the **alkane** family of hydrocarbons having only single bonds between carbon atoms.
Parallel chromatography	Two or more independent chromatographic separations occurring concurrently in the same analyzer, each using a separate sample injector, column system, and detector. Mostly, the several sample injectors inject the same process sample and the different column trains concurrently separate and measure different analytes, thereby reducing analysis time and complexity. Alternatively, the injections are phased to provide a more frequent analysis, or the injected samples are even from different process streams.
Paramagnetism	A form of magnetism induced in some materials by an external magnetic field. In analysis, the paramagnetism of oxygen is a prime example. The movement of oxygen when exposed to a magnetic field is the principle employed by many process oxygen analyzers.
Partial peak	An incomplete peak on the chromatogram that has lost some of its molecules, usually due to column switching.
Partial pressure	The portion of the absolute pressure of a gas that is due to a specific compound, often expressed in kilopascals (kPa). The total pressure of a mixture of gases is the sum of the partial pressures of its constituents.
Partition	An obsolete technical term now superseded by **distribution** in gas chromatography but retained to describe the similar process in liquid chromatography.
Partition isotherm	A graphical plot at specified temperature of analyte concentration in the liquid phase versus analyte concentration in the gas phase. Any deviation from a straight-line relationship results in an asymmetric peak shape.

Pascal (Pa)	The SI derived unit of **pressure** (*P*). A force of one newton acting on a surface of one square meter: Pa = N/m^2
Pascal-second (Pa · s)	The SI derived unit of dynamic viscosity (η or μ). To convert from centipoise (cP): 1 Pa · s = 1000 cP.
PDD	See **pulsed discharge detector**.
Peak	Literally, the shape of the detector response to an eluting component, when recorded on the chromatogram. Colloquially, a compact group of component molecules migrating through the column.
Peak dispersion	The broadening of a peak due to its molecules moving farther apart.
Peak gate	The time between two preset scheduled events bracketing the expected retention time of a single named peak or a group of peaks.
Peak, normal	See **normal peak**.
Peak, partial	See **partial peak**.
Peak picker	An analog pneumatic or electronic device used in early PGCs to capture the height of a peak so a computer could read its value. Alternatively, the peak picker would transfer the held value to a long-term memory to output a continuous trend signal.
Peak tail	A trail of component molecules following a peak that distorts the peak and may form an extended baseline offset declining exponentially. May be due to slow sample injection, non-linearities in the phase equilibria, adsorption of component molecules on surfaces, or entrapment of molecules at unpurged cavities in the walls of the flow path.
Peak width at base (w_b)	The width of a chromatogram peak in distance or time units, measured between the intersections of the triangulated peak sides with the extended baseline.
Peak width at half height (w_h)	The width of a chromatogram peak in distance or time units, measured parallel to the baseline at half the peak height.
PEEK	Polyether ether ketone. A thermoplastic polymer having good mechanical properties and chemical resistance at high temperatures.
Pentane C_5H_{12} nC_5' $M = 72.15$ g/mol	Also called n-pentane. A straight-chain paraffin having five carbon atoms in its molecule. Often seen on chromatograms of light hydrocarbons.
1-Pentanol $C_5H_{11}OH$ $M = 88.150$ g/mol	Also called n-pentyl alcohol. Used in the production of artificial flavors and as a solvent.
2-Pentanone $C_5H_{10}O$ $M = 86.13$ g/mol	Also called methyl propyl ketone. Used as a solvent. Selected as the third **McReynolds probe**.
Percent (%)	A concentration of one part in one hundred. The percent sign is numerically equal to 0.01.
Perfect resolution	Defined as a resolution of 1.5, which is just adequate for measuring equal symmetric peaks. More resolution is necessary for adjacent peaks of different size, particularly when the smaller peak elutes after the larger one.
Permanent dipole	See **dipole, permanent**.
Permeation	The act of gas molecules passing through an apparently solid material like the walls of polymer tubing. Don't use polymer tubing for ppm gas samples.

	May also refer to gas leaks at threaded or compression joints in stainless steel tube assemblies. As in **diffusion**, the rate of permeation is proportional to the partial pressure differential. Increasing the sample pressure will not stop its contamination by permeation of gases from the atmosphere or flare line.
Phase ratio (β)	The phase ratio is the ratio of gas-to-liquid volume in a column. In a capillary column, it's a function of the internal diameter of the tubing and the thickness of the liquid film. These two column variables are the equivalent of the **liquid loading** in a packed column.
Phenyl [—C_6H_5]	A functional group attached to an organic molecule, structurally a **benzene ring** that has lost one hydrogen.
Phosphorus P M = 30.974 g/mol	The chemical element with atomic number (Z) = 15. As a pure solid, it exists as two allotropic forms: red and white. Phosphorus is a highly reactive element and a very common component of minerals and living things, but not often encountered in PGC.
Photoionization detector (PID)	A photoionization detector employs ultraviolet photons from a discharge lamp to ionize the analyte molecules. The ions and electrons formed then migrate to the charged electrodes, as they do in an FID.
Photometry	The science and art of measuring analytes by their absorption or emission of electromagnetic radiation.
Photomultiplier	A highly sensitive electronic device to detect photons of light and generate a proportional electric current.
Photon adsorption	Most chemical compounds absorb photons of certain characteristic wavelengths, allowing measurement of their concentration by detectors tuned to those wavelengths.
Photon emission	Some chemical compounds absorb photon or heat energy and then emit photons of certain characteristic wavelengths, allowing measurement of their concentration by detectors tuned to those wavelengths.
Phthalate	An **ester** of phthalic acid: $HO_2C \cdot C_6H_4 \cdot CO_2H$
Physical property	A unique property of a molecule that can indicate its concentration in the sample, such as the paramagnetism of oxygen or the thermal conductivity of hydrogen.
Physical property analyzer	An instrument that measures a quality of the process sample other than its chemical composition. PGCs can calculate a few physical property measurements such as heating value or boiling point.
PID	See **photoionization detector**.
Piston valve	See **spool valve**.
Plate	A theoretical concept equivalent to the physical space within the column needed to create one equilibrium between the solute molecules in the gas phase and those in the liquid phase.
Plate height (H)	The length of column required to generate one plate.
Plate number (N)	The effective number of equilibria exhibited by a component peak.
Plate theory	The original theory (by analogy with liquid distillation columns) that separation was due to successive equilibria of solute molecules between a mobile gas phase and a stationary liquid phase. See **rate theory**.

PLOT column	The stationary phase in a porous-layer open-tubular column is a very thin layer of solid material coated on the inside wall of the tube to selectively adsorb sample molecules from the sample gas.
Plug flow	An idealized flow regime where all the molecules are moving at the same velocity, that is, at the same speed in the same direction.
Plug injection	An ideal sample **injection profile** having the minimum width.
Plunger valve	A valve designed to inject liquid samples. The sample volume is a hole or circumferential groove in a rod. Liquid sample fills the volume in a cold part of the valve. Actuation drives the rod into a hot zone typically within the column oven where the sample rapidly vaporizes.
Pneumatic detector	A PGC detector that uses a flow orifice and capillary tube in series to generate a pressure signal. The pressure signal is due to changes in gas viscosity and density when a component peak enters the detector.
Pneumatic signal (3–15 psig)	The standard pneumatic analog data-transmission signal for process control.
Poisson distribution	The probability distribution that closely models the outcome of the **theoretical plate** theory of peak formation in a chromatographic column. For very low plate numbers, the Poisson distribution predicts asymmetric peaks but these become symmetrical and Gaussian in shape at the higher plate numbers found in typical columns.
Polar	Any solute or stationary phase whose molecules have an uneven distribution of electrons within, creating an internal **dipole**.
Polar column	A column with a polar stationary phase that retains solutes in order of their boiling point plus additional retention due to their polarity.
Polarizability	The ability of an external force to disrupt the even electron distribution within a nonpolar molecule, thereby creating an internal **dipole**.
Polarization	Disruption of the even electron distribution within a nonpolar molecule by an external force, thereby creating an internal **dipole**.
Polyimide	A lightweight, heat-resistant, chemical-resistant, and flexible plastic used for electrical and thermal insulation and for the manufacture of molded parts.
Polymer	A substance having long-chain molecules of indeterminate length comprising repetitive structural elements due to the repeated attachment of monomer segments. The simplest polymer is polyethylene with the chain structure: $[-CH_2-CH_2-]_n$.
Porous polymer	A synthetic solid stationary phase. The original porous polymers developed by Hollis were small polystyrene beads with varying degrees of polymer crosslinking. Hollis, O.L. (1966). Separation of gaseous mixtures using porous polyaromatic polymer beads. *Analytical Chemistry* 1966382, 309–316 (February 1, 1966). 10.1021/ac60234a38
Pound mass (lb)	The American unit of mass or weight, now defined as exactly 0.45359237 kg.
Poundal (pdl)	An American unit of **force**: the force required to give a mass of one pound an acceleration of one foot-per-second-per-second.
Pound force (lbf)	An American unit of **force**: the gravitational force exerted on a mass of one pound on the surface of Earth.
Power	The rate of energy use in joules per second (J/s) or watts (W).

ppb	A concentration of one part in a billion (1×10^{-9}).
ppm	A concentration of one part in a million (1×10^{-6}).
ppt	A concentration of one part in a trillion (1×10^{-12}).
Precision	Also called repeatability. See **uncertainty interval**.
Pressure (P): $P = \dfrac{F}{A}$	Defined as the **force** (F) exerted by a fluid on a unit area (A) of the walls of its container, in newtons-per-square-meter. The **SI unit** of pressure is the pascal (Pa), where: Pa = N/m^2 The American unit of pressure is the pound-per-square-inch (psi).
Pressure, absolute	The total pressure exerted by a fluid.
Pressure balance switching	See **live tee**.
Pressure, gauge	The fluid pressure in excess of atmospheric pressure at the location and time of measurement.
Prime cause of everything	A troubleshooting mantra that recognizes valve actions as the proximate cause of every peak, spike, bump, or step in an otherwise flat and smooth baseline.
Prime rule	See **housekeeping rule**.
Priority interrupt	An older data-acquisition protocol that sends a signal to instruct a host computer to interrupt routine processing and read new data available at the PGC.
Process analyzer	Colloquially, any quality-measuring instrument (QMI).
Probe (to test liquid phases)	One of a set of standard analytes used to compare the polarity of liquid phases. The probes most often used to measure **McReynolds constants** are: benzene, 1-butanol, 2-pentanone, nitropropane, and pyridine.
Program	A sequence of timed events and calculations that control the operation of a process gas chromatograph.
Programmer	A device, usually electromechanical, that controlled the sequence of operations in early process gas chromatographs. Superseded by the microprocessor-based control unit.
Propadiene C_3H_4 C_3'''' $M = 40.065$ g/mol	Also called allene. The smallest **diene** with the carbon skeleton: C=C=C. Occurs on chromatograms of light olefin process streams.
Propane C_3H_8 C_3' $M = 44.097$ g/mol	A hydrocarbon gas. The third member of the paraffins homologous series, with the straight carbon skeleton: C—C—C. Frequently present in petroleum gases.
1-Propanol C_3H_7OH $M = 60.096$ g/mol	Also called n-propyl alcohol. Used as a solvent and disinfectant.
2-Propanone CH_3—CO—CH_3 $M = 58.08$ g/mol	Also called acetone or methyl-methyl-ketone. Smallest molecule in the ketone family. A flammable, volatile solvent used to clean tubing.
Propene C_3H_6 C_3'' $M = 42.081$ g/mol	Also called propylene. A hydrocarbon gas, the second member of the olefins homologous series with the straight carbon skeleton: C=C—C. An important monomer for producing polypropylene.
Propyl group [—C_3H_7]	An attachment to an organic molecule comprising three carbon atom and seven hydrogen atoms. Not an independent chemical compound.
Propylene	See **propene**.

Propyne C_3H_4 C_3''' $M = 40.064$ g/mol	Also called methyl acetylene. A reactive hydrocarbon gas, the second member of the acetylenes homologous series with the straight carbon skeleton: C≡C–C. Often a desired analyte in light olefin gases.
Pulsed discharge detector (PDD)	The PDD uses a pulsed electric discharge in helium, or helium doped with another noble gas, to raise helium atoms to an energetic state while emitting ultraviolet photons. The detector is popular in PGCs as it can mimic other ionization detectors without using a radionuclide.
Pyridine C_5H_5N $M = 79.102$ g/mol	An intermediate in the industrial production of herbicides. Selected as the fifth **McReynolds probe**.

Q

Qualitative analysis	To determine what chemical substances are present in a sample.
Quality-measuring instrument (QMI)	An instrument to measure a quality of a process fluid as distinct from its current condition. The category includes analytical instruments measuring composition and non-analytical instruments measuring a physical property of the fluid, but both types are "process analyzers."
Quantitative analysis	To determine the amount of selected chemical substances present in a sample.
Quantum cascade laser (QCL)	A wide-range tunable laser that produces a high optical power output at ambient temperature.

R

Radial diffusion	The spreading of peak molecules across the radius of the column tube due to the **laminar flow** profile of the carrier gas flow in the column. The band of component molecules in the fast-moving central flow is in contact with a lower concentration of molecules near the walls causing radial movement and band spreading.
Radical	In organic chemistry, a radical is an atom or molecule that has at least one unpaired valence electron, often due to the loss of a hydrogen atom. This enables the radical to attach to another atom or molecule.
Raman spectroscopy	A type of spectroscopy used in chemistry to identify molecules and study chemical bonding. The vibrational frequencies of a molecule are specific to its chemical structure and bonding. Industrial Raman spectrometers can identify and quantify the various types of molecules present in a sample.
Ramp (peak)	A term sometimes used for a **remnant peak**, particularly when it includes a slice from the rapidly declining tail of the major component, which gives the remnant a triangular ramp-like shape.
Ramp (temperature)	The gradual increase of column oven temperature during a temperature-programmed analysis.
Random error	Unpredictable variation in a measurement due to uncontrollable variations in environment or procedure. Limits the **precision** of a measurement.
Range	See **measurement range**.
Rate-of-arrival detector	A detector that outputs a signal proportional to the arrival rate of analyte molecules into the sensor. Often called a **mass-flow detector**.
Rate theory	A theory that disallows the formation of equilibria and sees peak width as a cumulation of multiple random effects, the statistical variances of which are additive.

Read the chromatogram	A mnemonic to remind troubleshooters that all chromatographic faults are discernable in the chromatogram record if they know where to look.
Reference chromatogram	A stored chromatogram that's trusted to represent perfect (or at least adequate) PGC performance. It's a good idea to save two reference chromatograms, one on calibration sample and one on process sample.
Reference peak	A selected peak that is always present on the chromatogram and used to predict the location of other peaks. The processor predicts the position of another analyte by using the fixed ratio between the retention times of analyte peak and reference peak.
Regrouping column system	A system that reverses the carrier gas flow in one of the columns to recombine a set of separated peaks. Generally, this term is not used for **backflush systems**. It's focused more on column systems that regroup a set of analytes (such as all the C_4's) for measurement as a composite peak.
Regrouping function	See **regrouping column system**.
Relative response factor (RRF)	The detector response factor of an analyte relative to a standard substance like benzene, measured under the same operating conditions. Retrieve RRFs from the literature. The ratio of two RRFs gives their relative sensitivity on the specified detector and may allow the calibration of an analyte without using an external standard.
Relative retention	The ratio of the adjusted retention times of two peaks on a chromatogram. Formally called their **separation factor**.
Remnant peak	Sometimes called a **ramp**. A distorted peak on the chromatogram due to a small slice of major component captured together with the analyte(s) during a heartcut valve action. The top of a remnant retains the shape of the major tail, but its vertical sides soften as it migrates through the second column, adopting the shape of half peaks.
Repeatability	See **precision**.
Resolution (R_s)	The degree to which the peak areas of two adjacent peaks are separate from each other, defined as the separation between the peaks divided by their average base width.
Response factor	A factor obtained during calibration by dividing the analyte peak area by its known concentration. To calculate analyte concentration, the analyst divides each peak area by its response factor. See **calibration factor**.
Response time	The total response time of a PGC is the sum of the delays in the sampling system plus the analysis time.
Retention factor (k)	The retention factor of a peak is the ratio of its adjusted retention time to the holdup time of the column.
Retention index (I)	A way to express the retention time of any solute relative to the retention times of two nearby straight-chain paraffins on the same column. The n-paraffins are assigned index values of 100n, where n is their number of carbon atoms. A solute with an index of 560 will elute between n-pentane and n-hexane. Since the index is a ratio of two peak retention times, it tends to be independent of the column operating conditions. Also called the Kováts Index.
Retention time (t_R)	The elapsed time to a component peak apex, measured from the instant of sample injection, being the average time that those component molecules spend in the

	column. Also, equal to the sum of the holdup time and the adjusted retention time of that component.
Retention volume (V_R)	The retention time of a peak expressed in volume units and equal to the volume of carrier gas needed to elute that peak from the column.
Root cause	The fundamental and original cause of a malfunction as distinct from the proximate and observable cause.
Rotary valve	A chromatographic valve, typically having two positions. When actuated, a rotor turns on a stator to make the desired connections.

S

Sample capacity	The maximum amount of an analyte that will produce an undistorted peak shape. It's usually measured with a solute of similar polarity as the liquid phase and might be a lot less for a dissimilar solute. Each analyte has its own sample capacity for a particular set of conditions.
Sample conditioning	The process of modifying the condition of the process sample to match the conditions required by the analyzer. For a PGC, it may include temperature and pressure control, removal of solid or liquid matter that might damage or contaminate the injector or columns, and maintaining a constant flow rate.
Sample flow	As applied to a PGC, the flow of sample fluid passing through the sample inject valve. Liquid flow is commonly set at 30 mL/min but might be as low as 10 mL/min. Gas flow is much higher, often 100–150 mL/min.
Sample injector	A means to inject reproducible samples of the process fluid into the carrier gas stream.
Sample loop	An external loop of tubing on a gas sample injector valve that provides the desired sample volume.
Sample splitter	A device for injecting extremely small samples into capillary columns. The carrier gas containing the injected sample is split into two flows, one entering the column and the other (containing most of the sample) going to vent.
Sample volume	The precisely dispensed volume of sample gas or liquid injected into a chromatographic column.
SCOT column	The stationary phase in a support-coated open-tubular column is a liquid coated on very fine support particles in a uniform layer on the inner wall of the tube.
Sebacate	An ester of sebacic acid: $HO_2C \cdot (CH_2)_8 \cdot CO_2H$
second (s)	The SI base unit of time (t).
Second retention mechanism	The notion that peak molecules are always subject to additional delay due to adsorption at active sites on contact surfaces. This is likely the main cause of the **normal peak** shape, slightly asymmetric.
Selectivity	The ability of a stationary phase to separate two solutes, indicated by their **separation factor**.
Self-sharpening peak	This is an asymmetric peak from a stationary phase with a severely curved adsorption or partition isotherm. Then, every elevation of the peak is traveling at a different speed, faster (or slower) than the base. By this mechanism, the top of the peak tends to become sharp with an almost vertical front (or rear). The other side forms an even ramp, so the peak is almost triangular.
Semi-diffusion	An arrangement of passages in a thermal conductivity detector that has the thermal elements partly in and partly out of the direct flow path of the carrier gas.

Separation (*S*)	The distance or time between the apexes of two adjacent peaks on a chromatogram. Separation does not indicate the degree of overlap between peaks as it doesn't account for peak width. See **resolution**.
Separation factor (*α*)	The ratio of adjusted retention times between a specified later peak and a specified earlier peak on an isothermal chromatogram. The ratio is usually constant on a given liquid phase and independent of column operating conditions, provided those conditions are constant during analysis.
Signal noise	Minor random variation in detector output signal caused by a myriad of small electronic, fluidic, or environmental instabilities that are uncontrollable.
Silica gel SiO_2	An active solid capable of adsorbing gas molecules, previously used as a solid stationary phase.
Silicon Si $M = 28.085$ g/mol	The chemical element with atomic number (Z) = 14. A nonmetal with chemical properties similar to carbon.
Siloxane	Chemically, an organic compound containing the Si—O functional group. Colloquial, for dimethylsiloxane that when polymerized yields poly(dimethylsiloxane), the least polar of the silicone oils.
Silicon coating	An effective way of minimizing the attraction between a polar solute and the tube wall. The silicon layer acts as a barrier to prevent the solute from touching the metal.
Silicone oil	A **siloxane** polymer. All silicone oils are polysiloxanes.
SI unit	An international standard unit of measure approved and certified by the Système international d'unités in France. All SI units are based upon seven invariant base units and are directly related to other SI units by the equations of physics. Therefore, they need no conversion factors.
Skew	Of a peak: see **asymmetry ratio**.
Sliced peak	A peak cut by column switching and having a piece missing. The vertical cut of the valve action may be smoothed by passing through a column, but it's often visible as a peak front rising faster than its back, or vice versa. Keep records of calibration factors to detect a newly sliced peak.
Slide valve	A two-position chromatographic valve that uses a sliding plate to switch the connections between ports. Slide valves may use an electric or pneumatic actuator and are functionally equivalent to rotary or diaphragm valves.
Slope detect	A software method of determining when a peak is about to emerge from a column by measuring the rate of change of a chromatogram signal. Typically, the processor recognizes the start of a peak when the rate of change exceeds a preset threshold value. A similar procedure determines when the peak has finished.
Solenoid valve	A two-position valve operated electromagnetically by an electric current in a wire coil. Solenoid valves for ac power have a copper ring that may be in contact with the fluid. Copper in contact with acetylenes can form an explosive acetylide, so don't use ac solenoid valves with hydrocarbon samples.
Solid support	A granular inert solid such as **Celite** on which is coated the liquid phase.
Solubility	As used here, solubility refers to the percentage of component molecules dissolved in the liquid phase at equilibrium. This usage of solubility incorporates the volume

	ratio of the two phases, so a change in phase volume ratio would also change the observed solubility. A more technical approach using the concentration of molecules in the phases is independent of phase ratio. See **distribution coefficient**.
Solute	The substance that dissolves in a solvent. Mostly synonymous with **component** in gas–liquid chromatography.
Solute polarity	The strength of a permanent or induced dipole within the solute molecules. The retention of the solute is a function of both solute- and liquid-phase polarities.
Solvation	The act of dissolving a solute in a solvent.
Solvent	In the theory of GC, this is the liquid stationary phase that dissolves the solutes. It should not be confused with the solvent used in laboratory GC to dissolve the sample before analysis.
Solvent peak	A large peak at the beginning of a laboratory chromatogram due to the solvent used to dissolve the sample before injection. Not applicable to PGC chromatograms.
Span	See **measurement span**.
Spatial separation	The distance separation of components within a column, for which the doubling rule does not apply.
Specific heat (c)	Also called the specific heat capacity. Simply stated, the specific heat of a substance is the amount of energy required to raise the temperature of a unit mass of the substance by one degree. The SI unit for specific heat is the joule-per-kelvin-per-kilogram J/(kg K). The specific heat of a substance may vary, sometimes substantially, depending on the starting temperature and pressure of the sample. For gases, the specific heat at constant pressure (c_P) is different than the specific heat at constant volume (c_V).
Spectrophotometer	An analytical instrument that splits the radiation absorbed or emitted by a sample into its spectral wavelengths and measures the intensity of certain characteristic wavelengths to determine the concentration of an analyte.
Spectroscopy	The science and art of measuring the intensity of emitted or absorbed radiation as a function of wavelength, particularly when applied to the identification and analytical measurement of chemical compounds.
Spike	An instantaneous deflection of the baseline with little observed duration. Due to fine solids in the detector, pressure perturbations in the detector vent, electromagnetic interference (radiated or carried in the power supply), or electronic malfunction. Spikes may occur at random or in sync with chromatogram events – when diagnosing it helps to know which.
Split ratio	When using a split injection technique, the split ratio is the amount of an injected sample that enters the column versus the amount vented. In practice, it's simply the ratio between the carrier gas flow rate into the column and the carrier gas flow rate exiting the splitter vent.
Split injection	See **sample splitter**.
Splitter	See **sample splitter**.
Spool valve	An early PGC valve, now obsolete, used for gas sample injection and column switching. A rod carrying O-rings restrained in grooves slides in a tube to make the desired connections.

Spot sample	See **grab sample**.
Squalane $C_{30}H_{62}$ $M = 422.826$ g/mol	A nonpolar liquid phase that acts as a standard for zero polarity. However, squalane is too volatile for most PGC applications and is often superseded by a nonpolar silicone oil. Chemical name: 2,6,10,15,19,23-hexamethyltetracosane.
Standard deviation (σ)	A measure of the random variation in a measurement equal to the square root of its **variance**.
Standard sample	A contained fluid with composition known to the desired level of uncertainty.
Start backflush	A timed event in the Method that initiates the backflush function.
Start next analysis	An event in the Method that results in sample injection. Ideally, it's time zero on the analysis clock, but some PGCs perform other functions first, so the actual injection occurs several seconds into the analysis cycle.
Stationary phase	The active solid or immobilized liquid in a column that is responsible for chromatographic separation.
Step	A sustained change in baseline level, usually returning to the original level before the end of the cycle. Caused by a valve action that initiates a flow change or feeds contaminated carrier gas into a column. A step occurring after a short delay is probably a flow change, while contamination will take more time to reach the detector. In both cases, the delay to recover after valve return is the same as the initial delay.
Step stream	An instruction in the program or Method to stop the current sample flow and start sample flow from the next process stream scheduled for analysis.
Stream switching	The manual or automatic practice of sequencing samples for analysis from two or more process streams.
Stripper column	See **backflush**.
Stuttering technique	Two or more short backflush operations designed to keep components longer in the first column. After other components have progressed along the second column, the stuttered components are allowed in. Then a full backflush occurs.
Succinate	An ester of succinic acid: $HO_2C \cdot (CH_2)_2 \cdot CO_2H$
Sulfur S $M = 32.06$ g/mol	Also called sulphur. The chemical element with atomic number (Z) = 16. A yellow solid that forms several stable allotropes.
Surface deactivation	The passivation of a surface (particularly of a metal) by chemical treatments including electropolishing and silicon treatments to minimize the adsorption of polar gas molecules.
Systematic error	A constant bias from the correct measurement value due to a zero or span error, or because of a nonlinearity in the response curve.
System constants	A proposed modification of the first five McReynolds constants where each constant includes a solute factor and a liquid phase factor. Ideally, these ten variables would derive the retention factor of any solute on any liquid phase. At present, only of academic interest.

T

Tailing peak	Has two similar meanings. For an asymmetric peak, it indicates the peak falls slower than it rises. All normal peaks have this distortion to a minor degree, so this usage of "tailing" is not encouraged. The more important usage is to describe a

	peak that doesn't return all the way to the baseline but instead trails long behind the main body of the peak.
TCD	See **thermal conductivity detector**.
Temperature programming	Operating a column on a preprogrammed temperature schedule comprising one or more periods of running at a constant set temperature, one or more periods of ramping at a set rate to a higher temperature, and a final period of rapid cooling.
Temporal separation	The time separation of components on a chromatogram and subject to the doubling rule.
Thermal conductivity	The ability of a material to transfer heat energy between two points when a temperature differential exists between them.
Thermal conductivity detector (TCD)	The original and still common detector in process gas chromatographs, capable of detecting any substance in the carrier gas. It detects the change in heat loss from a heated element when analyte molecules enter the detector.
Thermistor	A temperature-sensitive metal-oxide resistor, typically glass coated. Used as a detector element in some TCDs.
Theoretical plate	See **plate**.
Timed event	An event such as a valve actuation or peak gate that must occur at a preset time after sample injection.
Timer	Any device that enables the occurrence of timed events.
Titanium Ti $M = 47.867$ g/mol	The chemical element with atomic number $(Z) = 22$. A lightweight, strong, and corrosion-resistant metal.
Titrator	A wet-chemical analyzer that measures the volume of a standard reagent that reacts with the analyte in a fixed volume sample.
Toluene	See **methylbenzene**.
Trace analysis	The measurement of low concentrations. The term is mostly applied to the analysis of low ppm amounts of impurities in otherwise pure fluids.
Trap-and-hold column system	A PGC column system that traps selected peaks in a non-flowing column and releases them later for further separation.
Trend record	The historical record of a trend signal presented on a chart recorder trace.
Trend signal	A continuous measurement signal that changes after each new analysis to represent the latest analyte value.
Triangulating	A graphical method of characterizing a peak by drawing straight lines through the inflexion points at each side of the peak and extending them to intersect the projected baseline.
1,1,2-Trichloroethene	A halogenated solvent.
Trichloroethylene	See **1,1,2-trichloroethene**.
Trueness	The closeness of a measurement to the true value of the measurand. In standard terminology, accuracy is equal to trueness plus precision.
Tunable diode laser (TDL)	A sensitive photometric gas analysis technique. It's high sensitivity comes from using a powerful diode laser source tuned to exactly match the wavelength most strongly absorbed by the analyte.
Two-dimensional GC	A laboratory term for any separation achieved by two columns containing different stationary phases. Not used in PGC.

U

Ultraviolet — The portion of the electromagnetic spectrum with wavelength from 10 to 400 nm, shorter than that of visible light but longer than X-rays. Many process gas analyzers use ultraviolet absorption photometry, wherein the bonding electrons in analyte molecules strongly absorb ultraviolet energy at certain wavelengths that are characteristic of molecular structure.

Uncertainty interval — The range of measurement values believed to contain the true value of the measurand. The measurement must also state the **confidence level**.

Undecane
$C_{11}H_{24}$ nC_{11}
$M = 156.313$
— Also called n-undecane. A straight-chain paraffin having 11 carbon atoms in each molecule.

Unretained peak — See **air peak**.

Utilization of theoretical efficiency (UTE) — A measure of column efficiency used mainly with capillary columns. Expresses their measured plate number as a percentage of their theoretical maximum, a useful indicator of column condition. A good column will typically have a UTE of about 60–70 % when running at its optimum flow rate.

V

Vacancy peak — A negative peak appearing on the chromatogram at the retention time of a known component, due to the carrier gas containing more of that component than the injected sample does.

Valency — The number of chemical bonds exhibited by an element. For instance, carbon has a valency of 4.

Validation — A procedure for checking the calibration of an instrument by periodically measuring a sample of constant composition. Validation doesn't change the calibration of the instrument. The PGC compares the validation results statistically with a large number of previous validations to evaluate the need for recalibration or maintenance attention.

Validation fluid — The validation sample must be constant, so a large volume of fluid is necessary. However, the PGC doesn't need to know the concentration of the analytes because it will measure them. Sometimes the validation sample is simply a large cylinder filled with the process fluid.

Validation frequency — 10–20 prior validations are necessary for statistical analysis, so the PGC should automatically validate at least once daily to build up a collection of historical values.

Validation program — A series of regular analyses observations to gain confidence in the performance of a PGC. A program can adopt multiple techniques, including observing process data, statistical quality control, and comparison with other analyzers.

Valley point — The lowest point between adjacent peaks on the chromatogram.

Valve — Most process gas chromatographs use special chromatographic valves. They route the flow of carrier gas and component peaks into a selected column, detector, or vent.

Valve timing — The actuation and deactuation times of chromatographic valves are timed events specified in the analysis Method. To achieve the desired resolution of analyte peaks, most column systems rely on highly repeatable valve timing.

Van der Waals force — A generic term for intermolecular forces; includes dispersion, polar, and ionic forces.

Vapor pressure (of liquid)	The vapor pressure of a liquid is the equilibrium partial pressure of its vapor above the surface of that liquid. Liquids boil when their vapor pressure becomes equal to the surrounding pressure. Under that condition, the gas phase above the liquid is all vapor from the liquid.
Variable	A measurable property of an object or phenomenon that may change. Metrology is the science of evaluating a **continuous variable**.
Variance (σ^2)	A measure of the variability of a measurement due to random error. Estimated from a set of results by dividing the sum of the squares of the deviations from the mean value by the number of results -1.
Velocity (u) of carrier gas (m/s)	The velocity of the carrier gas at the column exit as calculated by dividing the flow rate by the free cross-sectional area of the column. See also, the **average velocity** of the carrier gas calculated by dividing the column length in meters by the **holdup time** in seconds.
Viscosity – dynamic (μ or η) (Pa s)	A measure of a fluid's resistance to flow due to internal friction within the moving fluid. A high-viscosity fluid resists motion because it has a lot of internal friction. The SI unit of viscosity is the pascal-second (Pa s). The more common unit is the centipoise (cP), equal to the millipascal-second (mPa s).
Viscosity – kinematic (ν) (m²/s)	The ratio of the viscosity of a fluid to its density. The SI unit of kinematic viscosity is the square-meter-per-second (m²/s) but rarely used. The common unit is the stokes (St) or centistokes (cSt) where: 1 St = 1 cm²/s
Voltage-to-frequency converter	A device that converts an analog voltage signal to a proportional frequency signal. As used in chromatographs, the output frequency was a square wave so the peak integrator could count the pulses.
Volume (V)	The size of a three-dimensional space, measured in cubic-meters (m³) or cubic-feet (ft³).
Volume percent	A unit of concentration: the ratio of analyte volume to total volume of a sample, expressed as a percentage.

W

watt (W)	The SI derived unit of power (P). The definition of one watt is a power dissipation of one joule-per-second (J/s).
WCOT column	A wall-coated open-tubular column comprising a liquid stationary phase coated onto the walls of capillary tube often made of fused silica. While highly effective, the fused silica is fragile, so process gas chromatographs prefer metal capillary columns with silicon-treated internal walls.
Weight percent	A unit of concentration: the ratio of analyte weight to total weight of a sample, expressed as a percentage. Effectively equal to mass percent.
Wet chemistry	Methods of chemical analysis involving chemical reactions in aqueous or nonaqueous solution.
Wheatstone bridge	An electrical circuit for detecting a small change of resistance. Has four resistors connected in a series-parallel arrangement. When the resistances are equal, the voltages at the junction of each series pair are equal giving a differential of zero millivolts. If one or more of the resistances changes, the differential voltage rapidly varies from zero giving a sensitive detection of resistance change.
Width of peak	See **peak width at base** or **peak width at half height**.

X

Xenon
  Xe
  $M = 131.293$ g/mol
: The chemical element with atomic number (Z) = 54. A heavy, colorless, odorless, and monatomic gas, the fifth of six noble gases. Not used in PGC.

X-ray fluorescence (XRF)
: Exposing an element to high-energy X-rays or gamma rays may cause the atoms to **fluoresce**, emitting X-rays at wavelengths that are characteristic of that atom. For each element measured, a dedicated process instrument would measure the incident fluorescent energy at one selected wavelength. However, a laboratory instrument might scan a range of X-ray wavelengths and output a spectrum of peaks representing the concentration of many different elements.

m-Xylene
: See **1,3-dimethylbenzene**.

o-Xylene
: See **1,2-dimethylbenzene**.

p-Xylene
: See **1,4-dimethylbenzene**.

Z

Zeolite
: See **molecular sieves**.

Zero line
: As used herein, the zero line is the baseline for integration. The integrated peak area comprises the total area above the zero line. Thus, setting up an appropriate zero line is an important prerequisite to accurate measurement.

Index

Absolute pressure, *see* Pressure
Accuracy   180, 298–302, 319, 322
Acetylenes   341
Acid washed   50
Adipates   93
Adjusted retention time, *see* Retention
Adjusted retention volume, *see* Retention
Adsorbent solid, *see* Solid phases
Adsorption
  columns   103–109
  isotherm   104
  sites   206, 218, 223
Affinity   3, 5, 42–48, 49, 50, 62, 68–70, 74, 83, 85, 89, 91, 98–100, 104, 108, 127, 209, *see also* Mutual affinity
Agreed reference value (ARV)   302
Air peak   17–20, 26, 30, 129, 131, 133, 136–139, 150, 151, 164, 339, 350, *see also* Holdup time
Alpha (α), *see* Separation factor
Alarms   8, 301–302, 308
Analysis time   30, 46, 54–55, 62–67, 120, 128–142, 152, 159–173, 191, 230–231, 251, 260–261, 266, 290, 291, *see also* Cycle time
Analyte   2–10, 27, 49, 67–73, 103–105, 119, 181–185, 214–215
Analytical difficulty   283
Analyzer management system   302
Analyzer shelter   8
Apolane-87   100–101
Applet   224, 349
Argon
  as analyte   47, 84, 106–107, 109
  as carrier gas   156, 159, 349
Asymmetric peaks   205–225, *see also* Peak shape
  SCI-FILE   212–214
Asymmetry ratio   206, 214
Availability, *see* Reliability

Average pressure, *see* Carrier gas pressure
Average velocity, *see* Carrier gas velocity
Backflush
  column design   260–262
  diagnosis   350–351
  event timing   256–257, 262–269
  flow rate   251–252, 259–269
  full details   250–273
  with heartcut   284, 285, 286, 288
  housekeeping   231, 234, 244
  limitations   106, 259–260
  major tail cut   275
  to measure   252–254
  regrouping   252–253, 256, 257–260, 268–269
  settings   256–257
  stuttering   257
  theory   259–262
  valveless   254–256
  valve systems   251, 253, 258
  to vent   250, 251–252
Band   23–27, 155–156, 173, 208, 223
Bara and barg   240
Baseline
  anomaly   306, 308–310
  correction   4, 319
  disturbance   311, 322, 331–332, 341, 354
  flat and smooth   304–310, 359
Baseline symptoms
  cycling   310–311, 317–318, 323
  drift   310–315, 317, 323
  noise   56, 60, 63, 71–74, 92–93, 180, 185, 189, 195, 299–300, 305, 307, 310–311, 319–323, 340, 344
  wander   310–311, 315–316, 323
Base width, *see* Peak width
Bell curve, *see* Gaussian distribution

Process Gas Chromatography: Advanced Design and Troubleshooting, First Edition. Tony Waters.
© 2025 John Wiley & Sons Ltd. Published 2025 by John Wiley & Sons Ltd.

Index

Beta (β), *see* Phase ratio
Bladder 286
Boiling point
  elution order 44–46, 85–85, 230, 262
  sample injection 217
Bonded liquid phase, *see* Liquid phase
Boyle's law 43
Bumps on baseline 312, 333–334, 338, 352–355

Calibration, *see also* Validation
  alarm 299, 301, 341
  check 342–343
  error 11, 209, 254, 303, 330, 345, 349, 359
  frequency 7, 302
  inappropriate 297–298
  sample 302, 332, 345
Capacity factor, *see* Retention factor
Capillary columns
  comparison 64–74, 82–83, 92, 179, 232–235
  description 3, 6, 39–40, 41–42, 57–62, 96–97
  fused silica 58
  metal 59, 106
  performance 103, 134, 182, 192, 214, 218, 237, 312, 346–348
Carbon, *see* Solid phases
Carrier gas
  argon 156, 159, 349
  choice 171–172
  helium 2, 57, 59, 63, 66, 158, 159, 171–172, 207, 343, 344, 349
  hydrogen 2, 56, 57, 63, 158, 159, 171–172, 207, 225, 336–337, 341, 343, 344, 349
  mixed 349
  nitrogen 2, 63, 156, 159, 171, 349, 352
Carrier gas flow
  diagnosis 332, 340, 344, 349, 351–353, 356, 358
  measurement 126
  optimum 63, 149, 150–153, 162
  path 187–188, 192, 206, 237
  settings 63, 251, 259–260
Carrier gas pressure
  calculations 240–244, 262–269
  drop 52–54, 74, 127, 154, 156, 159, 171, 188, 237
  effect 260, 193, 243
  SCI-FILE 238–240
Carrier gas purifier 47
Carrier gas velocity

  effect 17, 134, 137, 139, 192, 237, 259, 260, 264
  calculations 19, 150–172, 238–242
  optimum 63, 70, 150–153
Celite, *see* Diatomaceous earth
Chart record 16, 17, 41, 221, 336
Chemical Names, *see in* Glossary 379–416
Chromatogram
  diagnosis 304, 329–364, 335
  display 4, 6, 26
  measurements 16–19, 24
Chromatogram diagnosis
  bumps 334, 352–353
  peaks 334, 338–352
  SCI-FILE 335–337
  spikes 334, 352–356
  steps 334, 356–358
Chromatographs, *see* Gas chromatographs
Chromatography, *see* Gas chromatography
Chromosorb 50
Column
  care 106
  codes 39, 83, 94
  conditioning 51, 54, 92, 313
  identification 83
  life 51, 64, 73, 92, 134, 351
  train, *see* Applet
Column names
  cutter 277
  guard 250
  vendor codes 39, 83, 89, 94
Column parameters
  diameter 40, 52, 54, 56, 59–70, 127, 135, 158–159, 171, 184, 239, 269
  flow rate, *see* Carrier gas flow
  length 24–27, 29–30, 32, 54, 56, 74, 99, 124–125, 133, 139, 142, 149–151, 154, 160–170, 237, 240–243, 258, 260–269, 278
  pressure, *see* Carrier gas pressure
  temperature 69, 92, 96, 100, 134–141, 157–158, 173, 214, 225, 312, 314, 331, 351
Column performance
  bleed 51, 60, 63, 64, 71, 72–74, 86–87, 90, 92, 134, 256, 309, 312, 314, 317, 340, 357
  efficiency 25, 28, 30–32, 52–53, 62–71, 82, 83, 88, 96, 121, 134, 137, 139, 144, 150–160, 180, 186, 213–214, 237, *see also* Golay and van Deemter

overload, see Feed volume and Sample capacity
time efficiency   170
Columns in series Chapters   9–11, *passim*
Column system design
    fail safe   252
    pressure effects, *see* Carrier gas pressure
Column tubing
    cleaning   52
    electropolishing   52
    fused silica   42, 52, 58–59, 74, 100, 234, 342
    glass   41, 52, 54, 57–59
    Hastelloy™   52
    silicon coating   52, 189, 235
    Teflon™   42, 52
Column types
    capillary   3, 6, 40, 41–42, 57–64, 64–74, 92, 96, 106, 124, 127, 134, 179, 321, 432, 438
    micropacked   41–42, 52, 54, 56–57, 66, 74, 98, 179, 235, 254, 277
    open tubular (WCOT)   41–42, 59–62
    packed   3, 32, 40–41, 49–56, 64–74, 127, 155, 157, 184, 224, 234–235, 314, 317
    porous layer (PLOT)   41, 48, 60, 106–109, 224, 322, 334
    support coated (SCOT)   60–61
Component   6, 85, 207, 342, 347
Comprehensive method (as in Laboratory GC)   234, 276
Compressibility factor   182, 241, 265, 374
Concentration detectors, *see* Detector types
Control chart   301–302
Copolymer   60, 88–94
Correlogram   233
Credibility   298
Critical pair   83, 120
Cutter column, *see* Column names
Cyano silicones, *see* Liquid phases
Cycle time   159, 230, 261–262, 266, 268, 336, *see also* analysis time
Cycling baseline, *see* Baseline symptoms

Dead legs   158, 206, 217, 223, *see also* Peak dispersion
Dead time, *see* Holdup time
Deans switch, *see* Live Tee
Debye forces, *see* Forces
Desorption   209–211, 212, 214, 223, 309
Detector
    damage   73

dispersion   189, 194, 196
function   3–8, 41, 52, 65, 72, 73, 179, 189, 192, 194–195, 217, 223, 225, 235, 304–310, 310–323
intercolumn (ITC)   256, 284
malfunction   226, 303, 307–310
response time   195–196
sensitivity   61, 71, 72, 73, 83, 92, 105, 159, 184, 185
troubleshooting   304–323, *see also* Baseline symptoms
volume   65, 158, 173, 179, 180, 195
Detectors
    electron capture (ECD)   232, 274
    flame ionization (FID)   131, 151, 189, 194, 195, 234, 274, 275, 307, 315, 317, 318, 321, 340
    flame photometric (FPD)   274, 309, 340
    mass spectrometric (MSD)   6, 90, 234
    thermal conductivity (TCD)   151, 189, 194–195, 274, 307, 309, 315, 317, 321, 340, 349, 355, 356
    vacuum ultraviolet (VUV)   234
Detector types
    concentration   189, 194, 307
    mass-flow   189, 194, 223, 307
Deuterium   47
Diagnosis, *see also* Troubleshooting
    SCI-FILE: On Diagnosis   335–337
Diatomaceous earth (Celite)   43, 49–50. 60
Diffusion, *see also* extracolumn broadening
    axial (longitudinal)   156, 192–196, 290, 357
    coefficient   156, 158, 159, 194
    eddy   155
    in gas phase   157–158, 159
    leak   314
    in liquid phase   156–157, 159, 173
    radial   158, 192, 193–196
Dimethyl silicone   89, 96, 97, 100
Dipole, *see* Forces
Dispersive forces, *see* Forces
Distribution
    constant   5–6, 103–104, 125, 128–131, 133, 181, 208–212, 216, 240
    definition   126
    SCI-FILE   125–128
Distribution column system   290–291
Doubling rule   63, 281, 334, 338–339
Drift, *see* Baseline symptoms
Dual column system   232, 234–235, 290, 345
Dynamic equilibrium, *see* Equilibrium

Eddy diffusion   155
Effective plate number   28
Electronegativity   44
Electron capture detector (ECD), *see* Detectors
Electronic pressure controller (EPC)   244, 252, 254, 277–278, 291, 314
Electronic pressure regulator (EPR), *see* EPC
Electropolishing   52
Equilibrium   5, 23, 25, 32, 104, 126–127, 154–157, 163, 208, 240
Event times, *see* Timed events
Exponentially modified Gaussian (EMG)   215
Extracolumn
   broadening   179–203 *passim*
   SCI-FILE   190–196
   variance   158, 190, 195
   volume   126, 180, 187, 254, 277
   worked examples   196–198

Failure   39, 252, 259, 297–299, 303–310
Fast analysis   54, 64, 67, 74, 83, 86, 159, 168, 170–173
Feed volume   67–68, 71–72, 183–187, *see also* Sample capacity
Film thickness   51, 60–62, 66, 70, 80, 127, 135, 137, 151, 157, 159, 172, 181, 223, 348
Flame ionization detector (FID), *see* Detectors
Flame photometric detector (FPD), *see* Detectors
Flow, *see* Carrier gas flow
Fluoro silicones   84, 90–91, 97, 111
Forces
   dispersion (London)   43–45
   generic (Van der Waals)   45
   induced dipole (Debye)   45
   ionic   45
   permanent dipole (Keesom)   44–45
Frontal chromatography   357
Fronting peak, *see* Peak shape
Fused silica, *see* Column tubing

Gas chromatographs
   comparing process and laboratory   235
   development of   232–235
   laboratory   6, 41, 82, 87, 232, 326, 332
   process   5–6, 8, 9, 87, 88, 119–120, 126, 154, 156, 158, 230, 233, 250, 332
Gas chromatography
   gas-liquid (GLC)   15, 49, 103

   gas-solid (GSC)   46, 103–104, 214
Gas compressibility, *see* Compressibility factor
Gaussian distribution   24
General elution problem   231
Golay   41, 155, 158, 192
Guard column, *see* Column types

HayeSep   84, 108, 109
Heartcut   274–296
   column system   276–279
   event timing   281, 284–286
   function   55, 219, 222–224, 232, 236, 244, 256, 274–276
   in laboratory work   234
   by Live Tee   277–278
   multiple cuts   282–284
   single cut   279–28
   by valve   276–277
Heat of solution or adsorption   213, 222
Heavies
   definition   86
   measure   252–254
   regroup peak   224, 252, 256, 259, 269, 313
   removal   48, 86, 119, 120, 231, 251, 252, 260–261, 262–269, 290–291
Height equivalent to a theoretical plate (HETP), *see* Plate height
Helium
   as analyte   47
   as carrier   2, 57, 59, 63, 66, 158, 159, 171–172, 207, 343, 344, 349
   purge   54
Holdup
   time   16–20, 28, 129, 131, 134, 137, 150–153, 154, 164, 173, 334–225, 339, 350, 357, *see also* Air peak
   volume   126–127
Homolog   338–339
Housekeeping, *see also* Backflush
   column   8, 55, 250, 291
   rule   231
Hydrogen
   as analyte   5, 46, 47, 51, 84, 107, 195, 348
   anomalous response   349
   bonding   45, 83, 84, 85–86, 93, 95
   as carrier   2, 56, 57, 63, 158, 159, 171–172, 336
   leak   335–337
   makeup gas   194
   reaction   207, 225, 341, 343, 344, 348

Index

Induced dipole, *see* Forces
Inflection points   24–25
Instrumentation symptoms, *see* Troubleshooting baselines
Integration
   error   195, 216, 299, 303, 305, 319, 320, 322, 330, 341
   forced   224
Intercolumn detector, *see* Detector
Injection
   capillary columns   65
   plug   180, 184–187, 191, 194, 196–197
   profile   26, 67, 68, 71, 158, 180–181, 190, *see also* Feed volume
   sample   2, 3
   split   68
   time mark   3, 16, 18, 19
   volume   70–71, 72, 74, 185–187
Ionic liquid phase, *see* Liquid phase
Isotherm
   adsorption   104–106, 212
   linear   182
   nonlinear   104–105, 181, 210, 216, 222, 223, 224
   partition   181, 182, 208–209

Keesom forces, *see* Forces
Kinetic theory of gases   43
KISS principle   283
Kováts Index, *see* Retention Index
Krypton   47

Laminar flow   158, 188, 192–193
Langmuir isotherm   104
Leading peak   26, 206–207, 211, *see also* Peak shape (fronting)
Lexicon of laboratory GC Methods   234
Limit of detection (LOD)   319
Liquid chromatography   6
Liquid loading   50, 51, 61–65, 83, 99, 127, 181, 185, 258, *see also* Phase ratio
Liquid phase
   bonded   3, 60, 64, 71, 73, 74, 90, 92, 232
   ionic   45, 84, 94
   polarity   43–46, 50, 59, 60, 70, 83–101, 108, 109, 218, 223, 224, 257, 260–262, 281, 341, 347
   selection   65, 67, 82, 85, 96, 100, 236
   selectivity   20, 82, 83, 89, 90, 91, 93, 95, 96, 100–101, 107–109, 128
   temperature limits   92, 93, 107
   vendor codes, *see* Column codes
   viscosity   157, 159, 171–173
Liquid phase performance
   bleed, *see* Column bleed
   optimization   16, 51, 123, 128–142
Liquid phases
   Carbowax™   94, 95, 97, 101, 107
   DEGS   93, 94, 97, 101
   dialkyl esters   92–93, 111
   FFAP   84, 94, 95
   making a choice   96–103
   nitrile ethers   95
   non-silicone   92–95
   PEG   94, 95
   polyesters   93, 94
   polyethers (polyols)   45, 95, 97
   polysiloxanes (silicone oils)   45, 60, 85, 87–92, 95–97, 101, 107
   silicone, cyano   84, 85, 90, 91, 94, 95, 97, 101
   silicone, fluoro   84, 90–91, 97
   silicone, methyl   60, 84, 87–90
   silicone, phenyl   84, 88–90
   squalane   85, 87, 89, 90, 94, 100–101
   TCEP   84, 85, 90, 91, 94, 95, 100–101, 107
Live tee   254–256, 277–278, 353
London forces, *see* Forces
Longitudinal diffusion, *see* Diffusion

Major component   68, 185, 222, 223, 232, 256, 273–291
Marker spikes   281, 332, 333
Mass-flow detectors, *see* Detector types
McReynolds constants   99–103, 108
McReynolds probes   100–101, 108
Mean   24–25
Measurement, *see also* Calibration
   accuracy   180, 298–299, 301–302, 319, 322
   confidence   298–299, 302–303
   precision (repeatability)   74, 180, 216, 298–302, 319–320, 241
   range   64, 71, 74, 284, 319
Megabore   54, 59, 66–67, 74, 182
Mesh size   52–54, 83, 269
Methane
   as air peak   131, 151
   as analyte   84, 107, 336
   as major   275, 279, 288

Method   48, 225, 262, 285, 306, 314, 332, 334–335, 346, 349,
    350, 352, 353, 355, 358
Micropacked columns, *see* Column types
Microprocessor   8, 234, 290
Migration rate   104, 154–155, 176, 211, 224, 266, 268
Mobile phase   2–3, 19, 23, 25, 158, 240, *see also* Carrier gas
Modulator   233
Molecular sieve   46, 47, 60, 64, 106, 107, 232, 250
Mutual affinity   42–43, 45
    SCI-FILE   43–45

Negative peaks, *see* Peak shape
Neon   47
Nitrogen
    as air peak   131, 151, 156, 335–337
    analyte   5, 46, 47, 84, 106, 107
    as carrier, *see* Carrier gas
    calibration mixture   256, 285, 345
    oxides   51, 107, 108, 109
    purge   54, 60, 340
Noise, *see* Baseline symptoms
Noise reduction   189, 195, 320
Nonpolar, *see* Liquid phase polarity
Normal curve, *see* Peak shape Gaussian
Normal distribution, *see* Peak shape Gaussian
Normal peak, *see* Peak shape
Number of theoretical plates, *see* Plate number

Obstruction factor   156, 159
Olefins   47, 84, 91, 93, 100, 107, 172, 341, 347
Open-tubular column, *see* Column types
Optimize
    analysis time   63, 159, 170–173
    column   10, 16, 19, 30, 63, 123, 149–173
    k-value   123, 129–131, 133, 134–142, 225, 159–170, 225
    SCI-FILE: On Rate Theory   153–159
    SCI-FILE: On Retention Factor   129–131
Optimum flow rate, *see* Carrier gas flow
Overlapping peaks, *see* Peak shape
Oxygen
    as analyte   5, 46, 47, 84, 106, 107, 109
    atoms   87, 89, 91, 93, 95
    in carrier gas   47, 48, 86, 92
    reactions   73

Packed columns, *see also* Column types
    factor   155–156, 159, 192, 239

packing   50, 54, 57, 106–108, 158, 243
Paraffins   4, 21, 43–45, 84, 86, 91, 95, 98–101, 339
Parallel chromatography   58, 120, 244, 257, 289, 290, 340
Parallel columns   231, 235
Partial peak, *see* Peak shape
Partition   181–182, 207–210, 214
    coefficient, *see* Distribution constant
    isotherm, *see* Isotherm
Parts-per-million (ppm)   51, 52, 62, 72, 219, 222, 223,
    274–291
Peak area measurement   4, 195, 209, 210, 214–216, 257,
    289, 305, 319
Peak height measurement   195, 206, 224, 257, 305, 319,
    340, 342, 348
Peak shape
    asymmetric (skewed)   24, 49, 56, 105, 126, 204–229
    bimodal (split)   207, 224, 347
    flat top   181, 224
    fronting (leading)   68, 105, 181, 205–211, 224
    Gaussian   22, 24, 103, 154, 186, 191, 204, 207, 212–214,
        218, 305, 334, 341, 349
    merged (overlapping)   218–224, 233, 240, 341, 349, 351
    negative   346, 349
    normal (real)   205, 207, 208, 211, 220
    offscale   216–218, 221, 223, 275, 288, 340
    partial   222, 224, 261
    remnant   224, 276–288, 334, 344, 351, 357
    self-sharpening   209, 215
    tailing   60, 62, 68, 74, 105, 181, 194, 204–205, 209,
        216–218, 221–223, 275, 331, 340, 341, 346–347
    triangular   288, 305, 341
Peak width
    at base   16, 17, 21, 24–29, 31, 121, 136–141, 183, 190,
        195–197, 212–218, 263–264, 305
    at half-height   17, 21–24, 31
Performance data   97, 125, 166, 298, 301
Performance triangle   63
Permanent dipole, *see* Forces
Permanent gases, *see* Separation
Phase ratio (β)   61–63, 69–70, 127, 131, 133–135, 137, 141
Phenyl silicones, *see* Liquid phases
Phthalates, *see* Liquid phases
Plate height
    in distance units   15, 25–28, 32, 65–70, 96, 124–125, 130,
        138–141, 152–172, 181, 190
    theoretical minimum   53–54, 59, 62, 64, 195
    in time units   186–187, 193

Plate number   15, 23, 25, 28, 30–32, 40, 53, 56, 64–66, 69–72, 82, 88, 96, 121–125, 127, 130, 133, 134, 138–142, 154, 163–170, 182–187, 212–214, 222
   effective   28
Plate theory
   SCI-FILE: On Plate Theory   24–28
PLOT columns, *see* Column types
Plug flow   188, 192, 197–198
Plug injection, *see* Injection
Poisson distribution   212, 222
Polar column, *see* Liquid phase
Polarity
   scale   83, 85, 89, 90, 91, 94, 95
Polarizable   44–46, 84, 89, 91, 93
Polyesters   84, 93, 95, 97
Polyimide   58
Polyols   84, 94, 95, 97
Polysiloxanes   87, 89, 90
Porapak™   48, 108, 109, 250, 279
Power region   129–142
ppm, *see* Parts-per-million
Precision (repeatability)   216, 298–302, 319–320
Pressure
   absolute   237, 240
   average   127, 238, 182, 238–244, 258
   effect on solute solubility   240
   in series columns   237–244, 262–269
Pressure drop
   calculations   265–269
   in columns   52, 53, 54, 74, 127, 154, 156, 159, 171, 188, 243
   nonlinearity   237, 238
   SCI-FILE: On Pressure Drop   238–240
   in tubes   188, 193
Prime cause of everything   332, 333, 337
Prime objective   120
Probe
   McReynolds   100, 101, 108
   Rohrschnider   100
Process gas chromatographs
   development of   232, 233, 274
Processor, *see* Microprocessor
Programmer, *see* Microprocessor

Raschig rings   23
Ramp
   remnant peak, *see* Peak shape

temperature, *see* Temperature programming
Rate theory
   SCI-FILE   153–159
Real power   235, 291
Regroup   230, 231, 250, 252, 258, 259, 262, 269
Reliability   9, 60, 67, 93, 97, 171, 180, 257, 283, 298, 299
Remnant, *see* Peak shape
Resistance to mass transfer   51, 156–158, 172–173
Resolution
   adequate   119, 120–121
   of asymmetric peaks   215–222, 224
   on capillary columns   65–67, 69, 71–72, 134, 182–198
   and column length   27, 29–30, 133
   definition   21–23
   equation   121–122, 123–131
   improving   51, 57, 59, 62, 63, 83
   introducing   11, 15, 26, 64
   optimizing   63, 134–142, 150, 154, 160–171
   on packed columns   74
   perfect   22
   and plate number   26, 31–32, 40, 47, 53, 73
   and retention factor   130, 134–142
   SCI-FILE: On Resolution   121–122
Response factor   384
Retention
   data   70, 98, 126
   database   236, 301–302
   factor (k)   15, 19, 51, 64, 69, 70, 72, 102, 122–134, 151, 155–159, 181, 192, 224, 235, 239
   index (Kováts)   98–102
   mechanism   42, 45, 87, 106
   mechanism (second)   209–211, 214, 221, 222
   optimum   130, 139
   SCI-FILE: *On Retention Data*   98–102
   SCI-FILE: *On Retention Factor*   129–131
   time (adjusted)   18–19, 24
   time (definition)   16, 24–27, 130
   volume   125–126, 182, 184–185, 193
Rohrschneider probes   100
Root cause of failure   222–225, 298, 301, 304–308, 310, 317, 334, 337–338

Saddles   23
Sample
   capacity   40, 51, 59–61, 63, 67, 68–74, 83, 181–187, 209, 274, 341, 348, *see also* Feed volume
   conditioning   8, 223

Sample (cont'd)
    gas   5, 55, 180, 182, 186, 218, 251, 253, 291, 337, 314, 339, 342, 343, 347
    flow rate   342
    injection, see Injection
    liquid   68, 186, 217, 223, 251, 253, 344, 347
    pressure   342
    size   55, 65, 67–68, 71–72, 182, 192, 209, 223–225, 384
    size calculations   185–187
    splitter   64, 68, 74, 223, 232, 255, 347–348
    volume   56, 62–64, 67–68, 71, 159, 173, 180–187, 192, 196, 205, 209, 211, 213, 217, 222, 224, 232, 253, 274, 340–348
SCI-FILE Subject Index   378
SCOT columns, see Column types
Sebacates   93
Second retention mechanism   209–211, 214, 221, 222
Selectivity   20, 82, 83, 89–93, 96, 100, 101, 107–109, 112
Self-sharpening, see Peak shape
Separation
    calculation   27, 263
    definition   21
    factor (α)   15, 20–21, 29, 30, 72, 82, 98, 122, 123–125, 127–130, 133–137, 140, 224
    linear   27
    logarithmic   27
Separation of
    acetylenes   107, 341
    alcohols   45, 52, 64, 91, 93, 95, 100, 107, 109, 339, 347
    aldehydes   84, 91, 95, 100, 106, 109
    aromatics   84, 91, 93, 100, 262
    carbon oxides   5, 84, 107, 109, 27
    cis/trans isomers   84, 91
    ethers   84, 100, 109
    ketones   84, 86, 91, 100, 109, 339
    noble gases   106
    olefins   91, 93, 100, 107, 172, 341, 347
    paraffins   4, 21, 43–45, 84, 86, 95, 98–101, 107, 172, 339
    permanent gases   46, 74, 84, 107, 108, 109
    sulfides   52, 84, 106, 107, 109
    water   47, 48, 84, 86, 106, 107, 108, 109
Shelter, see Analyzer shelter
Sigma, see Standard deviation
Signal processing   189, 195, 224, 308, 310
Silicon coating   52

Silicon oils, see Polysiloxanes
Siloxane polymers, see Polysiloxanes
Solid phase(s)   3, 17, 42, 46, 47, 51, 57, 103, 104, 135, 149, 151, 224, 231, 232, 334
    alumina   46, 60, 107, 108
    Bentone   107
    carbon molecular sieves   46, 48, 75, 84, 106, 107
    graphitized carbon   48, 49, 106, 107
    porous polymers   46–48, 50, 60, 84, 107, 108, 109, 210, 223, 281
    porous silica   107
    silica gel   46, 107
    zeolite molecular sieves   46, 48, 106, 107
Solid support   49, 60, 83, 96, 135
Solubility   5, 6, 101, 128, 151, 209, 222, 224, 285, see also Distribution coefficient
    limit   181–184, see also Sample capacity
    pressure effect   240
    temperature effect   69, 131, 133
Solute
    affinity   68, 85, 127
    concentration   127, 207–209, 211, 240
    definition   5
    polarity   83, 84, 85
Solvation   42, 102, 104
Solvent see also liquid phase
    affinity   85, 97, 154
    coating   50–52, 60, 73, 92, 312, 313
    definition   5
    peak   5
Span   301
Spatial separation   26, 213–214
Spikes
    baseline   300, 305, 307, 310, 318, 319, 322, 333, 334, 336, 338, 345, 352, 353, 361, see also Baseline symptoms
    marker   281, 322, 333
Spiking a sample   285, 286
Split injection, see Injection
Squalane, see Liquid phases
Standard deviation   24–25, 183, 190, 196–197, 213, 305, 319
Start backflush   264, 336
Start next analysis   264
Stationary phase, see also Liquid phase; Solid phase
    affinity   3, 42–43, 68, 103, 127

definition   2, 40, 46, 49
equilibrium   5, 23, 104, 207–209, 213, 240
full details   82–118
function   6, 15, 61
identification   83, 84
mass transfer   158
overload   181–182, 211, 274, 285
polarity   83, 341
retention   19, 62, 215, 239, 309
SCI-FILE: On Mutual Affinity   43–45
selection   96, 97, 120, 223, 224, 231, 236, 250, 260, 291
selectivity   20, 27, 82, 128
Statistical
  quality control   8, 298, 301–303
  theory   25, 26, 154, 183, 190, 319
Steps, *see* Baseline symptoms
Stuttering, *see* Backflush
Succinate   93, 94, 97, 101
Surface deactivation   49, 50, 52, 218, 343, 345
Support   41, 49–51, 60, 83, 96, 135, 214, 218, 223
Symptoms   206, 216, 218, 219, 300, 304, 310, 312, 329, 330, *see also* Baseline
System constants   102

Tailing   60, 62, 68, 74. 105, 181, 186, 194, 204–206, 209, 211–212, 216–218, 221–223, 274–276, 331, 340, 341, 346–347
Tau   25, *see also* Standard deviation
Temperature
  effects of   6, 66, 69, 96, 100, 105, 126, 157, 214, 317, 331
  maximum   73, 74, 87, 92
  optimum   134, 135, 137–141, 158, 173, 224, 351
  programming   6–7, 27, 41, 47, 56, 59, 64, 134, 231, 232, 234, 312, 314, 339
Temporal separation   26, 214
TCD, *see* Detectors
TCEP   84, 85, 90, 91, 94, 95, 100, 101, 107
Theoretical plates, *see* Plate theory
Thermal conductivity detector, *see* TCD
Thermal peak distortion   213, 222
Time
  delay   188, 195, 211, 334, 352, 356, 357
  domain   27, 212, 222
  Time efficiency   162, 166, 168, 170, 171

Timed events   8, 243, 264, 381
Trace analysis, *see* Parts-per-million
Trap-and-hold column system   289–290
Triangular peaks, *see* Peak shape–triangular
Triangulating a peak   16–17
Troubleshooting *see also* Diagnosis
  baselines   304, 305, 306, 312–314, 315–316, 317–318, 319–322, 323
  chromatograms   4, 11, 15, 217, 219, 304, 306, 329, 330, 332, 333, 335, 336, 352–359
  column performance   8, 16, 19, 21, 133, 153, 243, 284
  detector problems   306–307
  peak position   242, 335, 336, 338–339, 341, 345
  peak separation   351–352
  peak shape   222, 286, 336, 338, 341, 346–348, 349
  peak size   340–341, 342–345, 349
  sampling system   303–304
  SCI-FILE: On Diagnosis   335–337
  techniques   3, 297, 304, 307–310
Tubing, *see* Column tubing
Two-dimensional GC   234, 235

Unibeads   107
Unretained component   17, 150–151, 238–239, *see also* Air peak
UTE (utilization of theoretical efficiency)   64

Validation   8, 298–303, 330
  control chart   301–302
  frequency   301
  procedure   301
Valley   22, 185, 224, 263–266
Valve types, *see also* Live Tee
  diaphragm   244, 251, 253, 289, 291, 316
  rotary   244, 253, 276, 289, 291
  slider   336–337
Van Deemter
  curve   153, 155–58
  equation   155, 158
  SCI-FILE: *On Rate Theory*   153–159
Van der Waals, *see* Forces
Vapor pressure   60, 72–73, 85, 87–88, 92, 93, 312
Variance
  on column   25–28, 154, 158, 213
  extracolumn   190–195, 196–198

Variance (*cont'd*)
   SCI-FILE: *On Extracolumn Variance* 190–195
   validation 301, 303
Velocity, *see* Carrier gas velocity
Vendor codes, *see* Column names

Wander, *see* Baseline symptoms
Water
   as analyte 84, 106, 107, 108, 109, 232
   backflushed 47, 48, 250
   in carrier gas 47, 48, 106
   deactivation 105, 108
   flushing 52
WCOT columns, *see* Column types
Width at half height, *see* Peak width

Zeolite 46, 106